Theoretical Mechanics

Theoretical Mechanics

EUGENE J. SALETAN
ALAN H. CROMER

Physics Department
Northeastern University

JOHN WILEY & SONS, INC.
New York London Sydney Toronto

Library of Congress Catalogue Card Number: 78-161494

ISBN 0-471-74986-9

Printed in the United States of America.

10 9 8 7 6 5 4 3 2 1

To Peggy

Куда, куда вы удалились,
Весны моей златые дни ?

Preface

Classical mechanics, which began with Galileo and Newton, is the basis of subsequent developments in physics, and therefore a deep understanding of the fundamental structure of mechanics is necessary for an understanding of modern physics. The purpose of this book is to exhibit this structure in a way that will be most useful to the physics graduate student in his study of quantum mechanics and field theory. At the same time, emphasis is placed on current research in the field of theoretical mechanics itself, and the book brings the subject into contact with some of this research.

A number of innovations have been introduced into the mathematical treatment. The traditional use of infinitesimals has been replaced consistently by the unambiguous operations of differentiation and integration. The motion of rigid bodies is treated in terms of operations on vector space. Lagrange multipliers in the variational principle are introduced in terms of inner products. A unified set of coordinates on phase space replaces the usual p's and q's in much of the development of the Hamiltonian formalism and canonical transformations, a notational device that greatly simplifies the proof and understanding of several important theorems.

In the selection of material an emphasis has been placed on modern topics such as symmetry and conservation laws, perturbation theory, adiabatic invariance, and group theory. This sometimes has been at the expense of more traditional subjects. The central-force problem is not treated in great detail, for example, but it is used throughout the book to illustrate other topics.

The final section of each chapter consists entirely of examples, applications and extensions of the theory that is presented earlier in the chapter. This allows a smooth development of the theory and permits a more extensive treatment of some examples than otherwise would be possible. In each chapter the final section is to be studied in parallel with the rest of the chapter.

The problems form an integral part of the development of the theory. For pedagogic reasons the proofs of some subsidiary results are sometimes left as problems for the reader, and the text then refers to these problems.

In general a mixture of formal and practical problems has been included to help the student to develop both his understanding and calculational skills.

We are grateful to our teachers, our colleagues, and our students. Among them we particularly acknowledge contributions from Professors Valentine Bargmann, Richard Arnowitt, Marvin Friedman, Douglas G. Currie, F. A. E. Pirani, and students who are too numerous to mention. We also thank Helen Schneider and Gertrude Tang for typing parts of the manuscript.

Boston, Mass. **E. J. SALETAN**
June, 1971 **A. H. CROMER**

Contents

Theoretical Mechanics

CHAPTER I

Fundamentals of Mechanics

This chapter discusses some aspects of elementary mechanics. It is assumed that the reader knows that $\mathbf{F} = m\mathbf{a}$ and that he has worked many problems using this knowledge. For this reason we do not attempt to develop an exhaustive treatment of the fundamentals, but only to emphasize what we consider to be some of the subtleties.

I-1 ELEMENTARY KINEMATICS

Let us begin by attempting a quantitative characterization of motion. Since motion involves change in the location of an object, and this location in turn is specified by the position of every point composing it, we must first consider how to specify the position of a point. This may be done by giving its coordinates relative to a coordinate system, called the *reference system* or *frame*. If the frame is chosen to be rectangular Cartesian with axes X_1, X_2, and X_3, the coordinates x_1, x_2, x_3 of the point are equal to the projections onto the axes of the line joining the point and the origin, as shown in Fig. 1. The position may also be given by the radius vector $\mathbf{x}$ from the origin to the point. The components of $\mathbf{x}$ are the coordinates of the point. We write

$$\mathbf{x} = x_1\mathbf{e}_1 + x_2\mathbf{e}_2 + x_3\mathbf{e}_3 = \sum_{\alpha=1}^{3} x_\alpha \mathbf{e}_\alpha, \tag{1}$$

where $\mathbf{e}_\alpha$ is the unit vector in the α direction.

As the point moves, its position vector will change, and we thus need a parameter t to label the different positions, so that $\mathbf{x}$ becomes a function of t, or

$$\mathbf{x} = \mathbf{x}(t). \tag{2}$$

We require further that t have the property of increasing monotonically as $\mathbf{x}(t)$ runs through *successively later positions* of the point. This concept of successively later positions is an intuitive one depending on the ability to distinguish between *before* and *after*, or to order events. In other words, for any two positions of the particle we shall assume that there is no question as to which is the earlier one. This is all related, clearly, to the intuitive notion of

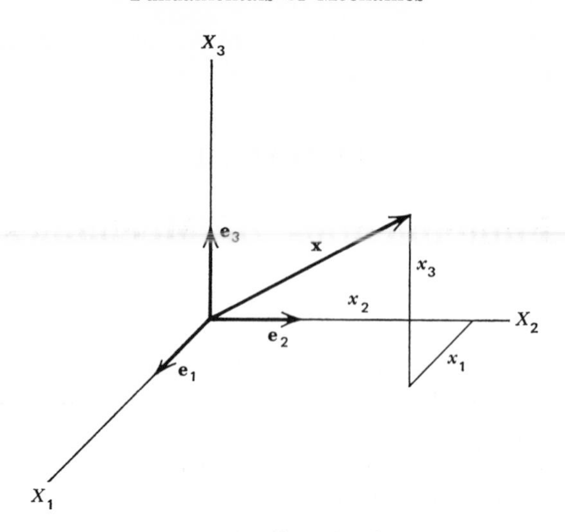

Figure 1. The vector **x** in a rectangular Cartesian frame.

time, and hence t is in some sense a quantification of this notion. In short, for any two values t_1 and t_2 of t such that $t_2 > t_1$, the point occupies the position $\mathbf{x}(t_2)$ at a time later than (after) it occupies $\mathbf{x}(t_1)$.

We shall not try to say what time is in this section, but in the next one, when we discuss dynamics. It is characteristic of kinematic statements, like the ones we make in this section, that they depend on time having only two properties: (a) t increases monotonically with successive positions, and (b) the first and second derivatives of **x** with respect to t exist.

Once a coordinate system is given, Eq. (2) can be represented by the set of three equations

$$x_\alpha = x_\alpha(t) \qquad (\alpha = 1, 2, 3). \tag{3}$$

But Eqs. (3) are only a representation of (2). Equation (2) gives the time dependence of the *physical vector* from the origin to the moving point, whereas (3) is a set of equations for the time dependence of the coordinates: the three functions appearing in (3) will depend on the particular coordinate system chosen. Once the frame is chosen, however, (3) is a set of three parametric equations for the *trajectory* of the particle (the curve it sweeps out in its motion) in which t appears as the parameter.

Perhaps it should be pointed out that we are assuming that space is Euclidean and has three dimensions. That is, three functions are needed to specify the position of a particle as a function of time, with no a priori restrictions on them. There is no relationship between these three functions which holds for all particle trajectories. Later we shall consider classes of trajectories restricted in one way or another, so-called *constrained motions*, but we shall assume nevertheless that space itself is three-dimensional in this dynamical sense.

The velocity $\mathbf{v}$ of the point is defined in terms of $\mathbf{x}$ and t, namely,

$$\mathbf{v}(t) = \dot{\mathbf{x}}(t) \equiv \frac{d\mathbf{x}}{dt}. \tag{4}$$

It is convenient to write this in terms of the distance s along the trajectory, measured from an arbitrary starting point. Mathematically, s is defined as

$$s(p_1) = \int_{p_0}^{p_1} \left[\sum_{\alpha=1}^{3} (dx_\alpha/dp)^2 \right]^{1/2} dp,$$

where p is any parameter along the trajectory which is monotonic between $\mathbf{x}(p_0)$ and $\mathbf{x}(p_1)$. Then in terms of s

$$\mathbf{v} = \frac{d\mathbf{x}}{ds}\frac{ds}{dt}.$$

But $d\mathbf{x}/ds$ is just the unit vector $\boldsymbol{\tau}$ tangent to the trajectory. To see this consider Fig. 2, which shows a section of a space curve. The tangent vector at

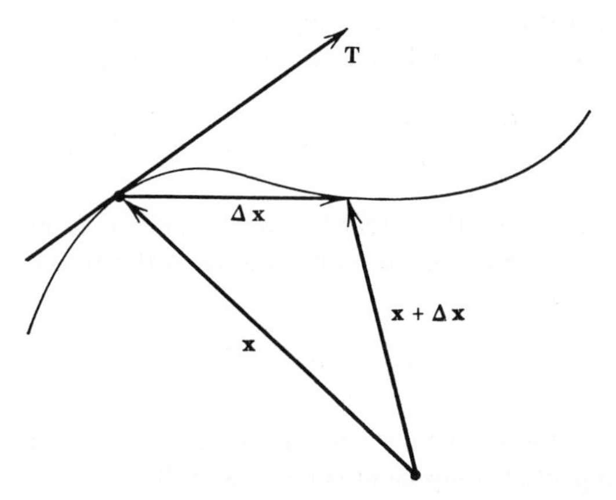

Figure 2. A vector **T** tangent to a space curve.

the point $\mathbf{x}$ on the curve is in the direction indicated by $\mathbf{T}$. The chord vector $\Delta\mathbf{x}$ between the points $\mathbf{x}$ and $\mathbf{x} + \Delta\mathbf{x}$ approaches parallelism to $\mathbf{T}$ in the limit as $\Delta\mathbf{x} \to 0$. In order to obtain a finite vector in this limit we must divide $\Delta\mathbf{x}$ by some scalar that goes to zero with $\Delta\mathbf{x}$. If we choose the arc-length Δs, we arrive at

$$\boldsymbol{\tau} = \lim_{\Delta s \to 0} \frac{\Delta\mathbf{x}}{\Delta s} = \frac{d\mathbf{x}}{ds},$$

which is of unit length and parallel to $\mathbf{T}$. Then we have

$$\mathbf{v} = \boldsymbol{\tau}\frac{ds}{dt} = \boldsymbol{\tau}v, \tag{5}$$

which says that $\mathbf{v}$ is everywhere tangent to the trajectory and equal in magnitude to the *speed* v along the trajectory.

The acceleration of the point is defined as the vector

$$\mathbf{a} = \dot{\mathbf{v}} = \ddot{\mathbf{x}} = \frac{d\mathbf{v}}{dt} = \frac{d^2\mathbf{x}}{dt^2}.$$

From Eq. (5) we have further that

$$\mathbf{a} = \frac{d}{dt}(\boldsymbol{\tau}v) = \frac{d\boldsymbol{\tau}}{dt}v + \boldsymbol{\tau}\frac{dv}{dt}.$$

Now $\boldsymbol{\tau}$ is a unit vector, so that $\boldsymbol{\tau}\cdot\boldsymbol{\tau} = 1$, and therefore

$$0 = \frac{d}{dt}(\boldsymbol{\tau}\cdot\boldsymbol{\tau}) = 2\boldsymbol{\tau}\cdot\frac{d\boldsymbol{\tau}}{dt}.$$

Thus $d\boldsymbol{\tau}/dt$ is perpendicular to $\boldsymbol{\tau}$, lying in a direction normal to the curve. Let $\mathbf{n}$ be the unit vector in this normal direction.[1] Then

$$\frac{d\boldsymbol{\tau}}{dt} = \left|\frac{d\boldsymbol{\tau}}{dt}\right|\mathbf{n} = \left|\frac{d\boldsymbol{\tau}}{ds}\frac{ds}{dt}\right|\mathbf{n} = \left|\frac{d\boldsymbol{\tau}}{ds}\right|v\mathbf{n}.$$

It is shown in Problem 1 that $|d\boldsymbol{\tau}/ds| = \rho^{-1}$ is the curvature of the trajectory at $\mathbf{x}$, where ρ is the radius of curvature. Inserting this into the expression for $\mathbf{a}$, we obtain

$$\mathbf{a} = \frac{v^2}{\rho}\mathbf{n} + \frac{d^2s}{dt^2}\boldsymbol{\tau}. \tag{6}$$

Thus the acceleration has components parallel and perpendicular to the trajectory. The tangential component is $\dot{v} = \ddot{s}$ and the normal component is the *centripetal* acceleration v^2/ρ.

We shall not go deeper into point kinematics, since further developments are probably better treated in conjunction with a knowledge of dynamics. Therefore we now turn from the mathematical treatment of motion to a discussion of the physical principles that determine the motion of point particles. This section on kinematics has established the language in which we now state the axioms lying at the basis of classical mechanics.

[1] Actually there are many normals to the curve at $\mathbf{x}$, but $\mathbf{n}$ is what is called the principal normal, lying in the instantaneous plane of the curve at $\mathbf{x}$.

I-2 PRINCIPLES OF DYNAMICS

Real objects are not points, yet we shall often use point kinematics to describe their motion. This is because small enough objects approximate points, and their motion can be described with some accuracy in this way. Moreover, as we shall see, results about extended objects can be derived from axioms about idealized point particles.

We start by introducing the notion of an *isolated particle*. By this we shall mean a sufficiently small physical object, removed sufficiently far from all other matter. The degree of isolation, that is, how small it must be and how far we must remove it, depends on the precision of measurements, and we shall henceforth make statements about isolated particles with the following understanding. They are true for measurements of length whose uncertainties are large compared to characteristic dimensions of the object, and for measurements of time (defined below) whose uncertainties are large compared to characteristic times of any changes within the object. Also the distances and times measured should be small compared to the distance from the nearest object and the time to get to it. When isolation is understood in this sense, the statements we make become more accurate for bodies more isolated. The laws of mechanics can then be formulated in two principles.

PRINCIPLE 1. There exist certain frames of reference, called *inertial*, with the following two properties. *Property A*. Every isolated particle moves in a straight line in such a frame. The unit of time can be defined by using the motion of any such isolated particle as a standard: equal lengths are marked off on its (straight-line) trajectory, and the time intervals in which the particle crosses successive marks are defined as equal intervals of time. It then follows by definition that the standard particle is moving at constant velocity. *Property B.* When the time is so defined every other isolated particle also moves at constant velocity in this frame. In other words the definition of time is independent of the particle by which it is defined.

Remarks. 1. The existence of a single inertial frame, as postulated in Principle 1, implies the existence of many more, all moving at constant velocity with respect to each other. *2.* Although it is not practical to measure time in terms of the motion of an isolated or, as we shall say, a *free* particle, it will be shown in Chapter V that the laws of motion derived from Principle 1 and Principle 2 (below) imply that the rotation of an isolated rigid body about a symmetry axis also takes place at a constant rate and can thus be used as a measure of time. In practice the Earth is usually used for this purpose, but corrections have to be introduced to take account of the fact that the Earth is neither completely isolated nor completely rigid. The most accurate modern standards of time are atomic in nature and are based on a long chain of reasoning stretching from the two principles and involving quantum as well as classical concepts. *3*. In terms of the usual statement of the laws of motion, Principle 1 is equivalent to Newton's First Law. *4*. Since inertial frames are defined in terms of isolated bodies, they cannot in general be extended indefinitely. They have, in other words, a local character, for if we attempt to extend them, the degree of isolation changes. Suppose, for instance, there exist two inertial frames very far apart. Then if they are extended until they intersect, it may turn out that a particle which is free in the first frame is not free in the second. Considerations such as these play a role in physics, but are not important in the present volume [Taylor (1966)].

PRINCIPLE 2. *A. Conservation of Momentum.* Consider two particles 1 and 2 isolated from all other matter, but not from each other, and observed from an inertial frame. In general they will not move at constant velocities: their interaction causes accelerations. Let $\mathbf{v}_i(t)$ be the velocity of particle i at time t, where $i = 1, 2$. Then there exists a positive constant μ_{12} and a constant vector $\mathbf{K}$ independent of the time such that

$$\mathbf{v}_1(t) + \mu_{12}\mathbf{v}_2(t) = \mathbf{K} \tag{7}$$

for all time t. Moreover, μ_{12} is independent of the inertial frame in which the motion is being observed and independent of the particular motion ($\mathbf{K}$ is not). That is, if the motion is interrupted and a new test is made, the same μ_{12} can be used: it depends only on particles 1 and 2, but not on the interaction. If a similar experiment is performed with particle 1 and a third particle 3, a similar result is obtained, and the same may be said for an experiment involving particles 2 and 3. We then have

$$\begin{aligned} \mathbf{v}_2(t) + \mu_{23}\mathbf{v}_3(t) &= \mathbf{L}, \\ \mathbf{v}_3(t) + \mu_{31}\mathbf{v}_1(t) &= \mathbf{M}. \end{aligned} \tag{8}$$

B. Existence of Mass. The μ_{ij} are related according to

$$\mu_{12}\mu_{23}\mu_{31} = 1. \tag{9}$$

It follows from Eq. (9) that there exist positive constants m_i, $i = 1, 2, 3$, such that Eqs. (7) and (8) can be written

$$\begin{aligned} m_1\mathbf{v}_1 + m_2\mathbf{v}_2 &= \mathbf{P}_{12}, \\ m_2\mathbf{v}_2 + m_3\mathbf{v}_3 &= \mathbf{P}_{23}, \\ m_3\mathbf{v}_3 + m_1\mathbf{v}_1 &= \mathbf{P}_{31}, \end{aligned} \tag{10}$$

where the $\mathbf{P}_{ij}$, called *momenta*, are constant vectors. It is left to the reader to prove that this is always possible. Note that the m_i are not unique, for it is their ratios, the μ_{ij}, which are determined by experiment and which are therefore unique. Given any set of m_i, another set can be obtained by multiplying them all by the same constant. What is done in practice is that some body is chosen as a standard (say 1 cm^3 of water at 4°C) and the m_i of all other bodies are related to it. The important thing about the m_i is that once such a standard is chosen there is just one number m_i associated with each body, independent of what other body it is interacting with. These m_i are, of course, called the *masses* of the bodies. In terms of the usual statement of the principles of mechanics, Principle 2 in the form of Eq. (10) states the conservation of momentum in a two-particle interaction.

Let us now take the time-derivative of Eqs. (10). We obtain

$$m_1\mathbf{a}_1 + m_2\mathbf{a}_2 = 0, \tag{11}$$

and similar expressions from the other two equations (we have used the fact that the vectors on the right-hand sides of those equations are time independent). This equation, or rather the one obtained by taking the time derivative of (7), could have been used as a starting point to define mass and to arrive at the same results. From it, of course, we could have derived (10). Thus they are equivalent. Equation (11) is perhaps more familiar as a starting point for classical mechanics, for it is essentially a restatement of Newton's Third Law of Motion. Indeed, let us define the *force* $\mathbf{F}$ acting on a particle by the equation

$$\mathbf{F} = m\mathbf{a} = m\,\frac{d^2\mathbf{x}}{dt^2}\,. \tag{12}$$

Then if $\mathbf{F}_{ij}$ is the force on particle i *due to* the presence of particle j, Eq. (11) reads

$$\mathbf{F}_{12} + \mathbf{F}_{21} = 0, \tag{13}$$

which is Newton's Third Law.

The two principles are equivalent to Newton's Three Laws of Motion, which, in their usual formulation, involve a number of logical difficulties which the present treatment tries to avoid. Thus for instance Newton's First Law, that a particle moves with constant velocity in the absence of an applied force, is incomplete without a definition of force and might at first seem to be but a special case of the Second Law. Actually it is an implicit statement of Principle 1 and was logically necessary for Newton's complete formulation of mechanics. Although such problems arise in a discussion of his formulation, we should remember that it is his formulation that lies at the basis of classical mechanics as we know it today. It is hard to believe that Newton was completely unaware of the inherent logical difficulties and unable to overcome them. The formulation of the laws of mechanics as we have given it is, in fact, merely an interpretation of Newton's Laws. This interpretation is due originally to Mach (1942). Our development is closely related to some work by Eisenbud (1958).

It is interesting that Eq. (12), Newton's Second Law, is now a definition of force, rather than a fundamental law. Why, one may ask, has this definition been so important in classical mechanics? The answer lies, as one may have expected, in its physical content. It is found empirically that in many physical situations it is $m\mathbf{a}$ that is known a priori rather than some other dynamical property: it is the force which can be specified independent of the mass, acceleration, or any other properties of the particle whose motion is being studied. (Strangely enough, in one of the most common cases of motion, namely a particle in a gravitational field, this is not true. In this case it is $\mathbf{a}$, not $m\mathbf{a}$, that is known a priori.) Moreover, forces in most cases satisfy a superposition principle; that is, the total force on a particle can be found by adding contributions from different agents, or from different systems of other particles. It is such properties that elevate Eq. (12) from a rather empty definition to a dynamical relationship. For an interesting discussion of this point see Feynman (1963).

Most of the problems we shall be dealing with will involve forces given as functions of position. Then Eq. (12) becomes the differential equation

$$\mathbf{F}(\mathbf{x}) = m\,\frac{d^2\mathbf{x}}{dt^2} = m\ddot{\mathbf{x}} \tag{14}$$

for the vector function $\mathbf{x}(t)$. This equation we shall then take as the basis of most of what follows in this volume, although we shall often speak also of forces that depend not only on $\mathbf{x}$, but also on $\dot{\mathbf{x}}$ and t. (In principle $\mathbf{F}$ could also be a function of higher time derivatives of $\mathbf{x}$.) In general, then, our problem will be to solve an equation such as (14) for $\mathbf{x}(t)$, thus obtaining what we shall call the *motion* of the particle.

From the principles of mechanics we may derive immediately some rather useful relations. First we define the linear *momentum* $\mathbf{p}$ of a single particle of mass m as

$$\mathbf{p} = m\mathbf{v}. \tag{15}$$

Since m is a constant, we may write Eq. (12) in the form

$$\mathbf{F} = \frac{d\mathbf{p}}{dt}, \tag{16}$$

and then if the external force on a particle is zero, it follows immediately that its momentum remains constant. As we have seen already, the total momentum in a two-particle interaction, that is, the sum of the momenta of the single particles, remains constant. It is this constancy of the momentum that leads one to single it out. At this point, incidentally, Eq. (16) is merely a restatement of (12), but this is because we have not discussed the possibility of a varying mass. In Section 4 we shall deal with masses that do not remain constant in time, and it will be seen then that the motion is correctly described by (16) rather than by (12).

Second, we define the *angular momentum* **L** of a particle about some *inertial point* (a point that moves at constant velocity in an inertial system) as

$$\mathbf{L} = \mathbf{x} \times \mathbf{p}, \tag{17}$$

where **x** is the vector from the inertial point to the particle. The derivative of **L** is then

$$\dot{\mathbf{L}} = \frac{d}{dt}(\mathbf{x} \times \mathbf{p}) = \mathbf{v} \times \mathbf{p} + \mathbf{x} \times \dot{\mathbf{p}}.$$

Now $\mathbf{v} \times \mathbf{p} = m(\mathbf{v} \times \mathbf{v}) = 0$, and then Eq. (8) leads to

$$\dot{\mathbf{L}} = \mathbf{x} \times \mathbf{F} = \mathbf{N}, \tag{18}$$

where we have defined the *torque* **N** about the inertial point. It is seen that the relation between torque and angular momentum is similar to that between force and momentum; in particular, if the applied torque is zero, the angular momentum is constant.

The third quantity we introduce is *energy*, but we shall devote the next section to it.

I-3 WORK AND ENERGY

In terms of the concept of force, the general problem of particle mechanics may now be stated as follows: given **F**, to find $\mathbf{x}(t)$, the general motion of the particle. This problem is trivially solved if **F** is known as a function of the time alone. By integrating Eq. (12) twice, we have

$$\mathbf{x}(t) = \mathbf{x}(t_0) + (t - t_0)\mathbf{v}(t_0) + \frac{1}{m}\int_{t_0}^{t} dt' \int_{t_0}^{t'} \mathbf{F}(t'')\, dt'', \tag{19}$$

where $\mathbf{x}(t_0)$ and $\mathbf{v}(t_0)$ are the position and velocity of the particle at the initial time t_0.

Unfortunately, however, the time dependence of **F** is almost never known a priori and can be found only after the problem is solved. In fact **F** is usually given explicitly as a function of **x**, **v**, t, and it is only after **x** and **v** are known as functions of the time that $\mathbf{F}(t) = \mathbf{F}(\mathbf{x}(t), \mathbf{v}(t), t)$ can be written down and Eq. (19) used. But this requires first solving the problem, so that (19) is useless.

Since **F** is often given as a function of **x** alone, on the other hand, integration over **x** is more reasonable. That this alternative procedure can be useful becomes evident on integrating **F** along the trajectory. We have

$$\begin{aligned}\int_{\mathbf{x}(t_0)}^{\mathbf{x}(t)} \mathbf{F} \cdot d\mathbf{x} &= \int_{t_0}^{t} \mathbf{F} \cdot \dot{\mathbf{x}}\, dt = m\int_{t_0}^{t} \frac{d^2\mathbf{x}}{dt^2} \cdot \frac{d\mathbf{x}}{dt}\, dt = \tfrac{1}{2}m\int_{t_0}^{t} \frac{d}{dt}(\dot{x}^2)\, dt \\ &= \tfrac{1}{2}m\,|\mathbf{v}(t)|^2 - \tfrac{1}{2}m\,|\mathbf{v}(t_0)|^2. \end{aligned} \tag{20}$$

We call $\frac{1}{2}mv^2$ the *kinetic energy* T of the particle, and we call the integral on the left-hand side the *work* W done on the particle along the trajectory. Then Eq. (20) gives the kinetic energy (and hence speed) of the particle at time t in terms of its initial kinetic energy (speed) and the work. The rate at which work is done on the particle is called the power, and is seen from the second expression in Eq. (20) to be given by

$$\dot{W}(t) = \frac{dW}{dt} = \mathbf{F} \cdot \dot{\mathbf{x}}. \tag{21}$$

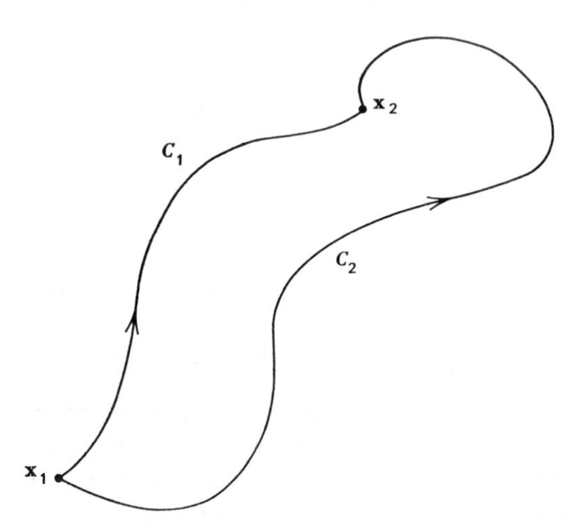

Figure 3. Two paths connecting the points $\mathbf{x}_1$ and $\mathbf{x}_2$.

The work is in general a function not only of the end points $\mathbf{x}(t_0)$ and $\mathbf{x}(t)$, but also of the trajectory; thus we have

$$T(t) = T(t_0) + W(\mathbf{x}_0, \mathbf{x}, C), \tag{22}$$

where C denotes the trajectory. In general (22) is therefore not very useful either, for W can be calculated only when C is known, and this again requires solving the problem at least partially. But for many of the forces occurring in nature W depends actually only on $\mathbf{x}$ and $\mathbf{x}_0$, not on the path C connecting them. A force that gives rise to such a W, that is, one whose line integral from one point to another is path-independent, is called *conservative*.

Because such forces are important, let us briefly study some of their properties. Let $\mathbf{F}$ be a conservative force, and let C_1 and C_2 be any two paths connecting two points $\mathbf{x}_1$ and $\mathbf{x}_2$, as shown in Fig. 3. Then by definition

$$\int_{C_1} \mathbf{F} \cdot d\mathbf{x} = \int_{C_2} \mathbf{F} \cdot d\mathbf{x}$$

or

$$\int_{C_1} \mathbf{F} \cdot d\mathbf{x} - \int_{C_2} \mathbf{F} \cdot d\mathbf{x} = \oint \mathbf{F} \cdot d\mathbf{x} = 0.$$

Thus if **F** is conservative, the integral $\oint \mathbf{F} \cdot d\mathbf{x}$ around *any* closed path is zero. Stokes's theorem can be used to transform the line integral into a surface integral giving

$$\oint \mathbf{F} \cdot d\mathbf{x} = \int_{\Sigma} (\nabla \times \mathbf{F}) \cdot d\mathbf{S}. \tag{23}$$

Here Σ is any surface bounded by the closed path of the line integral, and $d\mathbf{S}$ is the vector element of area on Σ. Since the left-hand side of Eq. (23) is zero for *any* closed path, the right-hand side must be zero for any surface Σ. This implies that

$$\nabla \times \mathbf{F} = 0 \tag{24}$$

is a necessary condition for **F** to be conservative. It is obvious that the argument can be reversed to show that (24) is also sufficient.

Given a force **F** which depends only on **x**, consider the differential equation

$$\mathbf{F}(\mathbf{x}) = -\nabla V(\mathbf{x}) \tag{25}$$

for a scalar function $V(\mathbf{x})$. From the vector identity $\nabla \times (\nabla V) = 0$, which holds for any V, it follows that if a solution V exists, **F** is conservative. The converse is also true: if **F** is conservative, then V exists. In fact, (24) is just the set of integrability conditions for Eq. (25). The function $V(\mathbf{x})$ is called the potential energy and is defined uniquely by Eq. (25) to within an arbitrary additive constant. The minus sign in (25) is for convenience, as we shall see.

Let us now return to (22). If **F** is conservative, W is independent of C, and we have

$$\begin{aligned} W(\mathbf{x}, \mathbf{x}_0) &= \int_{\mathbf{x}_0}^{\mathbf{x}} \mathbf{F} \cdot d\mathbf{x} = -\int_{\mathbf{x}_0}^{\mathbf{x}} \nabla V \cdot d\mathbf{x} \\ &= V(\mathbf{x}_0) - V(\mathbf{x}) \equiv V_0 - V. \end{aligned}$$

Equation (22) can then be rewritten in the form

$$V + T = V_0 + T_0, \tag{26}$$

which now says that in the motion of the particle the value of $V + T$ is everywhere and at all times the same as it was initially. We call $T + V$ the total (mechanical) energy E of the particle, and then (26) states the conservation of energy: if the force acting on a particle is conservative, the total energy E is constant.

Remarks. *1.* Since V is defined only up to an additive constant, so is E. This constant is usually chosen by specifying V to be zero at some arbitrary point. *2*. Neither Eq. (22) nor the more specialized Eq. (26) solve our original problem. All they tell us is that whatever $\mathbf{x}(t)$ is, that is, whatever the motion, the work done by $\mathbf{F}$ is the change in kinetic energy. We still must find the motion. However, this is a significant step on the way to the solution of the problem, as we shall see, for instance, in Section 5(a). In fact for many cases once we have stated this analytically for a particular problem, the problem is essentially solved. *3*. We call the reader's attention to the way energy is defined. It is largely *because* it is conserved that it is singled out for definition.[2]

I-4 SYSTEMS OF PARTICLES

We now wish to extend our discussion of dynamics to systems of several particles. Consider, therefore, N particles interacting with each other and acted on by external forces, each particle behaving according to the dynamical laws described in the previous section. Suppose, for instance, that known external forces are applied to the particles, but that the details of their interactions with each other are unknown. What can be said about their motion?

In order to say anything at all about their motion, we must extend Principles 1 and 2 to interactions of several particles. However the discussion can be simplified by assuming that the force on any particle is simply the vector sum of the forces "due to" all other particles present in the system. To find the total force on a particle, we find the force that would be exerted on it if only one other particle were present; we do this separately for each other particle until we have considered them all, and then we add these forces vectorially. Thus we are assuming that every force can be analyzed into two-particle forces. In addition there may be *external* forces on a particle, not due to particles in the system which is being analyzed.

Let $\mathbf{F}_i$ be the external force on the ith particle, whose mass is m_i, and let $\mathbf{F}_{ij}$ be the force exerted on the ith particle by the jth. Then $\mathbf{F}_{ii} = 0$ and, from Eq. (13), $\mathbf{F}_{ij} = -\mathbf{F}_{ji}$. The equation of motion of the ith particle is

$$m_i\ddot{\mathbf{x}}_i = \mathbf{F}_i + \sum_{j=1}^{N} \mathbf{F}_{ij}.$$

A simplification is obtained if this equation is summed with the equations of motion for all the other particles, that is, if it is summed over i. This yields

$$\sum_i m_i\ddot{\mathbf{x}}_i = \sum_i \mathbf{F}_i \equiv \mathbf{F}, \tag{27}$$

[2] There is a deeper meaning to energy. Whenever $T + V$ is not conserved, one tries to find some other physical quantity Q such that $T + V + Q$ is conserved. So far this has always been possible. Such Q's are nonmechanical (e.g., heat and radiation) and thus lead to an extension of the definition of energy and to the formulation of energy conservation in general. See Feynman (1963).

where the *total external force* **F** is the sum of only the external forces, since according to Eq. (13)[3]

$$\sum_i \sum_j \mathbf{F}_{ij} = \sum_j \sum_i \mathbf{F}_{ji} = -\sum_j \sum_i \mathbf{F}_{ij} = 0.$$

It is convenient to factor out the total mass $M = \sum m_i$ in Eq. (27). We have

$$\begin{aligned} \mathbf{F} &= M\left[\frac{1}{M}\sum_i m_i \ddot{\mathbf{x}}_i\right] \\ &= M\frac{d^2}{dt^2}\left[\frac{1}{M}\sum_i m_i \mathbf{x}_i\right] \\ &= M\ddot{\mathbf{X}}, \end{aligned} \tag{28}$$

where

$$\mathbf{X} = \frac{1}{M}\sum_i m_i \mathbf{x}_i \tag{29}$$

is defined as the *center of mass* of the system. This is the simplification we have been looking for, for what Eq. (28) says is that the center of mass moves as though the total mass were concentrated there and were acted upon by the total external force. This is the center-of-mass theorem.

The center of mass is a useful concept for many reasons. For instance the total momentum of the system of particles, that is the sum of the momenta, is seen to be

$$\mathbf{P} = \sum m_i \dot{\mathbf{x}}_i = M\dot{\mathbf{X}}. \tag{30}$$

Again it is as though the mass were all concentrated at **X**. Equation (28) can now be written

$$\mathbf{F} = \frac{d}{dt}\mathbf{P}.$$

It follows immediately that if the total external force vanishes, the total momentum is constant.[4]

Equation (28) is a very important result; for it allows us not only to combine smaller systems into larger ones and to analyze what may be called their overall behavior but also to do exactly the opposite, that is, to break up

[3] The first equality here is obtained simply by interchanging the roles of i and j in the sum; since both indices are summed, it does not matter how they are denoted.

[4] The present derivation is only valid for N-particle systems interacting through two-body forces, though the conclusion Eq. (28) is one of the most firmly established laws in physics. We could perhaps have tried to formulate something like Principle 2 for N-body forces, $N > 2$, by writing an equation similar to (7) for N velocities rather than for two. Then in the same sense that (7) is the statement of conservation of momentum for two-particle systems, this N-particle equation would state conservation of momentum for N-body forces and would thus be somewhat more general than the present result.

systems into smaller parts and to study the motion of these parts. It tells us in what sense the laws of motion apply to systems and subsystems and to what extent we can disregard their internal structure.

Unlike the momentum, the kinetic energy of the system of particles, defined as the sum of the individual kinetic energies, is not "the kinetic energy of the center of mass," that is, not just what it would be if all the mass were concentrated at the center of mass. It is found (see Problem 4) that

$$T = \tfrac{1}{2}M\dot{X}^2 + \sum_i \tfrac{1}{2}m_i\dot{y}_i^2. \tag{31}$$

where

$$\mathbf{y}_i = \mathbf{x}_i - \mathbf{X} \tag{32}$$

is the position of the ith particle relative to the center of mass. In addition to the kinetic energy of the center of mass, therefore, the total kinetic energy has a term which is the sum of the kinetic energies of all the particles relative to the center of mass.

The conservation of energy for a system of particles may be derived in the following way. In analogy with the one-particle case, we calculate the total work done on all the particles

$$\sum_i \int_{\mathbf{x}_{i0}}^{\mathbf{x}_{if}} \left(\mathbf{F}_i + \sum_j \mathbf{F}_{ij}\right) \cdot d\mathbf{x}_i = \sum_i \int_{\mathbf{x}_{i0}}^{\mathbf{x}_{if}} m_i\ddot{\mathbf{x}}_i \cdot d\mathbf{x}_i \tag{33}$$

in going from some initial configuration 0 to some final configuration f of the system. It is a simple matter to show that the right-hand side is just the change $T_f - T_0$ in the kinetic energy. As for the left-hand side, if each $\mathbf{F}_i$ is derivable from a potential V_i, then

$$\sum_i \int_{\mathbf{x}_{i0}}^{\mathbf{x}_{if}} \mathbf{F}_i \cdot d\mathbf{x}_i = \sum_i (V_{i0} - V_{if}) = V_{\text{ext},0} - V_{\text{ext},f}. \tag{34}$$

We write the second term on the left-hand side, taking the sum before the integral, in the form

$$\int_0^f \sum_{ij} \mathbf{F}_{ij} \cdot d\mathbf{x}_i = \int_0^f \sum_{ij} \mathbf{F}_{ji} \cdot d\mathbf{x}_j$$

by interchanging the roles of i and j in the sum. Now again by Eq. (13) $\mathbf{F}_{ij} = -\mathbf{F}_{ji}$, so that

$$\begin{aligned}\int_0^f \sum_{ij} \mathbf{F}_{ij} \cdot d\mathbf{x}_i &= -\int_0^f \sum_{ij} \mathbf{F}_{ij} \cdot d\mathbf{x}_j \\ &= \tfrac{1}{2}\int_0^f \sum_{ij} \mathbf{F}_{ij} \cdot d(\mathbf{x}_i - \mathbf{x}_j) \\ &= \tfrac{1}{2}\int_0^f \sum_{ij} \mathbf{F}_{ij} \cdot d\mathbf{x}_{ij}.\end{aligned}$$

Assume that $\mathbf{F}_{ij}$ is a function only of the relative positions $\mathbf{x}_{ij} = \mathbf{x}_i - \mathbf{x}_j$ of particles i and j, and that it is derivable from a potential function V_{ij} of this relative position. Then we have

$$\int_0^f \sum_{ij} \mathbf{F}_{ij} \cdot d\mathbf{x}_i = \tfrac{1}{2} \sum_{ij} (V_{ij,0} - V_{ij,f})$$

$$= V_{\text{int},0} - V_{\text{int},f}. \tag{35}$$

This last result is in fact the work it takes to move the system of particles from its initial to its final relative configuration (that is, disregarding the work done against external forces). This is most easily seen by referring each V_{ij} to infinite separation of the ith and jth particles (that is, by choosing the arbitrary constant in V_{ij} to be such that $\lim_{\mathbf{x}_{ij}\to\infty} V_{ij} = 0$) and then calculating the work done in assembling any particular configuration. If there is no external force, it takes no work to place the first particle in position, it takes V_{12} to bring up the second, it takes $V_{13} + V_{23}$ to bring up the third, and so forth. The final result is

$$\sum_{i<j} V_{ij} = \tfrac{1}{2} \sum_{ij} V_{ij} \equiv V_{\text{int}}.$$

We have used the fact that $V_{ij} = V_{ji}$ and $V_{ii} = 0$.

Remark. It is interesting that although $\mathbf{F}_{ij} = -\mathbf{F}_{ji}$, yet $V_{ij} = V_{ji}$. A careful verification of this by the reader may not be time wasted.

Collecting the results of Eqs. (33), (34), and (35), we have

$$(V_{\text{ext}} + V_{\text{int}} + T)_f = (V_{\text{ext}} + V_{\text{int}} + T)_0.$$

Note that this statement of energy conservation involves the internal as well as the external potential energy.

Let us now calculate the total angular momentum of a system of N particles. As we did for energy and momentum, we wish to state this in terms of the center of mass. The total angular momentum about the origin of our inertial reference system is

$$\mathbf{L} = \sum_i m_i \mathbf{x}_i \times \dot{\mathbf{x}}_i, \tag{36}$$

and that about the center of mass is

$$\mathbf{L}_c = \sum_i m_i \mathbf{y}_i \times \dot{\mathbf{y}}_i,$$

where $\mathbf{y}_i$ is defined by Eq. (32). We may then write

$$\mathbf{L} = \sum_i m_i (\mathbf{X} + \mathbf{y}_i) \times (\dot{\mathbf{X}} + \dot{\mathbf{y}}_i)$$

$$= \sum_i m_i \mathbf{X} \times \dot{\mathbf{X}} + \sum_i m_i \mathbf{y}_i \times \dot{\mathbf{y}}_i + \sum_i m_i \mathbf{y}_i \times \dot{\mathbf{X}}$$

$$+ \sum m_i \mathbf{X} \times \dot{\mathbf{y}}_i.$$

Now the last two terms vanish because they can be written

$$-\dot{\mathbf{X}} \times \sum_i m_i \mathbf{y}_i \qquad \text{and} \qquad \mathbf{X} \times \frac{d}{dt} \sum_i m_i \mathbf{y}_i,$$

and $\sum_i m_i \mathbf{y}_i = 0$ because it is, up to the factor M^{-1}, the position of the center of mass relative to the center of mass. Consequently

$$\mathbf{L} = \mathbf{L}_c + M\mathbf{X} \times \dot{\mathbf{X}}. \tag{37}$$

Thus the total angular momentum about the origin of an inertial system (and hence about any point moving at constant speed in an inertial frame) is

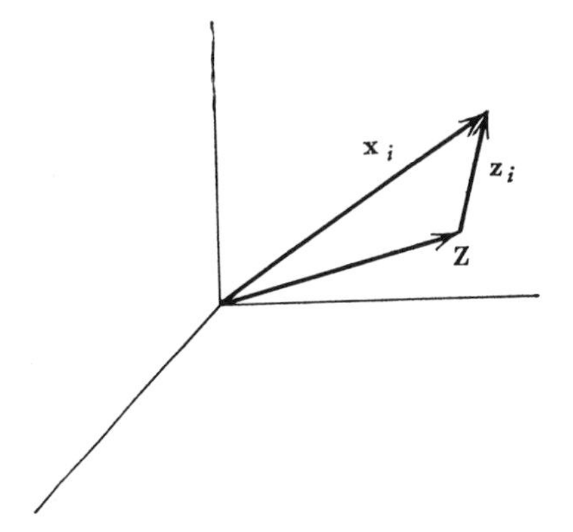

Figure 4. Relation between $\mathbf{Z}$, $\mathbf{z}_i$, and $\mathbf{x}_i$.

the sum of the angular momentum of the total mass as though concentrated at the center of mass and the angular momenta of all the particles about the center of mass.

In the one-particle case we have seen that the applied torque about the origin gives the rate of change of angular momentum about the origin. Let us see whether this applies to the many-particle system. We shall assume that the internal torques do not contribute to any change in the total angular momentum just as the internal forces do not contribute to the change in the total momentum (see Problem 5), and so we consider the external torque only. We have, for the total torque about an arbitrary (moving) point $\mathbf{Z}$,

$$\mathbf{N}_\mathbf{Z} = \sum_i \mathbf{z}_i \times \mathbf{F}_i = \sum_i m_i \mathbf{z}_i \times \ddot{\mathbf{x}}_i,$$

where $\mathbf{x}_i$ is the position of the ith particle in some inertial frame, and $\mathbf{z}_i = \mathbf{x}_i - \mathbf{Z}$ is its position relative to $\mathbf{Z}$, as shown in Fig. 4. Proceeding, we have

$$\mathbf{N}_\mathbf{Z} = \sum_i m_i \mathbf{z}_i \times (\ddot{\mathbf{z}}_i + \ddot{\mathbf{Z}}) = \frac{d}{dt} \sum m_i \mathbf{z}_i \times \dot{\mathbf{z}}_i + M\boldsymbol{\zeta} \times \ddot{\mathbf{Z}},$$

where $\boldsymbol{\zeta} = \mathbf{X} - \mathbf{Z}$ is the position of the center of mass relative to $\mathbf{Z}$. The first term in this equation is just $\dot{\mathbf{L}}_\mathbf{Z}$, the time rate of change of angular momentum about $\mathbf{Z}$, and therefore $\mathbf{N}_\mathbf{Z} = \dot{\mathbf{L}}_\mathbf{Z}$ if and only if $\boldsymbol{\zeta} \times \ddot{\mathbf{Z}} = 0$. This means $\ddot{\mathbf{Z}}$ must be parallel to the line connecting $\mathbf{Z}$ and the center of mass or, as special cases, $\ddot{\mathbf{Z}} = 0$ or $\boldsymbol{\zeta} = 0$. Of these, $\ddot{\mathbf{Z}} = 0$ implies that $\mathbf{Z}$ is moving at constant velocity relative to the origin of the inertial frame and so defines another inertial frame; $\boldsymbol{\zeta} = 0$ implies that $\mathbf{Z}$ is the center of mass of the system. Thus

$$\mathbf{N}_\mathbf{Z} = \dot{\mathbf{L}}_\mathbf{Z} \tag{38}$$

if (but not only if) $\mathbf{Z}$ is an inertial (nonaccelerating) point or the center of mass, even if the center of mass is accelerating. In the single-particle case we considered torques only about fixed (or inertial) points, so this problem did not occur.

The results we have obtained so far can be extended from discrete to continuous systems by replacing the discrete variable i by some continuous parameter. In the process sums will be replaced by integrals. For instance, consider a distribution of mass parametrized by a single variable λ running from 0 to some fixed value Λ. It is no longer possible, as with the discrete index i, to make $m(\lambda)$ the mass of the λth particle, for there are no discrete particles of nonzero mass. Instead, by $m(\lambda_0)$ we denote the sum of all the mass with λ between $\lambda = 0$ and $\lambda = \lambda_0$, so that $m(0) = 0$ and the total mass is $M = m(\Lambda)$. We assume that there exists a *mass density function* $\rho(\lambda)$ such that

$$m(\lambda_0) = \int_0^{\lambda_0} \rho(\lambda)\, d\lambda.$$

We shall sometimes write $\rho = dm/d\lambda$, for obvious reasons.

A single numerical variable λ will not do to parametrize most distributions, which are three dimensional. One needs three such variables or a single vector variable. Consider a distribution parametrized by the position vector $\mathbf{x}$ itself.[5] We then call the mass density function $\rho(\mathbf{x})$, and the mass in a small volume ΔV_i about some particular point $\mathbf{x}_i$ is then approximately $\rho(\mathbf{x}_i)\, \Delta V_i$. The total mass in the distribution is approximately the sum $\sum_i \rho(\mathbf{x}_i)\, \Delta V_i$ over a large set of points $\mathbf{x}_i$, where the ΔV_i cover the whole distribution without overlapping. The approximation becomes more accurate as more points $\mathbf{x}_i$ are taken and the ΔV_i are made smaller. In the limit of an infinite number of points and as the ΔV_i approach zero, provided such a limit exists, we

[5] This might lead to complications, for as the system moves the vector parameter associated with some moving part of it will change. In the present applications, however, this will not bother us.

obtain the integral

$$M = \int_V \rho(\mathbf{x})\, d^3x$$

over the total volume V of the distribution, where d^3x is the element of volume.

Let us calculate the total momentum in such a distribution. In a small neighborhood (volume) ΔV_i of $\mathbf{x}_i$ the mass, roughly $\rho(\mathbf{x}_i)\,\Delta V_i$, is moving roughly with velocity $\dot{\mathbf{x}}_i$. This is only roughly its velocity because different parts of ΔV_i have different velocities, but it becomes more nearly exact as ΔV_i gets smaller. The momentum of this small mass is then roughly $\Delta \mathbf{p}_i = \rho(\mathbf{x}_i)\dot{\mathbf{x}}_i\,\Delta V_i$. Now we sum over i as in the calculation of mass above, and go to the limit of small ΔV_i and many points $\mathbf{x}_i$. The result is

$$\mathbf{P} = \lim_{n\to 0} \sum_{i=0}^{n} \rho(\mathbf{x}_i)\dot{\mathbf{x}}_i\,\Delta V_i = \int_V \rho(\mathbf{x})\dot{\mathbf{x}}\, d^3x. \tag{39}$$

In Section 2 we promised to show that it is not $m\mathbf{a}$, but $d\mathbf{p}/dt$ which is equal to the force. We now turn to the demonstration. Consider a continuous mass distribution labeled with a one-dimensional parameter λ and acted on by an external force $\mathbf{F}$. We shall call such a distribution a *string*. Further, let $\mathbf{F}$ be acting not on the whole string but on a constantly changing part of it, specifically on the piece with $\lambda \leq \mu(t)$, where $\mu(t)$ is some differentiable function of the time. Assume also that the rest of the string is at rest, that is, that for $\lambda > \mu(t)$ the velocity $\partial\mathbf{x}(\lambda, t)/\partial t = 0$. Since for $\lambda \leq \mu(t)$ the velocity is not necessarily zero, as a function of λ the velocity may be discontinuous at $\lambda = \mu(t)$.

The mass of the string is $m(\Lambda)$ and constant, but there is no force on the piece with $\lambda > \mu(t)$. Thus we may think of the piece with $\lambda \leq \mu(t)$ as a small string with changing mass, with $\mathbf{F}$ acting only on this small string. The center-of-mass theorem, that is, Eq. (28), applies to the large string, and we then have

$$\mathbf{F} = \frac{d^2}{dt^2}\int_0^\Lambda \mathbf{x}(t, \lambda)\frac{dm}{d\lambda}\, d\lambda = \frac{d}{dt}\int_0^\Lambda \frac{\partial\mathbf{x}}{\partial t}\frac{dm}{d\lambda}\, d\lambda.$$

Since $\partial\mathbf{x}/\partial t$ may be discontinuous at $\lambda = \mu(t)$ we break up the integral into two parts. The integral from $\mu(t)$ to Λ vanishes, so that

$$\mathbf{F} = \frac{d}{dt}\int_0^{\mu(t)} \frac{\partial\mathbf{x}}{\partial t}\frac{dm}{d\lambda}\, d\lambda = \left[\frac{\partial\mathbf{x}}{\partial t}\frac{dm}{d\lambda}\right]_{\lambda=\mu(t)}\frac{d\mu}{dt} + \int_0^{\mu(t)} \frac{d^2\mathbf{x}}{\partial t^2}\frac{dm}{d\lambda}\, d\lambda. \tag{40}$$

The value of $\partial\mathbf{x}/\partial t$ in the first term is obtained by going to the limit as λ approaches μ from below, and is thus the velocity $\mathbf{v}$ at which the mass is moving as it joins the small string at $\lambda = \mu$. Further, $[dm/d\lambda]_{\lambda=\mu}\cdot d\mu/dt \equiv dm/dt$ is the rate at which the mass of the small string is increasing. The second term in (40) is just

$$m(\mu)\ddot{\mathbf{X}}(\mu)$$

where $\ddot{\mathbf{X}}(\mu)$ is the acceleration of the center of mass of the small string. We shall write Eq. (40), therefore, in the form

$$\mathbf{F} = \mathbf{v}\frac{dm}{dt} + m\frac{d\mathbf{V}}{dt}, \tag{41}$$

where $\mathbf{V} = \dot{\mathbf{X}}$. Note that $\mathbf{v} \neq \mathbf{V}$ in general, although if all parts of the small string are moving at the same velocity, as is often the case, $\mathbf{v}$ and $\mathbf{V}$ will be equal.

Let us now write down the rate of change of momentum of the small string. The momentum is changing first because new momentum in the form of new moving mass is being added to the string. This is proceeding at the rate $\mathbf{v}\, dm/dt$. Second, the mass already in the string is accelerating at the rate $d\mathbf{V}/dt$, so that its momentum changes at the rate $m\, d\mathbf{V}/dt$. Thus the right-hand side of (41) is the total rate of change of momentum, and we arrive at

$$\mathbf{F} = \frac{d\mathbf{p}}{dt}. \tag{42}$$

Does this treatment hold also for a single mass that is somehow increasing? This question is hard to answer, since we have no assumptions or axioms from which to start a discussion of such a varying mass. This treatment seems plausible, however, when we note the following. The acceleration of a particle acted on by some force can be calculated either (a) directly by Newton's Second Law or (b) by considering it to be in a larger fictitious system of particles of arbitrary (finite) mass such that the force on each additional fictitious particle is zero. By calculating the acceleration of the center of mass of the larger system and noting that none of the fictitious particles is accelerating, one can compute the acceleration of the original particle. The results obtained by (a) and (b) will be equal. It is left to the reader to verify this simple result. Now if the mass of a particle is changing somehow, we may consider it to be part of a larger fictitious system of constant total mass. Then the above result says that

$$\mathbf{F} = m\frac{d\mathbf{v}}{dt} + \mathbf{v}\frac{dm}{dt}.$$

All of this works because the momentum of the string for $\lambda > \mu(t)$ is zero. The reader may wish to check how (42) is altered if $v(\lambda)$ is nonzero for $\lambda > \mu(t)$.

I-5 EXAMPLES, APPLICATIONS, AND EXTENSIONS

(a) One-Dimensional Motion. The Energy Integral

A class of motions which is of particular interest is that in which a particle moves along a straight line under the influence of a force tangent to the line. Let the distance along the line be x and assume that the force F (we may dispense with vector notation in this one-dimensional case) is a function only of x. Then the problem can always be reduced to performing an integration in one dimension or, as is said, can be *reduced to quadrature*. Indeed, more complicated problems are often solved by putting them in the form of such a one-dimensional problem.

The equation of motion in this case may be written

$$m\ddot{x} = F(x). \tag{43}$$

If the force depends only on x, it can always be written as the derivative of another function which we shall call $-V$, so that the force is then what we have called conservative. In fact V is then (up to an arbitrary constant) the potential energy. We have already discussed conservative forces in three dimensions. In one dimension Eq. (26) becomes

$$\tfrac{1}{2}m^2\dot{x} + V(x) = E, \tag{44}$$

where E is the total energy of the system and is constant. The left-hand side of this equation is sometimes called the energy first integral.

Now Eq. (44) can be solved for $1/\dot{x} = dt/dx$ and immediately integrated to yield

$$t = \sqrt{\frac{m}{2}}\int\frac{dx}{\sqrt{E - V(x)}} + K, \tag{45}$$

where K is a constant to be evaluated from additional information. Let us call the indefinite integral $f(x, E)$, and suppose that we know the position $x(t_0) = x_0$ at some time t_0. Then clearly $K = t_0 - f(x_0, E)$, and (45) can be inverted to give

$$x = x(t - t_0, x_0, E).$$

It looks as though we have obtained three constants of integration t_0, x_0, E for a second-order differential equation. Actually E and K are the only constants of integration, but K must be calculated in terms of the initial information x_0, t_0. In practice one often takes $t_0 = 0$, so that only the energy E and initial position x_0 appear in the final expression.

Even if (45) cannot be integrated in terms of elementary functions, Eq. (44) contains a lot of information about the motion. Note first that Eq. (44)

does not specify the value of E, but says merely that no matter what the motion, the function on the left-hand side is a constant. By choosing different values of this constant E, we obtain different motions. But once E is given, it follows from (44) that the particle cannot move into a region in which the kinetic energy $E - V(x)$ is negative. It is then useful to plot $V(x)$ against x, as in Fig. 5, and to draw a horizontal line to represent the total energy E for any particular motion. Several examples are illustrated in Fig. 5.

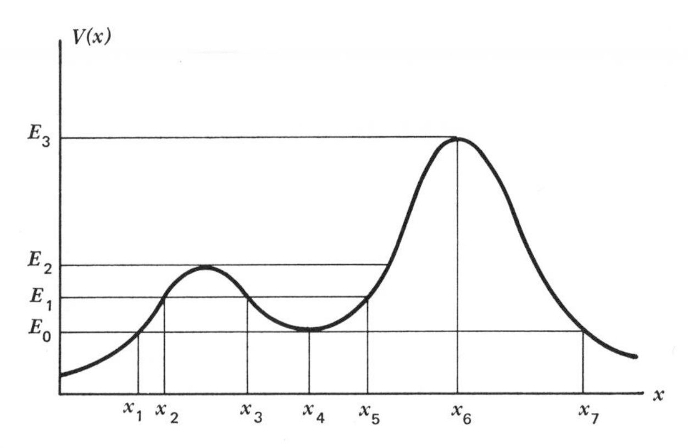

Figure 5. One-dimensional potential energy diagram. The curve indicates the potential energy $V(x)$.

If $E \leq E_0$, the particle cannot enter the region between x_1 and x_7. It makes no physical sense, for instance, to specify that the particle starts out at x_3 with total energy E_0. If, on the other hand, it is projected to the right from some point on the left-hand side of x_1 with total energy E_0, the particle will decelerate until it comes to rest at x_1. Then because $F = -dV/dx$ is negative, it will accelerate to the left, never again coming to rest. For total energy E_1 between E_0 and E_2 a new region opens up. In addition to motion similar to the previous case (except that it would come to rest at x_2), the particle can be trapped in the region between x_3 and x_5. In this case it will oscillate back and forth between the turning points x_3 and x_5. Its kinetic energy $E - V(x)$ will then be a maximum at x_4. For energies between E_2 and E_3 the motion will be bounded on the right (assuming that it starts from the region on the left-hand side of x_6) but not on the left. Finally, if $E \geq E_3$ the motion is unbounded.

Consider the case of total energy $E = E_1$ and the motion bounded on both sides (that is, by x_3 and x_5). The time to complete one round trip from x_3 to x_5 and back again is called the period P. Then simply by putting the limits of integration into Eq. (45) we have

$$P = 2\sqrt{\frac{m}{2}}\int_{x_3}^{x_5} \frac{dx}{\sqrt{E - V(x)}},$$

where we have used the fact that the time to go from x_3 to x_5 is equal to the time to go from x_5 to x_3. This formula is quite general for any oscillatory (or periodic) motion in any potential, the limits on the integral being the roots of the equation

$$E - V(x) = 0.$$

(b) Some Remarks about the Center-of-Mass Theorem

Equation (28), the center-of-mass theorem, is very important and its applications can be both useful and amusing. This is largely because in many systems it is relatively easy to say where the center of mass is located or at least to say how it is accelerating. Consider for instance a closed jar of mass m_J containing a mass m_A of air and a fly of mass m_F, the whole thing standing on a scale in a vacuum (so that we may ignore any buoyant force due to the atmosphere). Of course if the fly is on the wall of the jar, the scale reads the total weight $(m_J + m_A + m_F)g = W$ of the system. But suppose the fly is flying around inside the jar. What then will the scale read?

To answer the question we analyze the forces acting on the system. There is the total gravitational force acting downward and the normal force $\mathbf{N}$ exerted on the jar by the scale acting upward. The scale is constructed to read the force it exerts, so it will read just N. Now if X is the vertical position of the center of mass of the system and M is the total mass W/g, Eq. (28) gives

$$M\ddot{X} = N - W.$$

Thus N depends on the acceleration of the center of mass. We assume that the center of mass of the air remains fixed as does, naturally, that of the jar. It is then clear that if the fly does not accelerate in the vertical direction, $\ddot{X} = 0$ and hence $N = W$. Consequently without analyzing the details of how the weight of the fly is transferred to the scale we may conclude that so long as the fly is moving horizontally or without vertical acceleration, the scale records the entire weight.

Of course, the question of whether the scale takes account of the fly's weight is no different than whether it takes account of the weight of any particular molecule of the air. It is interesting to see in detail how the air, only a very small fraction of which is actually in contact with the jar, exerts a force on the jar and through it on the scale. We shall not, however, go into the details, but shall refer only to averaged out effects, such as pressure. Let the pressure on the bottom of the jar be P_0. Then the force which the air molecules exert on the bottom is σP_0, where σ is the area of the bottom. If the jar is a cylinder of height h, there is an upward force σP_h on the top, where P_h is the pressure at the top. The total vertical force due to the air is then $\sigma(P_h - P_0)$ which, according to the center-of-mass theorem, is the total weight of the air (since the center of mass of the air is not accelerating). Thus we have

$$P_h = P_0 - \frac{m_A g}{\sigma} . \tag{46}$$

From this condition and the equation of state of an ideal gas we can derive the variation in air density and pressure with height. Let $\rho(y)$ be the density of air at a height y from the bottom. Since the total mass is m_A we have

$$m_A = \sigma \int_0^h \rho(y)\, dy. \tag{47}$$

If air is assumed to be an ideal gas, the pressure is related to the density by

$$P = \frac{\rho k T}{m}, \tag{48}$$

where m is the average mass of an air molecule, T is the absolute temperature, and k is Boltzmann's constant. Substituting Eqs. (47) and (48) into Eq. (46), we obtain

$$\rho(h) = \rho(0) - \frac{g}{\sigma} \cdot \frac{m}{kT}\, \sigma \int_0^h \rho(y)\, dy,$$

which when differentiated with respect to h becomes

$$\frac{d\rho}{dh} = -\frac{mg}{kT}\, \rho(h).$$

The solution of this differential equation is

$$\rho = \rho_0 \exp\left\{-\frac{mgh}{kT}\right\}.$$

According to (48) the pressure varies in the same way, so that

$$P = P_0 \exp\{-mgh/kT\}.$$

This is the atmosphere equation, which gives the variation of pressure with height for an isothermal atmosphere.

(c) The Two-Body Problem

The simplest nontrivial system of particles consists of two particles acting on each other and acted on by external forces. Let $\mathbf{F}_1$ and $\mathbf{F}_2$ be the external forces on two masses m_1 and m_2, and let $\mathbf{F}_{12} = -\mathbf{F}_{21}$ be the force that m_2 exerts on m_1. We have seen in Eq. (28) that by adding the two equations of motion we arrive at

$$M\ddot{\mathbf{X}} = \mathbf{F}_1 + \mathbf{F}_2, \tag{49}$$

where $M = m_1 + m_2$, and $\mathbf{X} = (m_1\mathbf{x}_1 + m_2\mathbf{x}_2)/M$ is the center-of-mass position. However this equation tells us little about the detailed motion of the individual masses, and in order to see how they move we need one more equation of motion. Either one of the individual equations

$$\begin{aligned} m_1\ddot{\mathbf{x}}_1 &= \mathbf{F}_1 + \mathbf{F}_{12}, \\ m_2\ddot{\mathbf{x}}_2 &= \mathbf{F}_2 - \mathbf{F}_{12} \end{aligned} \tag{50}$$

will do, but it is usually more convenient to use a linarc combination of them. Multiply the first of Eqs. (50) by m_2/M and the second by m_1/M, and subtract. Since m_1, m_2, and M are always positive, the result is linearly independent of (49); it is

$$\mu\ddot{\mathbf{x}} = \mathbf{F}_{12} + \frac{1}{M}(m_2\mathbf{F}_1 - m_1\mathbf{F}_2). \tag{51}$$

where $\mathbf{x} = \mathbf{x}_1 - \mathbf{x}_2$ is the relative position vector, and $\mu = m_1m_2/M$ is called the *reduced mass*. By going over from Eqs. (50) to (49) and (51) we have restated the problem of finding the motion of a system of two particles in terms of the motion of their center of mass and their relative position.

The restatement we have obtained is particularly convenient in the typical problem for which $\mathbf{F}_{12}$ is a function of $\mathbf{x}$ and t (but not of $\mathbf{x}_1$ and $\mathbf{x}_2$ in any other combination) and in which there are no external forces. Then Eqs. (49) and (51) become

$$\begin{aligned} \ddot{\mathbf{X}} &= 0, \\ \mu\ddot{\mathbf{x}} &= \mathbf{F}_{12}(\mathbf{x}, t). \end{aligned} \tag{52}$$

The first of these equations is solved trivially, for what it tells us is that the center of mass moves with constant velocity. This means, incidentally, that there exists an inertial system in which the center of mass is at rest, and in fact that the center of mass may be chosen as the origin of this system, so that the solution of the first of Eqs. (52) becomes simply $\mathbf{X} = 0$. The second equation looks just like a one-particle equation for a particle of mass μ whose position vector is $\mathbf{x}$, and the entire problem can now be analyzed in one-particle terms. For instance if $\mathbf{F}_{12}$ is a conservative function of $\mathbf{x}$, that is if there exists a function $V(\mathbf{x})$ such that $\mathbf{F}_{12} = -\nabla V$, the total energy

$$E = \tfrac{1}{2}\mu\dot{x}^2 + V(\mathbf{x})$$

of the equivalent one-particle problem is conserved. We have mentioned that more complicated problems can often be reduced to equivalent one-dimensional ones. The procedure outlined here is often the first step in such a reduction.

Note, incidentally, that even if $\mathbf{F}_1$ and $\mathbf{F}_2$ are not zero, we still obtain the second of Eqs. (52) so long as $m_1\mathbf{F}_2 - m_2\mathbf{F}_1 = 0$. This is the case when the external forces are due entirely to a uniform gravitational field. The first of Eqs. (52) will also be relatively simple in this case, for the total external force will be constant. The situation becomes somewhat more complicated if the external gravitational field is nonuniform (see Problem 9).

(d) Changing Mass

Consider an idealized chain piled up at a point next to a hole in a table. Let the chain have mass ρ per unit length. If one end of the chain is allowed to slip through the hole, it will start to fall and pull the rest of the chain through. How does the lower end of the chain move as a function of time? We shall take as the initial conditions $x(0) = 0$, $\dot{x}(0) = 0$. Note that the mass in this problem is changing in just the sense of Eq. (41): there is an external force (gravity) acting on a varying amount of the total mass (the part that has already fallen through the hole). The rest of the mass (the piled up chain, all assumed to be located right at the edge of the hole) moves at constant velocity (zero) until it also is acted on by the external force.

In order to apply (41) to this case, we write $m = \rho x$, where x is the amount of the chain hanging off the table.[6] Then (41) becomes

$$F = \rho g x = \dot{x}(\rho \dot{x}) + \rho x \ddot{x}.$$

[This equation can also be obtained by using (42) with $p = mv = \rho x \dot{x}$.] If we divide through by $\rho \dot{x}$, we obtain

$$\frac{g x^2}{y} = \frac{dy}{dx},$$

where $y = x\dot{x}$. This is easily solved to give

$$\dot{x} = \sqrt{\tfrac{2}{3} g x},$$

in which we have used the condition that $x(0) = \dot{x}(0) = 0$. By taking the derivative with respect to time of this result, we find that the acceleration is constant and equal to $g/3$. It is seen that the mass constantly entering the system causes the acceleration to be less than g. Without the first term of (41), of course, this result would not have been obtained.

[6] Assume the links of the chain to be very short compared to any length entering the problem.

The energy of the system,

$$E = \tfrac{1}{2}m\dot{x}^2 - \tfrac{1}{2}mxg = -\tfrac{1}{6}\rho g x^2$$

is clearly not conserved, decreasing steadily as the chain falls. This is characteristic of problems like this in which each link is added to the system in an inelastic collision with the moving part of the chain. It is well known that if two masses collide and are constrained to move at the same velocity after the collision, then some mechanical energy is converted to thermal energy. Here this process is repeated over and over as each link falls off the table.

Problems

1. The curve $y = y(x)$ in two dimensions has curvature $y''(1 + y'^2)^{-3/2}$. Show that this is equal to $|d\boldsymbol{\tau}/ds|$, where $\boldsymbol{\tau}$ is the unit vector tangent to the curve, and s is the arclength measured from any point on the curve.
2. A charged particle moves in a constant electric field $\mathbf{E}$ which is at right angles to a constant magnetic field $\mathbf{B}$. Find the general motion and the general trajectory.
3. A particle moving in a uniform gravitational field experiences a retarding force $\mathbf{F} = -a\mathbf{v}$ proportional to its velocity vector. Find the general motion and show that if the particle starts from rest it reaches a terminal velocity whose magnitude is mg/a. Discuss the case in which the particle has an initial velocity greater than mg/a.
4. Express the total kinetic energy of a system of N particles in terms of their center of mass $\mathbf{X}$ and their positions $\mathbf{y}_i = \mathbf{x}_i - \mathbf{X}$ relative to the center of mass; that is, derive Eq. (31). Extend the result to a continuous system by using $\mathbf{x}$ as parameter, as in the derivation of (39).
5. Show explicitly that if each of the internal forces $\mathbf{F}_{ij}$ in an N-body system lies along the line connecting the ith and jth particles, the internal forces produce no net torque about a fixed point.
6. A particle of mass m_1 makes an elastic (energy conserving) collision with another particle of mass m_2. Before the collision m_1 has velocity $\mathbf{v}_1$ and m_2 is at rest relative to a certain inertial frame which we shall call the *laboratory system.* After the collision m_1 has velocity $\mathbf{u}_1$ making an angle θ with $\mathbf{v}_1$.
   (a) Find the magnitude of $\mathbf{u}_1$.
   (b) Relative to another inertial frame, called the *center-of-mass system,* the total linear momentum of the two-body system is zero. Find the velocity of the center-of-mass system relative to the laboratory system.
   (c) Find the velocities $\mathbf{v}_1'$, $\mathbf{v}_2'$, $\mathbf{u}_1'$, and $\mathbf{u}_2'$ of the two bodies before and after the collision in the center-of-mass system. Find the scattering angle θ' (the angle between $\mathbf{v}_1'$ and $\mathbf{u}_1'$) in terms of θ.
7. A shell of mass M is fired with an initial velocity whose horizontal and vertical components are v_1 and v_2. At the highest point of its trajectory the shell explodes into exactly two fragments of mass m_1 and m_2. The

explosion provides additional kinetic energy T, and the fragments separate initially in the horizontal direction. By using theorems on the motion of the center of mass, show that the two fragments strike the ground separated by the distance

$$\frac{v_2}{g}\left[2T\left(\frac{1}{m_1}+\frac{1}{m_2}\right)\right]^{1/2}.$$

8. Two masses m_1 and m_2 in a uniform gravitational field are connected by a spring of unstretched length b and spring constant k. The system is held by m_1 so that m_2 hangs down vertically, stretching the spring. At $t = 0$ both m_1 and m_2 are at rest and in equilibrium, and m_1 is released, so that the system starts to fall. Set up a suitable coordinate system and describe the subsequent motion of m_1 and m_2.
9. The Earth and the Moon form a two-body system interacting through their mutual gravitational attraction. In addition, each body is attracted by the gravitational field of the Sun, which in the sense of Section 5c is an external force. Take the Sun as the origin and write down the equations of motion for the center of mass $\mathbf{X}$ and the relative position $\mathbf{x}$ of the Earth-Moon system. Expand the resulting expressions in powers of x/X, the ratio of the magnitudes. Show that to lowest order in x/X the center of mass and relative position are uncoupled, but that in higher orders they are coupled because the Sun's gravitational force is not constant.
10. Show that a one-dimensional particle subject to the force $F = -kx^{2n+1}$, where n is an integer, will oscillate with a period proportional to A^{-n}, where A is the amplitude. Pay special attention to $n \leq 0$.
11. A particle of mass m moves along the x axis. Its potential energy at any point x is

$$V(x) = V_0x^2e^{-x^2},$$

where V_0 is a constant.

(a) What is the force on the particle?

(b) Find all the points on the x axis where the particle can be in equilibrium, that is, where the force is zero. Determine the value of the potential at each of these points and determine whether the equilibrium is stable or unstable, that is, whether in its immediate neighborhood the force is toward or away from the equilibrium point. Make a rough sketch of the potential and label the equilibrium points.

(c) Determine the maximum total energy E_0 which the particle can have and still execute bounded (i.e., periodic) motion. If $E < E_0$ is the motion necessarily bounded? Explain.

12. A particle of mass m moving in one dimension is subject to the force

$$F = -kx + \frac{a}{x^3}.$$

(a) Find the potential and make a sketch of it.

(b) Find the positions of equilibrium for the potential and show that the equilibria are stable.
(c) Find the angular frequency $\omega = 2\pi/P$ and period P for small oscillations about one of the equilibrium positions.
(d) For arbitrary E, find the values of the two turning points x_1 and x_2.
(e) From the energy integral and the values of x_1 and x_2 found in (d), obtain the period for arbitrary oscillations. Show that the period for arbitrary oscillations is equal to the period for small oscillations (see Chapter IV, Section 5b).

13. Consider a chain of finite individual links, each of length ε and mass $m = \mu\varepsilon$. The chain is piled up next to a hole on the table with n of the links hanging through and initially at rest. These start to fall and pull the rest of the chain through, each link starting the next one in a completely inelastic collision. Find the velocity of the end of the chain after k links have fallen through the hole. Show that in the limit as $\varepsilon \to 0$ this result goes over to the case discussed in Section 5(d) (but with different initial conditions).
14. For the following three fields of force draw the *lines of force* [Slater (1947), p. 50] and discuss whether or not a potential energy function exists. In all cases a is a constant.

(a) $F_1 = ax_2, \qquad F_2 = F_3 = 0.$

(b) $$F_\alpha = \frac{ax_\alpha}{x^3}, \qquad x = \sqrt{x_1^2 + x_2^2 + x_3^2}, \qquad \alpha = 1, 2, 3.$$

(c) $$F_1 = \frac{ax_2}{x^2}, \qquad F_2 = -\frac{ax_1}{x^2}, \qquad F_3 = 0.$$

15. A simple pendulum (mass m, length l) hangs vertically at rest in a uniform gravitational field. Its support is suddenly subjected to a constant acceleration **a** upward at a small angle α from the vertical. Describe the subsequent motion.

CHAPTER II

Lagrangian Formulation of Mechanics

The first chapter was intended as a review of the prerequisites for the rest of the book. In this second chapter we rewrite the general equations of motion in the Lagrangian form and then derive these equations from the variational principle. The resulting formulation is more useful for systems other than the simplest, and it has the additional advantage of logical conciseness. With its aid one can relate conservation laws to fundamental physical symmetries (Chapter III). Furthermore, it can be generalized to a large number of physical theories.

II-1 CONSTRAINTS

In the next two sections we shall put the equations of motion of a system of particles in a form which has become standard and which simplifies the analysis and solution in many special cases. In this form the equations are called Lagrange's equations. Before actually deriving them, however, we shall discuss motions which are restricted by certain conditions placed on them, for it is for such motion that the usefulness of Lagrange's equations first becomes apparent.

In a mechanical problem the motion of a system is often constrained by external agents applying forces that are initially unknown, although their effect on the motion is known. For instance, in the case of a body sliding on a curved surface under the influence of gravity, the gravitational force is known initially, but the force of the surface on the body is not. Nevertheless we know the effect of the surface, or rather of its force: it constrains the body to move on the specified two-dimensional surface. The force itself can be found in general only after the motion of the body has been determined.

There are several types of constraints. *Holonomic* constraints are constraints that can be expressed as K equations ($K < 3N$) of the form

$$f_I(\mathbf{x}_1, \ldots, \mathbf{x}_N, t) = 0, \qquad (I = 1, \ldots, K), \tag{1}$$

called the equations of constraint, connecting the position vectors of the N particles under consideration. The f_I are assumed to be continuous, differentiable functions of their variables, and their explicit time dependence allows for the possibility that the constraints vary with time in a known way, independent of the motion of the particles. In the case of a single particle constrained to move on a surface, for instance, the equation of constraint

$$f(\mathbf{x}, t) = 0 \tag{2}$$

is the equation of the surface at time t, and the time dependence allows for the possibility that the position or shape of the surface may vary with time. Furthermore, as we shall see in Chapter IV, we must assume that the constraints are stated in such a way that the equations $f_I = 0$ and $df_I/dt = 0$ impose $2K$ restrictions on the motion.[1] This can be put in a more precise mathematical way, but we leave that to Chapter IV.

General velocity-dependent constraints are constraints that can be expressed as K equations of the form

$$f_I(\mathbf{x}_1, \ldots, \mathbf{x}_N; \dot{\mathbf{x}}_1, \ldots, \dot{\mathbf{x}}_N, t) = 0, \qquad (I = 1, \ldots, K) \tag{3}$$

connecting both the positions and velocities of the particles. We shall discuss these in Chapter IV. Holonomic constraints are clearly a special case of general velocity-dependent constraints. Another special case is integrable velocity dependent constraints, which are a set of first-order differential equations in the $\mathbf{x}$'s that can, by integration, be put in holonomic form.

There are others, which do not fit into the class of general velocity-dependent constraints. For example, it may be necessary to express the constraints as K inequalities of the form

$$f_I(\mathbf{x}_1, \ldots, \mathbf{x}_N, t) > 0 \qquad (I = 1, \ldots, K)$$

as in the case of particles constrained to move inside a fixed region. There are no general methods for handling such constraints and so we shall restrict our discussion in this book to general velocity-dependent ones. In this chapter we treat only holonomic constraints.

Let us start this discussion by studying the motion of a particle restricted to move on the surface given by Eq. (2). The equation of motion of the particle is

$$m\ddot{\mathbf{x}} = \mathbf{F} + \mathbf{N}, \tag{4}$$

where the external applied force $\mathbf{F}$ is a known function of $\mathbf{x}$ and t (and perhaps of $\dot{\mathbf{x}}$), and $\mathbf{N}$ is the unknown force of constraint which the surface exerts on

[1] For instance a particle may be constrained to a plane either by the equation $\mathbf{a} \cdot \mathbf{x} = 0$ or by $(\mathbf{a} \cdot \mathbf{x})^2 = 0$, where $\mathbf{a}$ is a fixed vector. The time derivative of the first equation is $\mathbf{a} \cdot \dot{\mathbf{x}} = 0$, which restricts the velocity. The time derivative of the second is $2(\mathbf{a} \cdot \mathbf{x})(\mathbf{a} \cdot \dot{\mathbf{x}}) = 0$, which does not (since $\mathbf{a} \cdot \mathbf{x}$ is already known to vanish). Thus $(\mathbf{a} \cdot \mathbf{x})^2 = 0$ is not an acceptable holonomic constraint equation.

the particle. Now Eqs. (2) and (4) are only four equations in the six unknown components of $\mathbf{x}$ and $\mathbf{N}$, and so will not determine the motion. This can be seen physically from the fact the $\mathbf{N}$ can have any component parallel to the surface without altering the restriction implied by Eq. (2), that is, without causing the particle to move away from the surface.

In order to proceed, therefore, we restrict $\mathbf{N}$ to be normal to the surface. (We discuss the implications of this restriction later.) To write this analytically we use the fact that $\nabla f(\mathbf{x}, t)$ is also normal to the surface, so that

$$\mathbf{N} = \lambda \nabla f(\mathbf{x}, t), \tag{5}$$

where λ, a scalar function of t which depends on the motion, is still to be determined. The equations of motion now can be written

$$m\ddot{\mathbf{x}} = \mathbf{F} + \lambda \nabla f, \tag{6}$$

$$f(\mathbf{x}, t) = 0. \tag{7}$$

These are now four equations in four unknown functions of the time: $\lambda(t)$ and three components of $\mathbf{x}(t)$.

To solve (6) and (7) it would be convenient to eliminate $\lambda(t)$. This is easily done, since $\lambda \nabla f$ is normal to the surface, by taking only those components of Eq. (6) that are tangent to the surface. Let us therefore take the cross product of (6) with ∇f. Then we have

$$(m\ddot{\mathbf{x}} - \mathbf{F}) \times \nabla f = 0. \tag{8}$$

This equation tells us that $m\ddot{\mathbf{x}} - \mathbf{F}$ is perpendicular to the surface (parallel to ∇f), which only defines its direction. Thus it gives only two linearly independent relations among the three components of $m\ddot{\mathbf{x}} - \mathbf{F}$. The third equation needed to solve the problem is (7).

A second way to eliminate $\lambda(t)$ is the following. Let $\boldsymbol{\tau}$ be an arbitrary vector tangent to the surface at $\mathbf{x}$ at time t, i.e., restricted only by the condition

$$\boldsymbol{\tau} \cdot \nabla f = 0. \tag{9}$$

This is a single relation between the three components of $\boldsymbol{\tau}$, so that only two are independent. For instance τ_3 can be written in terms of τ_1, τ_2, and ∇f. The dot product of $\boldsymbol{\tau}$ with (6) gives

$$(m\ddot{\mathbf{x}} - \mathbf{F}) \cdot \boldsymbol{\tau} = 0, \tag{10}$$

which can be shown to be equivalent to (8) (indeed, it also says that $m\ddot{\mathbf{x}} - \mathbf{F}$ is perpendicular to the surface). Although this looks like only one equation, it is actually two, for two components of $\boldsymbol{\tau}$ can be chosen arbitrarily. If, as above, τ_1 and τ_2 are taken as the independent components, for instance, and τ_3 is expressed in terms of them, the coefficients of τ_1 and τ_2 must vanish separately in (10). Again a third equation is needed, and again (7) serves this purpose. After solving these equations to obtain $\mathbf{x}(t)$, one can use Eq. (5) to find $\mathbf{N}$ if the constraint force is desired. This second method is particularly useful when we try to generalize these considerations to a system of several particles.

Before generalizing these considerations to a system of N particles, let us consider the implications of Eq. (5) more fully. We assumed that $\mathbf{N}$ is perpendicular to the $f = 0$ surface in order to define the problem more specifically. This, as we shall now show, implies something about the energy of the system. Let us assume that $\mathbf{F}$ is derivable from a potential, so that $\mathbf{F} = -\nabla V(\mathbf{x}, t)$. Then

$$m\ddot{\mathbf{x}} \cdot \dot{\mathbf{x}} = \frac{d}{dt}(\tfrac{1}{2}m\dot{x}^2) = \mathbf{F} \cdot \dot{\mathbf{x}} + \lambda \nabla f \cdot \dot{\mathbf{x}}$$

$$= -\nabla V \cdot \dot{\mathbf{x}} + \lambda \nabla f \cdot \dot{\mathbf{x}}. \tag{11}$$

Now let $\mathbf{x}(t)$ be a solution of the equations of motion. Then since the particle remains on the surface, $f(\mathbf{x}(t), t)$ remains zero, and[2]

$$\frac{df}{dt} = \nabla f \cdot \dot{\mathbf{x}} + \frac{\partial f}{\partial t} = 0. \tag{12}$$

Also, we have

$$\frac{dV}{dt} = \nabla V \cdot \dot{\mathbf{x}} + \frac{\partial V}{\partial t}.$$

Inserting these into (11), we arrive at

$$\frac{dE}{dt} = \frac{d}{dt}[\tfrac{1}{2}m\dot{x}^2 + V] = \frac{\partial V}{\partial t} - \lambda\frac{\partial f}{\partial t}.$$

In other words, the total energy of the particle changes only if V and f are explicit functions of the time. If V is time independent, which is often the case,

$$\frac{dE}{dt} = -\lambda\frac{\partial f}{\partial t},$$

[2] We emphasize that the total time derivative has meaning only when applied to a function of t alone. Thus if we have some function $F(q, t)$, we must know $q(t)$ in order to calculate dF/dt. Then

$$\frac{d}{dt}F(q(t), t) = \frac{\partial F}{\partial q}\dot{q} + \frac{\partial F}{\partial t}.$$

Sometimes, however, this equation can be used to define a function of $q, \dot{q}, t$ in terms of $F(q, t)$. Let us call this the *formal derivative* of F with respect to t. It has the property that if some $q(t)$ is inserted into $F(q, t)$, the formal derivative becomes the actual derivative. Clearly similar remarks apply to higher derivatives and to functions of the form $F(q, \dot{q}, t)$.

or so long as **N** is perpendicular to the surface, the only change in the total energy comes from changes of the surface.[3]

This is not difficult to understand physically. If the surface is stationary, $\dot{\mathbf{x}}$ is always tangent to it, and $\mathbf{N} \cdot \dot{\mathbf{x}}$ or the rate at which **N** does work on the particle is zero. If the surface moves, however, $\dot{\mathbf{x}}$ need not be tangent to it, and $\mathbf{N} \cdot \dot{\mathbf{x}}$ need not be zero (see Fig. 6). All this involves the assumption that

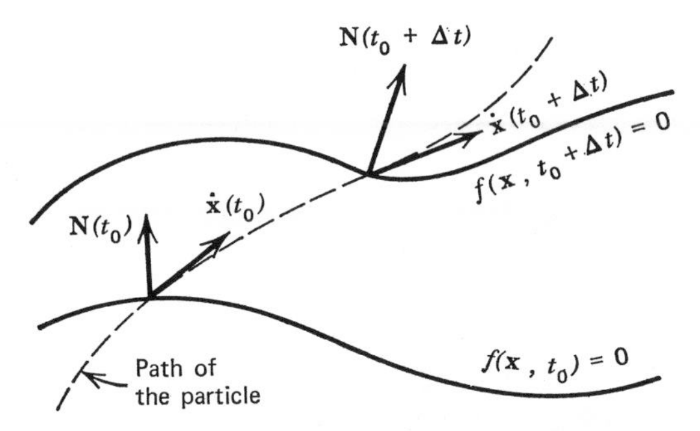

Figure 6. Motion with a time-dependent constraint. The constraint force **N** is always perpendicular to the constraint curve, but it need not be perpendicular to the particle path.

N has no component tangent to the $f = 0$ surface. For real surfaces **N** often has such components, ordinarily frictional in nature and causing dissipation of energy. Our surfaces must be more than just frictionless, however, since we also exclude nondissipative tangential forces. Surfaces which satisfy our assumption we call *smooth*.

Let us now generalize these considerations to a system of N particles with K independent holonomic equations of constraint given by Eq. (1). The equation of motion of the ith particle is

$$m_i \ddot{\mathbf{x}}_i = \mathbf{F}_i + \mathbf{N}_i, \tag{13}$$

where the applied force $\mathbf{F}_i$ is a known function of $\mathbf{x}_1, \ldots, \mathbf{x}_N, t$, and where $\mathbf{N}_i$ is the constraint force. The constraint equations do not specify the $\mathbf{N}_i$ uniquely, however, and so, as before, we add the assumption of smoothness, namely, the generalization of Eq. (5)

$$\mathbf{N}_i = \sum_{I=1}^{K} \lambda_I \nabla_i f_I, \tag{14}$$

[3] It is interesting that the converse is not true: $\partial f/\partial t = \alpha(dE/dt)$ does not imply that **N** must be normal to the surface, since **N** may deviate from the normal by any amount always perpendicular to $\dot{\mathbf{x}}$.

where ∇_i is the gradient with respect to $\mathbf{x}_i$, and the $\lambda_I(t)$ $(I = 1, \ldots, K)$ are K functions to be determined. (See Problem 1, which discusses this for a particle constrained to a curve.)

To compare this with the one-particle case let us calculate the rate at which the energy changes. We assume that each $\mathbf{F}_i$ is given in terms of some time-independent potential $V(\mathbf{x}_1, \ldots, \mathbf{x}_N)$ (though the restriction to time independence is not essential) by $\mathbf{F}_i = -\nabla_i V$. Then the rate of change of the energy is

$$\begin{aligned}\frac{dE}{dt} &= \frac{d}{dt}\left[\sum_{i=1}^{N} \tfrac{1}{2} m_i \dot{x}_i^2 + V\right] \\ &= \sum_i [m_i \ddot{\mathbf{x}}_i + \nabla_i V] \cdot \dot{\mathbf{x}}_i.\end{aligned}$$

Using (13) and (14), we obtain

$$\begin{aligned}\frac{dE}{dt} &= \sum_i [m_i \ddot{\mathbf{x}}_i - \mathbf{F}_i] \cdot \dot{\mathbf{x}}_i \\ &= \sum_i \mathbf{N}_i \cdot \dot{\mathbf{x}}_i \\ &= \sum_{i=1}^{N} \sum_{I=1}^{K} \lambda_I \nabla_i f_I \cdot \dot{\mathbf{x}}_i.\end{aligned}$$

Since the $\mathbf{x}_i$ must satisfy the equations of constraint at all times we have

$$\frac{df_I}{dt} = \sum_i \nabla_i f_I \cdot \dot{\mathbf{x}}_i + \frac{\partial f_I}{\partial t} = 0, \tag{15}$$

so that

$$\frac{dE}{dt} = -\sum_I \lambda_I \frac{\partial f_I}{\partial t}.$$

Thus as in the one-particle case, so long as V is time independent, E changes only if the f_I are explicit functions of the time, and so Eq. (14) implies that if the constraints are time independent, the forces of constraint do no work. In this case also the converse is not true: nonsmooth constraints can conserve energy. In other words, the existence of λ_I such that $\sum_I \lambda_I(\partial f_I/\partial t) = 0$ does not imply Eq. (14).

We can now proceed to obtain the equations of motion for the $\mathbf{x}_i$ from (13) and (14) in much the same way as we did by the second method [Eqs. (9) and (10)] in the one-particle case. For given $\mathbf{x}_i$ and t consider N arbitrary vectors $\boldsymbol{\tau}_i$ restricted only by the K conditions

$$\sum_{i=1}^{N} \nabla_i f_I \cdot \boldsymbol{\tau}_i = 0, \qquad (I = 1, \ldots, K). \tag{16}$$

We shall call these $\boldsymbol{\tau}_i$ *generalized tangent vectors.* Equation (16) gives K relations between the $3N$ components of the $\boldsymbol{\tau}_i$, so that only $3N - K$ of the components are independent. The dot product of $\boldsymbol{\tau}_i$ with the ith of equations (13), summed over i, gives, when we take account of (14) and (16),

$$\sum_i (m_i \ddot{\mathbf{x}}_i - \mathbf{F}_i) \cdot \boldsymbol{\tau}_i = 0. \tag{17}$$

Because there are $3N - K$ independent arbitrary components of the $\boldsymbol{\tau}_i$, this equation splits into $3N - K$ independent equation in the components of the $\mathbf{x}_i$: the coefficients of the $3N - K$ independent components of the $\boldsymbol{\tau}_i$ can be separately equated to zero once the dependent components have been eliminated using (16). Another K equations are needed to solve the problem, and Eqs. (1) serve this purpose. On solving the resulting set of $3N$ equations for the $\mathbf{x}_i(t)$, one can use (13) to find the $\mathbf{N}_i$ if the constraint forces are desired.

Equation (17) is sometimes called D'Alembert's principle. How to deal with it is not immediately evident, for to go from (16) to (17) is quite complicated. The problem is to find suitable $\boldsymbol{\tau}_i$ which will simplify the equations, and this is what we shall discuss in the next section. In any case, time-independent constraints are easily handled, for we may pick the $\boldsymbol{\tau}_i$ to be the $\dot{\mathbf{x}}_i$. Indeed, Eq. (15) shows that if $\partial f_I/\partial t = 0$ then $\sum_i \nabla_i f_I \cdot \dot{\mathbf{x}}_i = 0$ as required by (16). We shall see in the next section how to simplify the problem by choosing coordinates in which the constraint equations become trivial.

II-2 GENERALIZED COORDINATES: LAGRANGE'S EQUATIONS

In order to simplify the problem it is helpful to gain freedom by removing the restriction that we deal with Cartesian coordinate systems only. Let us therefore pick new *generalized coordinates* q_α, $\alpha = 1, \ldots, 3N$, such that the q_α can be expressed in terms of the components of the $\mathbf{x}_i$ and vice versa. In other words there exist relations of the form

$$\begin{aligned} q_\alpha &= q_\alpha(\mathbf{x}_1, \ldots, \mathbf{x}_N, t), \\ \mathbf{x}_i &= \mathbf{x}_i(q_1, \ldots, q_{3N}, t). \end{aligned} \tag{18}$$

The invertibility of these equations means that the Jacobian of the q_α with respect to the components of the $\mathbf{x}_i$ (and vice versa) must be nonzero. Also, we assume that these functions are continuous and twice continuously differentiable. Equations (18) can then be used to write the equations of motion in terms of the q_α. The point is to pick the q_α so that the equations of constraint are eliminated and the equations of motion become easier to solve in their new form. It is then easy to use (18) again to restate the solution, if desired, in terms of the $\mathbf{x}_i$.

When there are K independent constraint equations of the form of Eq. (1), it is particularly convenient to pick K of the q_α (we choose the last K) to depend on the $\mathbf{x}_i$ only through the f_I, that is to be functions only of the f_I. Thus we shall write

$$q_{n+I} = R_I(f_1, \ldots, f_K), \qquad I = 1, \ldots, K, \tag{19}$$

where $n = 3N - K$ is the number of *degrees of freedom* of the system. Often the R_I are just chosen to be the f_I themselves, but sometimes it is more convenient to choose them more generally, depending on how the constraints are written (for there is in general more than one way to write a constraint condition). Actually, in practice Eqs. (19) are seldom written out. Note in any case that if the q_{n+I} are to be independent, the Jacobian of the R_I with respect to the f_I must be nonzero, and therefore Eqs. (19) can be inverted to yield

$$f_I = f_I(q_{n+1}, \ldots, q_{n+K}). \tag{20}$$

With the choice of (19), the constraint equations (1) become

$$q_{n+I} = R_I(0, \ldots, 0); \tag{21}$$

that is, the last K of the q_α are constants independent of the time. This means that we have found the time dependence of these q_α. Having eliminated the problem of finding how these q_α depend on the time, we are left with a problem in n variables.

Thus we must now find the equations for the first n of the q_α. These are given implicitly by Eqs. (16) and (17). Recall that it was shown that as a consequence of (16), Eq. (17) is in reality a set of $n = 3N - K$ independent equations. The problem is now to choose the $\boldsymbol{\tau}_i$ so as to exhibit explicitly the n equations of motion for the q_α, $\alpha = 1, \ldots, n$.

Before proceeding it is convenient to introduce the *summation convention.* According to this convention, whenever we have a term with repeated indices, such as the i in $\mathbf{F}_i \cdot \boldsymbol{\tau}_i$ or the α in $\varepsilon_\alpha(\partial \mathbf{x}_i/\partial q_\alpha)$, the term is to be summed over the entire range of each repeated index. What this range is depends, of course, on the context. In the present discussion Latin indices such as i run from 1 to N and Greek indices such as α run from 1 to n. Thus, using the summation convention we have

$$\mathbf{F}_i \cdot \boldsymbol{\tau}_i \equiv \sum_{i=1}^{N} \mathbf{F}_i \cdot \boldsymbol{\tau}_i$$

and

$$\varepsilon_\alpha \frac{\partial \mathbf{x}_i}{\partial q_\alpha} \equiv \sum_{\alpha=1}^{n} \varepsilon_\alpha \frac{\partial \mathbf{x}_i}{\partial q_\alpha}.$$

It is seen that this is merely a shorthand which allows us to leave off the summation sign. Occasionally we shall not want to sum over a repeated index. In such cases we shall put one of the indices in parentheses or state that no sum is implied, or even perhaps do both. Thus in the expression $\lambda_{(I)}f_I$, for instance, no sum is implied over I. In addition, we shall occasionally have a triple index. To indicate a sum in this case we shall put *one* of the indices in parentheses. Thus

$$m_{(i)}\ddot{\mathbf{x}}_i \cdot \boldsymbol{\tau}_i \equiv \sum_{i=1}^{N} m_i\ddot{\mathbf{x}}_i \cdot \boldsymbol{\tau}_i.$$

We shall have to be quite careful to have only double indices when we want to imply summation, but it turns out that this is not difficult. Triple indices that are not summed and other special cases will be treated specially.

We now make the following assertion. Let ε_α, $\alpha = 1, \ldots, n$ be any set of n constants. Then the vectors

$$\boldsymbol{\tau}_i = \varepsilon_\alpha \frac{\partial \mathbf{x}_i}{\partial q_\alpha} \tag{22}$$

will satisfy (16). To prove this, we merely insert Eq. (22) into (16), obtaining

$$\nabla_i f_I \cdot \boldsymbol{\tau}_i = \varepsilon_\alpha \nabla_i f_I \cdot \frac{\partial \mathbf{x}_i}{\partial q_\alpha}.$$

(There is a sum over both i and α in this last expression.) Now from the rules of partial differentiation

$$\nabla_i f_I \cdot \frac{\partial \mathbf{x}_i}{\partial q_\alpha} \equiv \sum_{i=1}^{N} \nabla_i f_I \cdot \frac{\partial \mathbf{x}_i}{\partial q_\alpha} = \frac{\partial f_I}{\partial q_\alpha},$$

so that

$$\nabla_i f_I \cdot \boldsymbol{\tau}_i = \varepsilon_\alpha \frac{\partial f_I}{\partial q_\alpha} = 0.$$

The last equality follows because α runs only up to n, but not through $n + I$ $(I \geq 1)$, and the f_I depend, according to (20), only on the q_{n+I}. Thus the assertion is proven.

Let us now insert (22) into (17) to obtain the n equations of motion. We have

$$\varepsilon_\alpha(m_{(i)}\ddot{\mathbf{x}}_i - \mathbf{F}_i) \cdot \frac{\partial \mathbf{x}_i}{\partial q_\alpha} = 0.$$

Since the ε_α are arbitrary constants, the coefficient of each must be zero, and we may thus write

$$(m_{(i)}\ddot{\mathbf{x}}_i - \mathbf{F}_i) \cdot \frac{\partial \mathbf{x}_i}{\partial q_\alpha} = 0, \qquad \alpha = 1, \ldots, n. \tag{23}$$

When the $\mathbf{x}_i$ are written as functions of the q_α in accordance with Eq. (18), we obtain n equations of motion for the first n of the $q_\alpha(t)$. Their solution together with (21) is then a solution of the problem.

Let us then write (23) entirely in terms of the q_α. For this purpose we note that

$$m_{(i)}\ddot{\mathbf{x}}_i \cdot \frac{\partial \mathbf{x}_i}{\partial q_\alpha} = \frac{d}{dt}\left[m_{(i)}\dot{\mathbf{x}}_i \cdot \frac{\partial \mathbf{x}_i}{\partial q_\alpha}\right] - m_{(i)}\dot{\mathbf{x}}_i \cdot \frac{d}{dt}\frac{\partial \mathbf{x}_i}{\partial q_\alpha}. \tag{24}$$

Now we have

$$\mathbf{v}_i = \dot{\mathbf{x}}_i = \frac{d\mathbf{x}_i}{dt} = \frac{\partial \mathbf{x}_i}{\partial q_\alpha}\dot{q}_\alpha + \frac{\partial \mathbf{x}_i}{\partial t}, \tag{25}$$

so that

$$\frac{\partial \mathbf{v}_i}{\partial \dot{q}_\alpha} = \frac{\partial \mathbf{x}_i}{\partial q_\alpha}.$$

Furthermore,

$$\begin{aligned} \frac{d}{dt}\frac{\partial \mathbf{x}_i}{\partial q_\alpha} &= \frac{\partial^2 \mathbf{x}_i}{\partial q_\alpha \partial q_\beta}\dot{q}_\beta + \frac{\partial}{\partial t}\left(\frac{\partial \mathbf{x}_i}{\partial q_\alpha}\right) \\ &= \frac{\partial}{\partial q_\alpha}\left(\frac{\partial \mathbf{x}_i}{\partial q_\beta}\dot{q}_\beta + \frac{\partial \mathbf{x}_i}{\partial t}\right) = \frac{\partial \mathbf{v}_i}{\partial q_\alpha}. \end{aligned} \tag{25a}$$

By using (25) and (25*a*) we find that Eq. (24) can be written

$$\begin{aligned} m_{(i)}\ddot{\mathbf{x}}_i \cdot \frac{\partial \mathbf{x}_i}{\partial q_\alpha} &= \frac{d}{dt}\left[m_{(i)}\mathbf{v}_i \cdot \frac{\partial \mathbf{v}_i}{\partial \dot{q}_\alpha}\right] - m_{(i)}\mathbf{v}_i \cdot \frac{\partial \mathbf{v}_i}{\partial q_\alpha} \\ &= \frac{d}{dt}\frac{\partial T}{\partial \dot{q}_\alpha} - \frac{\partial T}{\partial q_\alpha}, \end{aligned}$$

where

$$T = \tfrac{1}{2}m_i v_i^2$$

is the total kinetic energy of the system. Then Eq. (23) becomes

$$\frac{d}{dt}\frac{\partial T}{\partial \dot{q}_\alpha} - \frac{\partial T}{\partial q_\alpha} - \mathbf{F}_i \cdot \frac{\partial \mathbf{x}_i}{\partial q_\alpha} = 0. \tag{26}$$

This is completely general, though it does not express the equations of motion in terms of the q_α alone. However if all the $\mathbf{F}_i$ are conservative, so that $\mathbf{F}_i = -\nabla_i V$, we can write

$$\mathbf{F}_i \cdot \frac{\partial \mathbf{x}_i}{\partial q_\alpha} = -\nabla_i V \cdot \frac{\partial \mathbf{x}_i}{\partial q_\alpha} = -\frac{\partial V}{\partial q_\alpha}$$

and so (26) becomes

$$\frac{d}{dt}\frac{\partial T}{\partial q_\alpha} - \frac{\partial T}{\partial q_\alpha} + \frac{\partial V}{\partial q_\alpha} = 0.$$

Since V is a function of the $\mathbf{x}_i$ and therefore only of the q_α and not the $\dot{q}_\alpha$, we have $\partial V/\partial \dot{q}_\alpha = 0$. Let us now define the *Lagrangian function* L of the system as

$$L = T - V, \tag{27}$$

and then in terms of this function the equations of motion become

$$\frac{d}{dt}\frac{\partial L}{\partial \dot{q}_\alpha} - \frac{\partial L}{\partial q_\alpha} = 0. \tag{28}$$

These are called *Lagrange's equations.* In them, the equations of motion are written entirely in terms of generalized coordinates, assuming, that is, that L is written in terms of the q_α and $\dot{q}_\alpha$. They hold also if there are no constraints, that is, if $K = 0$ and $n = 3N$.

Now these equations have been derived from the basic principles (from Newton's Laws) with L given by (27). On the other hand, Eq. (28) could be taken as fundamental. (Actually, in the next section we shall state mechanics in terms of the Lagrangian in still another way which we may also take as fundamental.) That is, we may assume that there exists a function $L(q, \dot{q}, t)$ of the q_α, $\dot{q}_\alpha$, and t such that Eqs. (28) are the correct equations of motion,[4] rather than assume that there exist vector functions $\mathbf{F}_i$ of the $\mathbf{x}_i$ such that $\mathbf{F}_i = m_{(i)}\ddot{\mathbf{x}}_i$ are the correct equations of motion. In either case the functions must be found before one can solve a problem (in fact in the Lagrangian formulation only one function is needed). In most cases these functions are known a priori, and when they are not, L is as easy to find as the $\mathbf{F}_i$.

Remarks. 1. We will see in Chapter III that L is greatly restricted by certain symmetry conditions, and so may be easier to find than the $\mathbf{F}_i$. *2.* One of the most important Lagrangians, that for a charged particle in an electromagnetic field, is not just $T - V$, where V is a scalar potential (see Section 4). *3.* One of the important properties of Lagrange's equations is that they are in a certain sense coordinate-system independent (see Problem 2). That is, Eq. (28) will look the same no matter what generalized coordinates are represented by the q_α.

In general the equations of motion are second-order differential equations for the n functions $q_\alpha(t)$, and thus involve $2n$ constants of integration. These constants are related to initial conditions imposed on the problem, such as the initial positions and velocities, or, say, the total energy or certain momenta. They may be determined in more complicated ways also, as for instance by giving two points on the trajectory.

It may be trivial, but is important to note that (28) is in no sense a set of differential equations for L. The Lagrangian is a given function with the

[4] In $L(q, \dot{q}, t)$ and other functions we will write q for the collection of q_α, and $\dot{q}$ for the collection of $\dot{q}_\alpha$. This convention will be used throughout.

aid of which we find the $q_\alpha(t)$ by Eqs. (28). In other words, (28) is an expression involving the $q_\alpha(t)$ and their derivatives [a *functional* of the $q_\alpha(t)$], which, when set equal to zero, becomes an equation for certain particular functions $q_\alpha(t)$. These define the possible physical paths. No other functions $q_\alpha(t)$ will cause the functional in (28) to vanish.

We wish to prove just one theorem concerning the form of the Lagrangian:

THEOREM. Let $L(q, \dot{q}, t)$ and $L'(q, \dot{q}, t)$ be two Lagrangians such that the equations of motion obtained from L are *exactly the same* as those obtained from L'. Then L and L' differ by the (formal) total time derivative of some function of the form $\Phi(q, t)$.

PROOF. We shall prove the theorem in only one dimension, leaving to Problem 8 the proof for n dimensions.

Let us write Lagrange's equations in the form

$$\frac{d}{dt}\frac{\partial L}{\partial \dot{q}} - \frac{\partial L}{\partial q} = \Lambda(\ddot{q}, \dot{q}, q, t) = 0,$$

$$\frac{d}{dt}\frac{\partial L'}{\partial \dot{q}} - \frac{\partial L'}{\partial q} = \Lambda'(\ddot{q}, \dot{q}, q, t) = 0.$$

Then to say that these two equations are *exactly the same* is to say that

$$\Lambda = \Lambda'$$

or that Λ' is the same function of $\ddot{q}$, $\dot{q}$, q, t as is Λ.

In detail this equation may be written

$$\Lambda - \Lambda' = \frac{\partial^2 \psi}{\partial \dot{q}^2}\ddot{q} + \frac{\partial^2 \psi}{\partial q\,\partial \dot{q}}\dot{q} + \frac{\partial^2 \psi}{\partial \dot{q}\,\partial t} - \frac{\partial \psi}{\partial q} = 0, \tag{29}$$

where $\psi = L - L'$. Since this is a functional equation, that is, true for all values of $\ddot{q}$, $\dot{q}$, q, t, and since L and L', and thus also ψ, are not functions of $\ddot{q}$, the coefficient of $\ddot{q}$ must vanish. Thus

$$\frac{\partial^2 \psi}{\partial \dot{q}^2} = 0,$$

which may be integrated to yield

$$\psi = \dot{q}F(q, t) + G(q, t).$$

Then (29) becomes

$$\frac{\partial F}{\partial t} - \frac{\partial G}{\partial q} = 0.$$

But this is just the condition[5] that there exists a function $\Phi(q, t)$ such that

[5] A fast way to see this is to think of q and t as Cartesian x and y coordinates in two dimensions, and of F and G as the x and y components of some vector **V**. Then what we have is $\nabla \times \mathbf{V} = 0$, or $\mathbf{V} = \nabla\Phi$, which is the desired result.

$F = \partial\Phi/\partial q$ and $G = \partial\Phi/\partial t$. Then, as asserted, we obtain

$$\psi = \dot{q}\frac{\partial\Phi}{\partial q} + \frac{\partial\Phi}{\partial t} = \frac{d\Phi}{dt}.$$

A number of applications of Lagrange's equation are illustrated in Section 4. In most practical cases the generalized coordinates needed to simplify the equations of constraint or otherwise to simplify the problem are apparent from inspection. These coordinates must then be related to the usual Cartesian coordinates and, once this is done, Eq. (25) can be used to find the kinetic energy in terms of the q_α and $\dot{q}_\alpha$. Inserting (25) into the expression for T we obtain

$$\begin{aligned} T = \tfrac{1}{2}m_{(i)}\mathbf{v}_i \cdot \mathbf{v}_i &= \tfrac{1}{2}m_{(i)}\left[\frac{\partial \mathbf{x}_i}{\partial q_\alpha}\dot{q}_\alpha + \frac{\partial \mathbf{x}_i}{\partial t}\right] \cdot \left[\frac{\partial \mathbf{x}_i}{\partial q_\beta}\dot{q}_\beta + \frac{\partial \mathbf{x}_i}{\partial t}\right] \\ &= a(q, t) + b_\alpha(q, t)\dot{q}_\alpha + c_{\alpha\beta}(q, t)\dot{q}_\alpha\dot{q}_\beta, \end{aligned}$$

where

$$\begin{aligned} a &= \tfrac{1}{2}m_{(i)}\frac{\partial \mathbf{x}_i}{\partial t}\cdot\frac{\partial \mathbf{x}_i}{\partial t} = \tfrac{1}{2}m_i\left|\frac{\partial \mathbf{x}_i}{\partial t}\right|^2, \\ b_\alpha &= m_{(i)}\frac{\partial \mathbf{x}_i}{\partial t}\cdot\frac{\partial \mathbf{x}_i}{\partial q_\alpha}, \\ c_{\alpha\beta} &= \tfrac{1}{2}m_{(i)}\frac{\partial \mathbf{x}_i}{\partial q_\alpha}\cdot\frac{\partial \mathbf{x}_i}{\partial q_\beta}. \end{aligned} \tag{30}$$

The coefficients a, b_α, and $c_{\alpha\beta}$ are not functions of the $\dot{q}_\alpha$, so that T is quadratic function of the $\dot{q}_\alpha$. Moreover, if the equations relating the q_α and $\mathbf{x}_i$ are time-independent then from (30) it is seen that a and b_α are zero and therefore that the kinetic energy is a homogeneous quadratic function of the $\dot{q}_\alpha$. This is, incidentally, often a useful property of the Lagrangian (see Problem 2 of Chapter III).

Remark. Lagrange's equations can be written

$$\frac{\partial^2 L}{\partial \dot{q}_\alpha\,\partial \dot{q}_\beta}\ddot{q}_\beta = G_\alpha(q, \dot{q}, t)$$

where the G_α are functions which are known if the Lagrangian is known. If the generalized accelerations $\ddot{q}_\beta$ are to be uniquely determined by the generalized coordinates and velocities q_β and $\dot{q}_\beta$, the determinant of the $\partial^2 L/\partial\dot{q}_\alpha\,\partial\dot{q}_\beta$ must not vanish. We shall always assume this to be so. There exist, however, ways of dealing with systems for which this is not true [Dirac (1964)].

II-3 HAMILTON'S PRINCIPLE

Let us pause now and take stock of where we have gone. Starting from the principles of mechanics (or Newton's Three Laws) in Section 2 of Chapter I, we have arrived at a new statement of the basic problem. Corresponding to a given mechanical system is a function of the q_α, $\dot{q}_\alpha$, and t, namely, the Lagrangian, with the aid of which one can write down a set of differential equations for the $q_\alpha(t)$. The solution of the problem is obtained by solving these equations with the initial conditions, thus predicting the state of the system for all later time. The concept of force has now taken on a secondary role. More important have become the kinetic and potential energies, for it is in terms of these that the Lagrangian is calculated. In fact, in Section 4 we have the example of a charged particle in an electromagnetic field in which the Lagrangian itself is of importance rather than even the kinetic or potential energies. In view of the primary importance of the Lagrangian, therefore, it is reasonable to state the principles of mechanics in terms of this function rather than in terms of the force function. In this section we shall reformulate Newton's Second Law entirely in terms of the Lagrangian, putting off further discussion of techniques for solving problems until a later time.

The many advantages of the Lagrangian reformulation will become apparent as we proceed; they extend, indeed, beyond the limits of classical mechanics. For our purposes at first, however, their main usefulness will be that we can work in generalized coordinates and so avoid the rigidity of Cartesian frames. Let us therefore consider the generalized position of a mechanical system as a point in an n-dimensional space with a coordinate system, each axis of which gives the value of a particular q_α. We call this *q-space* or *configuration space*. We represent configuration space as a plane[6] drawn at right angles to the time axis, and let us assume that we know that the system has certain coordinates $q(t_1)$ at some initial time t_1 and certain other coordinates $q(t_2)$ at some later time t_2. Then the problem is to find the trajectory on which the system moves in going from $q(t_1)$ to $q(t_2)$. [Recall that $q(t)$ stands for the collection of $q_\alpha(t)$.] Actually, we want to know more than the trajectory, for we want to know at which point on the trajectory the system is located at any time t between t_1 and t_2. Thus what we wish to find is the path that the system takes, indicated by the solid line in Fig. 7. We shall often call such a path the *motion*.

Another way to say this is as follows: of all possible paths $q(t)$ from $q(t_1)$ to $q(t_2)$, two of which are indicated by dotted lines in the figure, the system

[6] It is not always possible to represent q-space as a plane with rectilinear coordinates. For instance, if we choose latitude θ and longitude ϕ as coordinates on a sphere, $\theta = \pi/2$ and all ϕ correspond to one point. Thus the representation of Fig. 7 is meant to be only symbolic.

takes only one. On what basis is this one path chosen? The answer is that the system follows the motion, that is, the path $q(t)$ which is a solution of the equations of motion, Lagrange's equations. The main point of the present section is to show that this path has another important property which we shall first state rather roughly and perhaps inexactly.

Choose an arbitrary path $q(t)$ going from $q(t_1)$ to $q(t_2)$. Since along this chosen path the q_α and hence the $\dot{q}_\alpha$ are known as functions of the time, the

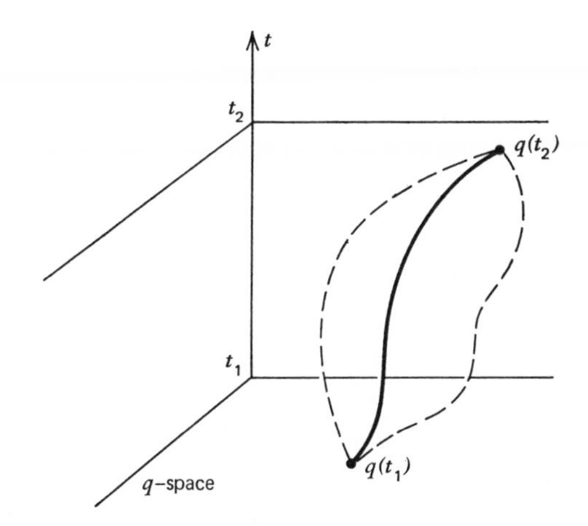

Figure 7. Possible motions for a system going from $q(t_1)$ to $q(t_2)$. The solid curve represents the physical motion.

Lagrangian can also be written as a function of the time on the path. On a different path, since the time dependence of the q_α will differ, the Lagrangian will be a different function of the time. Now for each possible path, form the integral

$$\int_{t_1}^{t_2} L(q(t), \dot{q}(t), t)\, dt, \tag{31}$$

which will have different values for different paths. *Then the path along which the mechanical system will move is an extremum of this integral.* From this statement, known as Hamilton's variational principle, we shall be able to derive Lagrange's equations.

To do so, let us quantify the statement, making it more precise. Consider, instead of all possible paths from $q(t_1)$ to $q(t_2)$, a family of them in which we identify each path by the value of a certain parameter ε. Let this family include the actual physical path that the system takes, and choose the parametrization so that $\varepsilon = 0$ on this path. Thus each path in the family is a function of t and is labeled by ε, so that we may write it in the form $q(t, \varepsilon)$ for fixed ε. Now let us specify further that $q(t, \varepsilon)$ be differentiable with respect to ε for fixed t. Such a family of paths will be called an ε-family. Since all paths start and end at $q(t_1)$ and $q(t_2)$, we have

$$\begin{aligned} q(t_1, \varepsilon) &= q(t_1, 0) = q(t_1) \equiv q_1, \\ q(t_2, \varepsilon) &= q(t_2, 0) = q(t_2) \equiv q_2, \end{aligned} \tag{32}$$

or $\partial q(t_1, \varepsilon)/\partial\varepsilon = \partial q(t_2, \varepsilon)/\partial\varepsilon = 0$. (Here $\partial q/\partial\varepsilon$ stands for the collection of the $\partial q_\alpha/\partial\varepsilon$.) For different paths the integral in Eq. (31) has different values, so we may write

$$S(\varepsilon) = \int_{t_1}^{t_2} L(q(t, \varepsilon), \dot{q}(t, \varepsilon), t)\, dt = \int_{t_1}^{t_2} L(t, \varepsilon)\, dt.$$

Then an accurate statement of the variational principle is that for every ε-family of paths satisfying the conditions we have named above, $S(0)$ is an *extremum*, that is,

$$\left.\frac{dS}{d\varepsilon}\right|_{\varepsilon=0} = \left[\frac{d}{d\varepsilon}\int_{t_1}^{t_2} L\, dt\right]_{\varepsilon=0} = 0. \tag{33}$$

This is Hamilton's principle: the time integral of the Lagrangian is an extremum on the actual physical motion. Since the initial and final times t_1 and t_2 are arbitrary, this is, as we shall see, a general formulation of Newton's Second Law. What we must now show is that Lagrange's equations follow.

Since the limits on the integral in (33) are fixed, we have

$$\frac{dS}{d\varepsilon} = \int_{t_1}^{t_2} \frac{\partial L}{\partial\varepsilon}\, dt. \tag{34}$$

The partial derivative is

$$\frac{\partial L}{\partial\varepsilon} = \frac{\partial L}{\partial q_\alpha}\frac{\partial q_\alpha}{\partial\varepsilon} + \frac{\partial L}{\partial\dot{q}_\alpha}\frac{\partial\dot{q}_\alpha}{\partial\varepsilon}. \tag{35}$$

There is now a notational difficulty. Since q_α and $\dot{q}_\alpha$ (the velocity along a path of the ε-family) are functions of t and ε, so is any function $F(q, \dot{q}, t)$, for example, L or $\partial L/\partial\dot{q}_\alpha$. Sometimes we want to take the time derivative of such a function holding q and $\dot{q}$ fixed. This we write $\partial F/\partial t$. Sometimes we want to take the time derivative along a path of the ε-family, that is, holding ε fixed. This we write dF/dt or $\dot{F}$. Thus for the present d/dt is a partial derivative (holding ε fixed), but a different one from $\partial/\partial t$. Now $\dot{q}_\alpha$ is $dq_\alpha(t, \varepsilon)/dt$ in just this sense, and hence we can change the order of differentiation in $\partial\dot{q}_\alpha/\partial\varepsilon$: that is, $\partial\dot{q}_\alpha/\partial\varepsilon = d(\partial q_\alpha/\partial\varepsilon)/dt$. We may therefore write the second term in (35) in the form

$$\frac{\partial L}{\partial \dot{q}_\alpha}\frac{\partial \dot{q}_\alpha}{\partial \varepsilon} = \frac{\partial L}{\partial \dot{q}_\alpha}\frac{d}{dt}\frac{\partial q_\alpha}{\partial \varepsilon} = \frac{d}{dt}\left[\frac{\partial L}{\partial \dot{q}_\alpha}\frac{\partial q_\alpha}{\partial \varepsilon}\right] - \left[\frac{d}{dt}\frac{\partial L}{\partial \dot{q}_\alpha}\right]\frac{\partial q_\alpha}{\partial \varepsilon}.$$

On inserting this into (35) we arrive at

$$\frac{\partial L}{\partial \varepsilon} = \left[\frac{\partial L}{\partial q_\alpha} - \frac{d}{dt}\frac{\partial L}{\partial \dot{q}_\alpha}\right]\frac{\partial q_\alpha}{\partial \varepsilon} + \frac{d}{dt}\left[\frac{\partial L}{\partial \dot{q}_\alpha}\frac{\partial q_\alpha}{\partial \varepsilon}\right]. \tag{36}$$

Later we shall use this result to calculate variations of L with respect to a parameter for paths which do not all start and end at the same point. For this reason, we emphasize the fact that the calculation of (36) is so far independent of this restriction.

Now we insert (36) into (34) in order to obtain an expression for (34). This yields[7]

$$0 = \frac{dS}{d\varepsilon} = \int_{t_1}^{t_2}\left[\frac{\partial L}{\partial q_\alpha} - \frac{d}{dt}\frac{\partial L}{\partial \dot{q}_\alpha}\right]\frac{\partial q_\alpha}{\partial \varepsilon}\,dt + \int_{t_1}^{t_2}\frac{d}{dt}\left[\frac{\partial L}{\partial \dot{q}_\alpha}\frac{\partial q_\alpha}{\partial \varepsilon}\right]dt. \tag{37}$$

The second integral is trivially obtained: it is

$$\left.\frac{\partial L}{\partial \dot{q}_\alpha}\frac{\partial q_\alpha}{\partial \varepsilon}\right|_{t_1}^{t_2} = 0,$$

vanishing according to (32). Thus the first integral alone must equal zero. Since this is true for every family of paths, the $\partial q_\alpha/\partial \varepsilon$ are arbitrary functions of the time (except that they must be zero at the end points). It then follows[8] that the other factor in the integrand must vanish for each α, and we have

$$\frac{d}{dt}\frac{\partial L}{\partial \dot{q}_\alpha} - \frac{\partial L}{\partial q_\alpha} = 0, \tag{38}$$

which are Lagrange's equations.

Similar equations were first obtained by Euler in the more general variational problem, and Eqs. (38) are therefore sometimes called the Euler-Lagrange equations. An interesting discussion of the variational principle is found in Feynman (1965), Chapter 19.

[7] All quantities are to be evaluated at $\varepsilon = 0$, although we leave off any notation indicating this. The results we obtain will consequently be for the $q_\alpha(t, 0)$, giving the physical motion. Note that although the $q_\alpha(t, 0)$ are certain fixed functions (still to be found of course), the $\partial q_\alpha/\partial \varepsilon$ are arbitrary functions of the time for $\varepsilon = 0$, and can be changed by choosing different ε-families of paths.

[8] R. Courant, *Differential and Integral Calculus*, Vol. II, p. 501, Interscience, New York, 1950.

At the end of Chapter IV we will show how constraints can be included in the variational principle without eliminating them by a clever choice of generalized coordinates. It will be shown that if the constraints can be written as $K < n$ equations of the form

$$f_I(q, t) = 0, \qquad I = 1, \ldots, K,$$

then L can be replaced by the new Lagrangian

$$\mathscr{L}(q, \lambda, \dot{q}, t) = L(q, \dot{q}, t) - \lambda_I f_I(q, t),$$

where the λ_I are K new variables (called Lagrange multipliers). The Euler-Lagrange equations obtained from $\mathscr{L}$ in the $n + K$ degrees of freedom $q_1, \ldots, q_n, \lambda_1, \ldots, \lambda_K$ then yield the correct equations of motion for the $q_\alpha(t)$.

II-4 EXAMPLES, APPLICATIONS, AND EXTENSIONS

(a) Particle Constrained to the Surface of a Sphere

Consider a particle fixed at one end of a massless rod whose other end is freely pivoted from a stationary point. If the external force is due to a uniform gravitational field, this system is called a *spherical pendulum.* In general, it may be thought of as consisting of a single particle constrained to remain on the surface of a sphere whose radius a is the length of the rod.

Let us start by treating this problem by the formal method we have discussed in developing the theory of constrained systems. It will be seen that this treatment is somewhat cumbersome. As is characteristic of a large class of constrained motions, the Lagrangian provides a simpler approach, which we shall demonstrate after we have completed the formal treatment. The equation of constraint is

$$f(\mathbf{x}, t) = x_\alpha x_\alpha - a^2 = 0, \tag{39}$$

where x_α is the α component of the position vector $\mathbf{x}$. Let $\mathbf{F}$ be the external force on the particle. Then Eq. (17) becomes

$$(m\ddot{\mathbf{x}} - \mathbf{F}) \cdot \boldsymbol{\tau} = (m\ddot{x}_\alpha - F_\alpha)\tau_\alpha = 0. \tag{40}$$

Here $\boldsymbol{\tau}$ is any vector tangent to the constraint surface, that is, any vector such that

$$\boldsymbol{\tau} \cdot \nabla f = 2x_\alpha \tau_\alpha = 0.$$

By solving this for τ_3 and inserting the expression so obtained into (40) we arrive at

$$\left[m\ddot{x}_1 - F_1 - \frac{x_1}{x_3}(m\ddot{x}_3 - F_3)\right]\tau_1 + \left[m\ddot{x}_2 - F_2 - \frac{x_2}{x_3}(m\ddot{x}_3 - F_3)\right]\tau_2 = 0,$$

where $\mathbf{F}$ is the external force on the particle. In this last expression, τ_1 and τ_2 are arbitrary, so that each of the expressions in brackets can be set equal to zero separately. We then obtain two equations of motion:

$$\begin{aligned} m(x_3\ddot{x}_1 - x_1\ddot{x}_3) &= x_3F_1 - x_1F_3 \\ m(x_3\ddot{x}_2 - x_2\ddot{x}_3) &= x_3F_2 - x_2F_3. \end{aligned} \tag{41}$$

Although this appears to involve three variables, any one of them, say x_3, can be eliminated by using the equation of constraint (39).

To treat this problem by the Lagrangian method, we would write down the Lagrangian function

$$L = \tfrac{1}{2}m\dot{x}^2 - V(\mathbf{x}), \tag{42}$$

where $V(\mathbf{x})$ is the potential due to the external force. We now pick appropriate generalized coordinates. These are obviously spherical polar coordinates, and then the Lagrangian becomes

$$L = \tfrac{1}{2}m(\dot{r}^2 + r^2\dot{\theta}^2 + r^2\dot{\phi}^2 \sin^2 \theta) - V(r, \theta, \phi). \tag{43}$$

The constraint equation becomes

$$r^2 - a^2 = 0,$$

whose solution, because r is restricted to nonnegative values, is $r = a$. Then if r is treated as a constant in (43), the problem reduces to one in two dimensions immediately. Of course, the same result could have been achieved by transforming from Cartesian to spherical coordinates in (41) and (39) (see Problem 7).

(b) Central Force Motion

We now turn to a system of two particles acted on by zero external force, but acting on each other through a force depending only on their separation and directed along the line joining them. This kind of force is called central. A large number of physical problems involves central force motion, since both the gravitational and electrostatic forces between point particles are central. These problems include such phenomena as the motion of planets around the sun, satellites around the earth, electrons around nuclei, and the scattering of elementary particles. (In the last two cases, of course, quantum rather than classical methods must be used.) The central force problem is of interest also because it is one of the few meaningful problems of mechanics that can be solved explicitly, that is, reduced to quadrature.

In Chapter I [Section 5(c)] we saw that if two bodies interact through a force which depends only on their separation, their motion in the absence of external forces can be found in terms of an equivalent one-particle problem. Since in central force motion, in addition, the potential depends only on the magnitude of the separation, Eq. (52) of Chapter I can be derived from the Lagrangian (see Problem 12)

$$L = \tfrac{1}{2}\mu\dot{x}^2 - V(x), \tag{44}$$

where x is the magnitude of $\mathbf{x}$, and $\mu = m_1 m_2 (m_1 + m_2)^{-1}$ is the reduced mass. By solving Eq. (52) of Chapter I we find the motion of the relative position vector, and when this is added to the center-of-mass motion (constant velocity), the problem is solved.

We now proceed with the equivalent one-particle problem. Since the torque vanishes (the force is along the position vector $\mathbf{x}$), the angular momentum $\mathbf{L} = \mu\mathbf{x} \times \dot{\mathbf{x}}$ is constant. This fact provides a great simplication: we have the identity

$$\mathbf{x} \cdot \mathbf{L} = 0,$$

where $\mathbf{L}$ is a constant vector. In other words, $\mathbf{x}$ lies in the plane perpendicular to $\mathbf{L}$, and in this plane we need only two variables to specify the position vector. The dimension drops from three to two. If $\mathbf{L} = 0$, incidentally, $\mathbf{x} \cdot \mathbf{L} = 0$ does not restrict $\mathbf{x}$ to a plane. But then $\mathbf{x}$ and $\dot{\mathbf{x}}$ are always parallel, which means that the particle moves on a straight line.

The resulting two-dimensional one-particle problem is treated as follows. Let x, θ be polar coordinates in the plane perpendicular to $\mathbf{L}$. (In the final statement of the solution the third dimension can be added by choosing a coordinate system with the 3-axis along $\mathbf{L}$; then x, θ, $x_3 = 0$ become cylindrical polar coordinates.) The Lagrangian may then be written

$$L = \tfrac{1}{2}\mu(\dot{x}^2 + x^2\dot{\theta}^2) - V(x),$$

and the Euler-Lagrange equations become

$$\frac{d}{dt}\left(\frac{\partial L}{\partial \dot{\theta}}\right) = \frac{d}{dt}(\mu x^2 \dot{\theta}) = 0, \tag{45}$$

$$\frac{d}{dt}\left(\frac{\partial L}{\partial \dot{x}}\right) - \frac{\partial L}{\partial x} = \mu\ddot{x} - \mu x\dot{\theta}^2 + \frac{dV}{dx} = 0. \tag{46}$$

Equation (45) is immediately integrated to yield

$$\mu x^2 \dot{\theta} = l, \tag{47}$$

where l is a constant. It is clear, in fact, that l is the magnitude of the angular momentum. By using (47) to eliminate $\dot{\theta}$ from (46) we arrive at

$$\mu\ddot{x} - \frac{l^2}{\mu x^3} + \frac{dV}{dx} = 0, \tag{48}$$

a single ordinary second-order differential equation. Equation (48) may be integrated once when we note that it looks just like the typical one-dimensional problem [see Eq. (43) of Chapter I, Section 5(a)]. The main difference is that $l^2/(\mu x^3) - dV/dx$, which corresponds to the $F(x)$ of Chapter I, now depends on the initial conditions. At any rate, if we write an equivalent one-dimensional potential

$$\mathscr{V}(x) = \frac{l^2}{2\mu x^2} + V(x), \tag{49}$$

the equation can be solved by the usual one-dimensional methods. In particular, the energy of the equivalent one-dimensional problem

$$\begin{aligned} E &= \tfrac{1}{2}\mu\dot{x}^2 + \frac{l^2}{2\mu x^2} + V \\ &= \tfrac{1}{2}\mu\dot{x}^2 + \tfrac{1}{2}\mu x^2\dot{\theta}^2 + V \end{aligned}$$

is constant.

The solution of (48) as given in Chapter I, Eq. (45) is

$$t = \sqrt{\frac{\mu}{2}} \int_{x_0}^{x} \frac{dx}{\sqrt{E - \mathscr{V}(x)}}, \tag{50}$$

which can in principle always be integrated and inverted to give $x(t)$. Even before integrating, however, much can be learned from a graph of $\mathscr{V}(x)$. For instance, whether or not the motion is bounded can be seen by inspection of such a graph. Finally, when $x(t)$ is known, (47) can be integrated to give $\theta(t)$.

One is often interested in the trajectory or orbit, that is, in $x(\theta)$, rather than in the detailed time dependence of x and θ. This can, of course, be found from $x(t)$ and $\theta(t)$ by eliminating t, but it can also be found more directly. According to (47)

$$\dot{x} = \frac{dx}{dt} = \frac{dx}{d\theta}\dot{\theta} = \frac{l}{\mu x^2}\frac{dx}{d\theta}.$$

This may be inserted into the equation defining E to yield

$$\frac{l^2}{2\mu x^4}\left(\frac{dx}{d\theta}\right)^2 + \frac{l^2}{2\mu x^2} + V = E,$$

whose solution is (we set $\theta_0 = 0$)

$$\theta = \sqrt{\frac{l^2}{2\mu}} \int_{x_0}^{x} \frac{dx}{x^2\sqrt{E - V(x) - l^2/(2\mu x^2)}}. \tag{51}$$

Once $V(x)$ is given, this equation can be used to find the equation of the orbit. We shall not go into the details of central-force motion, leaving much to the Problems. Many of the standard works discuss central potentials exhaustively.

(c) Rotating Coordinate Systems

The Lagrangian method has been developed from the principles of mechanics, and these principles apply only in inertial systems. Thus the Lagrangian function L must always be calculated in an inertial system. But once L has been found, it can be transformed to any other system, not necessarily inertial, and the Euler-Lagrange equations can be used to write the equations of motion for the coordinates of a particle in that system (see Problem 2).

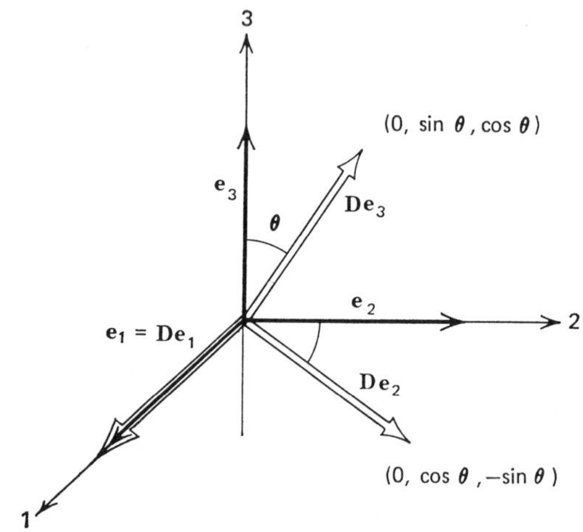

Figure 8. A stationary and a rotating coordinate system with a common X_3 axis.

Consider in particular the transformation to a rotating system. Such motion is important, for instance, because the Earth is rotating, and thus a frame fixed on the Earth is not inertial. We take two Cartesian frames with a common origin and 3-axis, as in Fig. 8. Let one of these, with axes X_1', X_2', $X_3' = X_3$, be rotating with constant angular velocity ω with respect to the other (inertial) system. Then if the coordinates of a particle in the inertial system are x_1, x_2, x_3, its coordinates x_1', x_2', x_3' in the rotating system are related to these by

$$\begin{aligned} x_1 &= x_1' \cos \omega t - x_2' \sin \omega t, \\ x_2 &= x_1' \sin \omega t + x_2' \cos \omega t, \\ x_3 &= x_3'. \end{aligned} \tag{52}$$

The kinetic energy in the inertial system is $T = \frac{1}{2} m \dot{x}_\alpha \dot{x}_\alpha$. To write it in terms of the coordinates in the rotating system we differentiate (52) with respect to the time and insert the expressions obtained for the $\dot{x}_\alpha$ into the formula for T. The result is

$$T = \tfrac{1}{2} m[\dot{x}_\alpha' \dot{x}_\alpha' + \omega^2(x_1'^2 + x_2'^2) - 2\omega(\dot{x}_1' x_2' - \dot{x}_2' x_1')].$$

Now let $V(\mathbf{x}')$ be the potential, also expressed in terms of the x'_α rather than the x_α. Then by writing $L = T - V$ in the usual way and taking the Euler-Lagrange equations with respect to the primed variables, we arrive at

$$\begin{aligned} m\ddot{x}'_1 - m\omega^2 x'_1 - 2m\omega\dot{x}'_2 &= -\partial V/\partial x'_1, \\ m\ddot{x}'_2 - m\omega^2 x'_2 + 2m\omega\dot{x}'_1 &= -\partial V/\partial x'_2, \\ m\ddot{x}'_3 &= -\partial V/\partial x'_3. \end{aligned}$$

Note that this can be written in the usual vector notation if ω is given vector properties: we let $\boldsymbol{\omega}$ be the vector of magnitude ω which points along the 3-axis. Then the equations of motion become

$$m\ddot{\mathbf{x}} = -2m\boldsymbol{\omega} \times \dot{\mathbf{x}} - m\boldsymbol{\omega} \times (\boldsymbol{\omega} \times \mathbf{x}) - \nabla V, \tag{53}$$

where we have dropped the primes.

Remarks. *1.* If $V = 0$, that is, if there is no force applied to the particle, it nevertheless seems to a stubborn observer in the rotating system that the particle is accelerating. If he insists on believing that he is in an inertial system, he must interpret this acceleration as being caused by certain forces. The first term in (53), which is velocity dependent, is called the *Coriolis force*, and the second is the well-known *centrifugal force*. In this sense, these forces are sometimes called *fictitious*. They can be useful in describing motion in a rotating system, but they are often incorrectly used to describe rotating particles in inertial systems. *2.* In problems involving time-dependent constraints, equations similar to (52) are used. The procedure in such problems is also quite similar (see, for instance, Problem 5).

(d) The Lagrangian for a Particle in an Electromagnetic Field

As we have mentioned, even if the forces depend on the velocity, it is often possible to find a Lagrangian that will yield the correct equations of motion, although it is not always a simple matter to find it. Particularly important is the case of a charged particle in an external electromagnetic field.

The force on a charged particle in an electromagnetic field is

$$\mathbf{F} = e\left(\mathbf{E} + \frac{1}{c}\mathbf{v} \times \mathbf{B}\right), \tag{54}$$

where e is the charge on the particle, $\mathbf{E}(\mathbf{x}, t)$ is the electric field, and $\mathbf{B}(\mathbf{x}, t)$ is the magnetic field. In general $\mathbf{E}$ and $\mathbf{B}$ can be written in terms of the scalar and vector potentials $\mathbf{A}(\mathbf{x}, t)$ and $\Phi(\mathbf{x}, t)$:

$$\begin{aligned} \mathbf{E} &= -\nabla\Phi - \frac{1}{c}\frac{\partial \mathbf{A}}{\partial t}, \\ \mathbf{B} &= \nabla \times \mathbf{A}. \end{aligned}$$

On inserting this and $\mathbf{F} = m\mathbf{a}$ into (54), we obtain

$$m\ddot{\mathbf{x}} = -e\nabla\Phi - \frac{e}{c}\frac{\partial \mathbf{A}}{\partial t} + \frac{e}{c}\dot{\mathbf{x}} \times (\nabla \times \mathbf{A}). \tag{55}$$

We shall try to find the desired Lagrangian by writing down the free-particle Lagrangian and adding a term which will give the velocity-dependent force of Eq. (55). It is relatively easy to see that part of this additional term should be $-e\Phi$, for this is just what is needed when $\mathbf{B} = 0$ and therefore when $\mathbf{A}$ may be taken to vanish. We then put

$$L = \tfrac{1}{2}m\dot{x}^2 - e\Phi + f(\mathbf{x}, \dot{\mathbf{x}}, t).$$

In order to find f we now calculate the αth Euler–Lagrange equation:

$$m\ddot{x}_\alpha = -e\frac{\partial \Phi}{\partial x_\alpha} + \frac{\partial f}{\partial x_\alpha} - \frac{\partial^2 f}{\partial \dot{x}_\alpha\, \partial x_\beta}\dot{x}_\beta - \frac{\partial^2 f}{\partial \dot{x}_\alpha\, \partial \dot{x}_\beta}\ddot{x}_\beta - \frac{\partial^2 f}{\partial \dot{x}_\alpha\, \partial t}. \tag{56}$$

The right-hand side has a term depending on the acceleration, which is absent from Eq. (55), so that $\partial^2 f/\partial \dot{x}_\alpha \partial \dot{x}_\beta = 0$. A particular solution of this equation is

$$f = \dot{x}_\beta G_\beta(x, t). \tag{57}$$

This is inserted back onto (56) and the equation so obtained is compared with the αth component of (55). After some trivial subtractions we arrive at

$$\dot{x}_\beta\left[\frac{\partial G_\beta}{\partial x_\alpha} - \frac{\partial G_\alpha}{\partial x_\beta}\right] - \frac{\partial G_\alpha}{\partial t} = \frac{e}{c}\dot{x}_\beta\left[\frac{\partial A_\beta}{\partial x_\alpha} - \frac{\partial A_\alpha}{\partial x_\beta}\right] - \frac{e}{c}\frac{\partial A_\alpha}{\partial t}.$$

It is obvious now what must be done. Setting $G_\alpha = A_\alpha e/c$, or $\mathbf{G} = \mathbf{A}e/c$, we finally obtain

$$L = \tfrac{1}{2}m\dot{x}^2 - e\Phi + \frac{e}{c}\dot{\mathbf{x}} \cdot \mathbf{A}. \tag{58}$$

Remarks. 1. It is interesting to see how gauge invariance of the electromagnetic field is reflected in the Lagrangian. For this purpose see Problem 10. *2.* The interaction energy of a charged particle with an external electromagnetic field is just $U = e\Phi - (e/c)\dot{\mathbf{x}} \cdot \mathbf{A}$, so that the Lagrangian we have obtained has the same form, in a slightly generalized sense, as in the velocity-independent case: $L = T - U$.

Problems

1. Consider a space curve given by two equations of the form

$$f_I(\mathbf{x}, t) = 0, \qquad I = 1, 2.$$

A particle is constrained to move along this curve. Show that in this case Eq. (14) results if the forces of constraint are required to be perpendicular to the curve. It may be helpful to think of the two constraint equations as defining a curve through the intersection of two surfaces.

2. One set of generalized coordinates is as good as another. Let q_α and q'_α be two sets of generalized coordinates related by n equations of the form

$$q'_\alpha = q'_\alpha(q, t), \qquad \alpha = 1, 2, \ldots, n.$$

The Lagrangian $T - V$ can be written in terms of either of the two coordinates. Let these two forms be called L and L', so that

$$L'(q', \dot{q}', t) = L(q, \dot{q}, t)$$

Show that if L satisfies the Euler-Lagrange equations in terms of the q_α then L' satisfies the Euler-Lagrange equations in terms of the q'_α.

3. A mass m_1 hangs from a spring of natural length a and spring constant k. Let x measure the displacement of m_1 from a. From m_1 hangs a massless rod of length l at the end of which is a second mass, m_2. The spring is constrained to move vertically and the rod is constrained to a vertical plane. Let θ be the angle the rod makes with the vertical. (a) Write the Lagrangian in terms of x and θ. (b) Determine the equations of motion for x and θ. (c) Check these equations by noting what happens if it is assumed that $\theta = \dot{\theta} = \ddot{\theta} = 0$ at all times.

4. A mass m is attached to a weightless string wrapped around a circular cylinder of radius b and is constrained to move in a plane perpendicular to the cylinder. The mass is given an initial velocity v_0, and the string wraps itself around the cylinder. Show that the angular velocity of the mass about the center of the cylinder is given by

$$\dot{\theta} = \frac{v_0}{r^2}\sqrt{r^2 - b^2},$$

where r is the radial distance from the center of the cylinder to the mass. Is angular momentum conserved?

5. A particle is constrained to move on a vertical circle of radius R in a uniform gravitational field. The circle rotates about the vertical axis with constant angular velocity ω. Write the (time-dependent) constraint equations. Obtain the equations of motion and discuss the motion. Find the equilibrium position of the mass on the circle.

6. A particle moves under gravity along a cycloid whose equation is

$$x = a\cos^{-1}\left(\frac{a - y}{a}\right) + \sqrt{2ay - y^2}.$$

Obtain and solve the equation of motion. Show that no matter where on the cycloid the particle starts at time $t = 0$, it will reach the bottom at the same time.

7. Obtain the Euler-Lagrange equations for the spherical pendulum from the Lagrangian of Eq. (43). Show that these equations are just those of (41) written in spherical coordinates. Show that the angular momentum about the vertical axis remains constant.

8. Consider two Lagrangians $L(q, \dot{q}, t)$ and $L'(q, \dot{q}, t)$ in the same n generalized coordinates $q_1, \ldots, q_n$. Prove that L and L' give exactly the same equations of motion if *and only if* L and L' differ by the formal time derivative of some function of the form $\Phi(q, t)$.
9. Show that the two one-dimensional Lagrangians $L_1 = (q + \dot{q})^2$ and $L_2 = q^2 + \dot{q}^2$ give the same equations of motion, and verify that their difference satisfies the criterion of Problem 8.
10. When the potential and vector potential of an electromagnetic field are subjected to a gauge transformation

$$\mathbf{A} \to \mathbf{A}' = \mathbf{A} + \nabla\Lambda$$

$$\Phi \to \Phi' = \Phi - \frac{1}{c}\frac{\partial\Lambda}{\partial t},$$

the electromagnetic field they describe does not change. If, however, $\mathbf{A}$ is replaced by $\mathbf{A}'$ and Φ by Φ', the Lagrangian of Eq. (52) will change.
(a) Show that the new Lagrangian will yield the same equations of motion.
(b) Show that if for (57) we took a more general solution of $\partial^2 f/\partial\dot{x}_\alpha\,\partial\dot{x}_\beta = 0$, the Lagrangian would automatically be determined only up to a gauge transformation.
(c) Discuss the relation of this to Problem 8.
11. Consider the Lagrangian $L = \frac{1}{2}m\dot{q}^2 - kq$. Obtain the motion $q(t)$. Construct some function $p(t)$ that has the same values as $q(t)$ at arbitrary times t_0 and t_1, but which differs in general elsewhere. Verify explicitly that $\int_{t_0}^{t_1} L\,dt$ is less when $q = q(t)$ than when $q = p(t)$.
12. Consider a two-body system with no external forces and such that $\mathbf{F}_{12} = -\mathbf{F}_{21}$ depends only on the relative separation vector $\mathbf{x}$ (not just on x) of the two bodies. Show that the Lagrangian splits into two unrelated parts, thus deriving a Lagrangian of the form of (44) for the equivalent one-body problem. Show that if $\mathbf{F}_{12}$ is directed along the relative position vector, but its magnitude depends (sufficiently nicely) only on x (not on $\mathbf{x}$), then there always exists a potential $V(x)$.
13. Prove Kepler's Second Law of planetary motion for a general central force: the radius vector of a particle moving in a central potential sweeps out equal areas in equal times.
14. Find the general orbit of a particle moving in the potential $V = -k/x$, where k is a positive constant. This is the Kepler problem. Discuss in particular the orbits for $E < 0$ and $E > 0$, where E is the total energy. Show that if $E < 0$ the orbit is an ellipse with one of its foci at the center of force, and that $E = -2a/k$, where a is the semimajor axis of the ellipse. Show further that in this case the period is $2\pi\sqrt{m/k}\,a^{3/2}$. (This is Kepler's Third Law of planetary motion.)
15. Find the general orbit for a particle moving in the modified gravitational potential $V = -k/x + \beta/x^2$. Show that if $k > 0$ and $|\beta| \ll \frac{1}{2}ka(1 - e^2)$, the orbit is an ellipse whose major axis precesses slowly with an angular velocity equal to $2\pi\beta[\tau(1 - e^2)ka]^{-1}$, where a is the semimajor axis of

the ellipse, e is the eccentricity, and τ is the period of motion on the ellipse.

16. Determine the central law of force for which all orbits are ellipses with the attracting center located at the center of the ellipse. Find the period of the orbit and show that it is the same for all orbits. Compare with Chapter I, Problem 12.
17. Find the general orbit for a particle moving in the central potential $V = -k/x^2$, where $k > 0$. What are the physically possible values of E? Discuss the two cases $l > \sqrt{2mk}$ and $l < \sqrt{2mk}$, where l is the angular momentum.
18. A problem of constrained motion can be thought of as the limit of an unconstrained problem in which the potential function gets more and more sharply peaked. Consider a particle moving in the central potential.

$$V(x) = k\left[\frac{1}{\alpha} - \frac{\alpha}{(x - R)^2 + \alpha^2}\right].$$

Draw a rough sketch of V for different values of α. For given energy and angular momentum obtain an expression for the maximum and minimum values of x. Find the force on the particle. Now go to the limit as $\alpha \to 0$, and show that the motion in the limit is the same as the motion for a problem with the appropriate constraint. Show also that in the limit the force on the particle has the properties we have assumed for the force of constraint.

19. A particle is constrained to move on the surface of a paraboloid of revolution about the 3-axis and opening in the positive 3-direction. There is a uniform gravitational field in the negative 3-direction. Obtain the Euler-Lagrange equations. Show that each motion is contained between a greatest and least value of x_3. Show that the projection of the radius vector on the (1, 2) plane sweeps out equal areas in equal times.

CHAPTER III

Symmetry and Conservation

In the formulation of the principles of mechanics in Chapter I the conservation of momentum was taken as a fundamental postulate. Other conservation laws are almost equally important, and we wish now to discuss conservation in some generality. This discussion will lead to interesting and important results also intimately connected with the structure of mechanics, results that shed light on much of physics. We shall see how symmetry is connected with conservation, and shall eventually be able to state the principles of mechanics largely in terms of some natural assumptions about the structure of space. This new statement will not contradict, of course, or even extend Newton's laws, but it is a kind that is easily generalized from particle dynamics to field theory and that is often used at the frontier of physical knowledge.

III-1 CONSTANTS OF THE MOTION

Consider a system of n degrees of freedom described by a Lagrangian $L(q, \dot{q}, t)$ (recall that q stands for the set $q_1, \ldots, q_n$). The function $\partial L/\partial \dot{q}_\alpha$ of q, $\dot{q}$, and t is called the *generalized momentum conjugate* to q_α, and is usually denoted by p_α. In terms of it the n Euler-Lagrange equations can be written in the form of $2n$ first-order equations:

$$p_\alpha = \frac{\partial L}{\partial \dot{q}_\alpha},$$
$$\dot{p}_\alpha = \frac{\partial L}{\partial q_\alpha}. \tag{1}$$

Now suppose that one of the generalized coordinates, say q_α, does not appear explicitly in L. (L must still depend on $\dot{q}_\alpha$, since if it depended on neither q_α nor $\dot{q}_\alpha$ there would be one less degree of freedom.) Then $\partial L/\partial q_\alpha = 0$ and from (1) we have $\dot{p}_\alpha = 0$ which can be immediately integrated to give

$$p_\alpha = \frac{\partial L}{\partial \dot{q}_\alpha} = P_\alpha, \tag{2}$$

where P_α is a constant. Thus we have obtained a partial solution of the equations of motion. Although we do not know the possible motions or even the possible trajectories, we do know that along every possible motion the function $p_\alpha(q, \dot{q}, t)$ remains constant.

Any such function F of $q, \dot{q}$, and t that remains constant along every possible motion will be called a *constant of the motion*. Thus we see that to every coordinate that does not appear explicitly in the Lagrangian, called a *cyclic coordinate*, there corresponds a constant of the motion, namely, its conjugate momentum.

Because it is sometimes useful to distinguish between constants of the motion that do and do not depend explicitly on the time, we shall call *conserved quantities* those which do not. If L does not depend explicitly on the time, the generalized momentum conjugate to a cyclic variable is then conserved. Moreover, even when L does depend on time, this time dependence is often in a term which does not involve $\dot{q}_\alpha$, so that such a momentum is still a conserved quantity.

Now let $F_1, F_2, \ldots, F_m$ be constants of the motion. Then any function of the form $G(F_1, F_2, \ldots, F_m)$ is also a constant of the motion. But such an additional constant of the motion is not particularly interesting, for the equation $G =$ const implies no more (in fact, usually less) about the motion than do the m equations $F_i =$ const, $i = 1, \ldots, m$. We shall therefore look only for independent constants of motion, that is functions $F_1, \ldots, F_m$ of $q, \dot{q}, t$ such that none of the F_i can be expressed as a function simply of the others. One is naturally led to ask whether there is a maximum number of such independent constants of the motion and, if so, how many. This question is answered by the following.

THEOREM. For a system with n degrees of freedom, there are exactly $2n$ independent constants of the motion.

The number of degrees of freedom of a system has been defined as the number of unconstrained variables q_α in the system. But in order to prove this theorem we must go a bit deeper into this concept. Let us assume that the equations of motion (the Euler-Lagrange equations) have been integrated to yield n equations of the form

$$q_\alpha = q_\alpha(C_1, C_2, \ldots, C_{2n}, t), \tag{3}$$

where the C_i are $2n$ constants arising in the integration of the n second-order differential equations. Each choice of the C_i constitutes a different time dependence of the q_α, and hence a different *particular motion*. Now consider any function $f(q, \dot{q}, t)$ which does not depend explicitly on the C_i and insert the expressions for q and $\dot{q}$ in terms of the C_i and t from Eqs. (3) and their derivatives. We write

$$f(q(C, t), \dot{q}(C, t), t) = G(C, t),$$

where C stands for the collection of the C_i. Now G can not be independent of the C_i no matter what f is [except for the trivial case of $f = f(t)$], for if it were, this equation would constitute a constraint, in general velocity dependent. That is, it would be a relation between the q_α, $\dot{q}_\alpha$, and t which is valid for all motions independent of the particular motion. That would mean that one (or some) of the q_α or $\dot{q}_\alpha$ would be known as a function of time as soon as the others had been found, which implies that the generalized coordinates do not vary independently.

For instance, consider a body falling in a uniform gravitational field. If one solves for the time in terms of the horizontal coordinate x_1 and inserts this expression into the equation for the vertical coordinate x_2, one obtains an equation connecting x_1 and x_2 (the equation of the parabolic trajectory). But the actual connection between x_1 and x_2 depends on the particular motion through the coordinates of the starting point and the components of the initial velocity. Nor is there any function of x_1, x_2, and t whose value is independent of such initial data, for if there were, we would solve for the vertical motion in terms of the horizontal quite generally.

Thus to say that the number of degrees of freedom is just n is to say that for every $f(q, \dot{q}, t)$ and for at least one i

$$\frac{\partial G}{\partial C_i} = B_i \neq 0. \tag{4}$$

Remark. In the last Remark of Section 2, Chapter II, we note that the determinant of the $\partial^2 L/\partial \dot{q}_\alpha \partial \dot{q}_\beta$ must not vanish. If it did, this again would give a relation among the q_α, $\dot{q}_\alpha$, and t, and could also be looked on as a constraint.

We now return to the proof of the theorem.

PROOF. For every function $f(q, \dot{q}, t)$ we have

$$\frac{\partial G}{\partial C_i} = \frac{\partial f}{\partial q_\alpha}\frac{\partial q_\alpha}{\partial C_i} + \frac{\partial f}{\partial \dot{q}_\alpha}\frac{\partial \dot{q}_\alpha}{\partial C_i} = B_i, \qquad i = 1, \ldots, 2n,$$

which is a set of $2n$ algebraic linear inhomogeneous (because at least one of the B_i fails to vanish) equations in the $2n$ quantities $\partial f/\partial q_\alpha$, $\partial f/\partial \dot{q}_\alpha$. Since f may be chosen arbitrarily, we know that in general the $\partial f/\partial q_\alpha$, $\partial f/\partial \dot{q}_\alpha$ exist, and hence these equations must in general have a unique solution. The necessary and sufficient condition that there exists a unique solution to a set of inhomogeneous linear equations is that the determinant of the coefficients is not zero (see Chapter IV, Section 3). In this case these coefficients are the $2n \times 2n = 4n^2$ quantities $\partial q_\alpha/\partial C_i$, $\partial \dot{q}_\alpha/\partial C_i$ ($i = 1, \ldots, 2n$ and $\alpha = 1, \ldots, n$), so that their determinant must be nonzero. Let us write this in the form

$$\left| \frac{\partial(q, \dot{q})}{\partial C} \right| \neq 0.$$

This quantity is also the Jacobian of the transformation [see Courant (1950), Vol. II, p. 155]

$$\begin{aligned} q_\alpha &= q_\alpha(C, t), \\ \dot{q}_\alpha &= \dot{q}_\alpha(C, t) \end{aligned} \tag{5}$$

from the $2n$ quantities q, $\dot{q}$ to C. Since this Jacobian is nonzero, which is the necessary and sufficient condition that Eqs. (5) can be inverted, the C_i can be expressed as functions of q and $\dot{q}$:

$$C_i = C_i(q, \dot{q}, t). \tag{6}$$

Each of these $2n$ functions of q, $\dot{q}$, and t is a constant of the motion since C_i is a constant. The actual value of the constant, however, depends on the particular motion, that is on the particular time dependence of q. These functions are clearly independent since, if they were not, Eq. (6) could not be inverted to yield (5). Finally, there are no other independent constants of the motion. For if $K(q, \dot{q}, t)$ is a constant of the motion we can always use Eq. (5) to write

$$K(q, \dot{q}, t) = A(C, t).$$

But since K is a constant of the motion

$$\frac{dK}{dt} = \frac{dA}{dt} = \frac{\partial A}{\partial t} = 0,$$

and consequently A is independent of t, or $K(q, \dot{q}, t) = A(C)$ is functionally dependent on the C_i. This completes the proof of the theorem: for a system with n degrees of freedom there are exactly $2n$ independent constants of the motion.

For any particular motion the C_i can be calculated at any time t, but in most cases they must be calculated from a given set of initial values of q_i and $\dot{q}_i$. Then we write (6) at time $t = 0$, and this tells us the values of the C_i for all time. If the C_i are expressed in terms of the initial values $q_\alpha(0) = Q_\alpha$ and $\dot{q}_\alpha(0) = \dot{Q}_\alpha$, Eq. (5) can also be written in terms of these initial values, and then we have

$$\begin{aligned} q_\alpha &= q_\alpha(Q, \dot{Q}, t), \\ \dot{q}_\alpha &= \dot{q}_\alpha(Q, \dot{Q}, t). \end{aligned}$$

The equations connecting the C_i and q, $\dot{q}$ are invertible at all times, so, in particular, the equations connecting the C_i and Q, $\dot{Q}$ are invertible. Therefore we can write Q, $\dot{Q}$ through the C_i as functions of q, $\dot{q}$, and this means that the Q_α and $\dot{Q}_\alpha$ form a set of valid constants of the motion, and we may now see the physical meaning of their maximum number $2n$: this is exactly the number of physical initial conditions at one's control. Once the initial coordinates and velocities are given, the motion is determined, and every constant of the motion is determined by the initial coordinates and velocities.

We now see also that the problem of finding all the constants of the motion is exactly the problem of integrating the Euler-Lagrange equations. Indeed, when we have found $2n$ independent constants of the motion $F_1, \ldots, F_{2n}$ we can invert the equations

$$F_i = F_i(q, \dot{q}, t)$$

to obtain

$$q_\alpha = q_\alpha(F, t),$$

$$\dot{q}_\alpha = \dot{q}_\alpha(F, t),$$

which then give us the motion (in fact we need only those for the q_α). The particular values of the constants F_i can be calculated from the initial conditions or whatever other information is given in order to solve the problem. We have seen that in some problems some of the constants of the motion are trivially found. This is because some of the Euler-Lagrange equations are trivially integrated. Now every time we find a constant we reduce the complexity of the problem, so it is evident that any tricks that exist for finding such constants are quite useful. They are, it turns out, more than useful, for they teach us much about dynamical theories and the structure of mechanics. In the next section we shall begin one of a series of generalizations of the discussion involving cyclic coordinates and conserved momenta.

III-2 TRANSFORMATIONS, SYMMETRIES, AND CONSERVATION LAWS

As we have seen in Chapter II (Problem 2), the Euler-Lagrange equations look the same in two different systems of generalized coordinates. Thus whenever we have used a certain set q to describe a system, we can transform to a new set q' by n transformation equations of the form

$$q'_\alpha = q'_\alpha(q, t). \tag{7}$$

We shall call such transformations *point* transformations to distinguish them from others of the form $q'_\alpha = q'_\alpha(q, \dot{q}, t)$. The coordinates q and q' describe the same physical point in configuration space, so that the Lagrangian will have the same value whether it is expressed in terms of q or q', although it will in general be a different function of q' than it is of q. These two functions are related by the equation

$$L(q, \dot{q}, t) = L'(q', \dot{q}', t) = L'\left(q'(q, t), \frac{d}{dt}[q'(q, t)], t\right). \tag{8}$$

Of course if q and q' are to be equivalent in the sense that either will do to describe the system, (7) must be invertible to give

$$q_\alpha = q_\alpha(q', t). \tag{9}$$

It is often more convenient to write these equations in the more symbolic form

$$\begin{aligned} q' &= Rq, \\ q &= Sq', \end{aligned} \tag{10}$$

where R and S signify the operations in Eqs. (7) and (9). We call R and S (and similar mathematical entities) *operators*, and we say that (the set) q' results from applying the operator R to (the set) q. Note that R and S are in general time-dependent operators; that is to say, they give different sets at different times or, in terms of the functional notation of Eqs. (7) and (9), they define different functions of q and q' respectively at different times. In general these operators will not be linear (see Chapter IV).

The reason we introduce the operator notation is that an operator algebra is easily set up. For instance consider the inversion of (7). We can write q' as a function of q' itself and of t by inserting (9) into (7), and then, of course, this function must be just q':

$$q'_\alpha = q'_\alpha(q(q', t), t) \equiv q'_\alpha.$$

In the operator notation this equation is

$$q' = R(Sq') \equiv q'. \tag{11}$$

Now let R and T be any two operators of the kind we are talking about and consider a third coordinate system $q'' = Tq'$. Expressing q' in terms of q we have

$$q'' = Tq' = T(Rq).$$

We shall define the operator TR, called the *product* of T and R by the equation

$$(TR)q \equiv TRq \equiv T(Rq).$$

That is, TR is the operator which has the same effect on q as is obtained by first transforming from q to q' by R and then transforming from q' to q'' by T. With this definition, (11) becomes

$$q' = RSq'.$$

Thus RS is a trivial operator, namely, the "no-transformation-at-all" operator. We shall call this the *identity* or *unit operator* and write

$$RS = \mathbb{1}.$$

In exactly the same way we could have established that $SR = \mathbb{1}$. We shall henceforth write $S = R^{-1}$ and $R = S^{-1}$, in an obvious notation. Thus the invertibility of the first of Eqs. (10) can be stated either in the form of the second, or merely by asserting that R^{-1} exists.

Now suppose that in the Lagrangian $L(q, \dot{q}, t)$ none of the q_α are cyclic, so that there is no obvious conserved momentum. There may, however, exist a transformation to a new set of generalized coordinates q' such that the new Lagrangian $L'(q', \dot{q}', t)$ is cyclic in one of the q'_α. If, for instance, q'_β does not appear explicitly in L', then, as we saw in Section 1, the momentum conjugate to q'_β is conserved, or

$$p'_\beta = \frac{\partial L'}{\partial \dot{q}'_\beta} = P_\beta, \tag{12}$$

where P_β is a constant. In terms of the original coordinates this can be written as

$$\frac{\partial L'}{\partial \dot{q}'_\beta} = \frac{\partial L}{\partial \dot{q}_\alpha} \frac{\partial \dot{q}_\alpha}{\partial \dot{q}'_\beta} = P_\beta \tag{13}$$

since q is independent $\dot{q}'$ (and hence L depends on $\dot{q}'$ only through $\dot{q}$). From Eq. (8) we have

$$\dot{q}_\alpha = \frac{\partial q_\alpha}{\partial q'_\gamma} \dot{q}'_\gamma + \frac{\partial q_\alpha}{\partial t},$$

so that

$$\frac{\partial \dot{q}_\alpha}{\partial \dot{q}'_\beta} = \frac{\partial q_\alpha}{\partial q'_\beta}.$$

Thus (13) can be written in terms of the transformation equations:

$$\frac{\partial L}{\partial \dot{q}_\alpha} \frac{\partial q_\alpha}{\partial q'_\beta} = P_\beta. \tag{14}$$

This equation is a statement in terms of the original variables that the momentum conjugate to one of the transformed variables is conserved. It gives a constant of the motion which is not immediately evident in the original variables. Needless to say, we are left with the problem of how to determine the coordinate transformation that will lead to such a cyclic coordinate.

Remark. Although p'_β may be a conserved quantity (i.e., time independent) in one of the coordinate systems, it will not necessarily be so in both; it will certainly be a constant of the motion in both, however.

It will be easier to find a coordinate system in which one of the coordinates is cyclic if we first look not at single transformations but at certain families of them. Consider, therefore, a set of transformations each one of which is labeled by the value of a certain parameter ε. Then for given ε we may write Eq. (7) in the form

$$q'_\alpha = q'_\alpha(q, t; \varepsilon), \tag{15}$$

and its inverse Eq. (9) in the form

$$q_\alpha = q_\alpha(q', t; \varepsilon). \tag{16}$$

We shall assume that the transformation functions depend on ε continuously and differentiably, and we shall always choose the parametrization so that $\varepsilon = 0$ corresponds to the identity transformation. In general we will be talking not about points q, but paths $q(t)$. Then q' will have a rather complicated time dependence, part coming from the time dependence of the transformation equations (15) and part from $q(t)$. We will often abbreviate the expression $q'(q, t; \varepsilon)$, writing

$$q'(q, t; \varepsilon) = q'(t, \varepsilon) = q'(t).$$

Then since $\varepsilon = 0$ corresponds to the identity transformation, we have

$$q'(q, t; 0) = q'(t, 0) = q(t).$$

Let us call such a set of transformations an ε-family. With this notation, Eqs. (15) and (16) can be written in the operator form,

$$q'(t, \varepsilon) = R(\varepsilon)\, q(t), \tag{17}$$

$$q(t) = S(\varepsilon)\, q'(t, \varepsilon), \tag{18}$$

where the ε-dependence of R and $S = R^{-1}$ is indicated, and $R(0) = S(0) = \mathbb{1}$. Since part of the time dependence of these equations comes from the explicit time dependence of $q(t)$ and $q'(t, \varepsilon)$ and part comes from the time dependence of $R(\varepsilon)$ and $S(\varepsilon)$, it follows that $q'(t, \varepsilon)$ may be time independent even if q is not. Finally, we shall write $dq'/dt = \dot{q}'(t, \varepsilon)$ for the partial time derivatives of $q'(t, \varepsilon)$ with ε fixed, as described at Eq. (35) of Chapter II.

For each value of ε there is a different transformation and hence from Eq. (8), a new Lagrangian function. We label the new Lagrangians L_ε (the subscript ε varies) rather than L', and then from Eqs. (8) and (17) we arrive at

$$L(q, \dot{q}, t) = L_\varepsilon(q', \dot{q}', t) = L_\varepsilon\left(R(\varepsilon)q, \frac{d}{dt}\,[R(\varepsilon)q], t\right).$$

By using Eqs. (17) and (18) we can rewrite this in the form

$$L_\varepsilon(q', \dot{q}', t) = L\left(S(\varepsilon)q', \frac{d}{dt}\,[S(\varepsilon)q'], t\right).$$

This equation is actually the definition of L_ε, and is thus an identity in $q'(t, \varepsilon)$: it is true for whatever set of functions is used for $q'(t, \varepsilon)$. Let $Q(t)$

be some arbitrary set of functions (which do not depend on ε). Then the equation

$$L_\varepsilon(Q, \dot{Q}, t) = L\left(S(\varepsilon)Q, \frac{d}{dt}[S(\varepsilon)Q], t\right)$$

tells us explicitly how to calculate the functional form of the new Lagrangian L_ε if we know L and the transformation $S(\varepsilon)$ from the new to the old coordinates. In particular, let us take $Q(t)$ to be some path in terms of the original coordinates, that is, let us write $Q(t) = q(t)$. Further, define

$$S(\varepsilon)q(t) = \bar{q}(t, \varepsilon). \tag{19}$$

Then we arrive at

$$L_\varepsilon(q, \dot{q}, t) = L(\bar{q}(t, \varepsilon), \dot{\bar{q}}(t, \varepsilon), t), \tag{20}$$

where (recall that d/dt is the partial derivative with ε fixed)

$$\dot{\bar{q}}(t, \varepsilon) = \frac{d}{dt}[S(\varepsilon)q(t)].$$

To understand this a little better, we can operate on both sides of (19) with $R(\varepsilon)$. Then, since $R(\varepsilon)S(\varepsilon) = \mathbb{1}$, we get

$$q = R(\varepsilon)\bar{q}(t, \varepsilon).$$

Comparing this with Eq. (17), we see that $\bar{q}(t, \varepsilon)$ is the set of coordinates which is transformed into q by $R(\varepsilon)$, whereas $q'(t, \varepsilon)$ is the set into which q is transformed by $R(\varepsilon)$. Clearly we have $\bar{q}(t, 0) = q(t)$.

Equation (20) shows that the ε-dependence of L_ε comes entirely from the ε-dependence of $\bar{q}(t, \varepsilon)$ and $\dot{\bar{q}}(t, \varepsilon)$. We have already come across such ε-dependence of a Lagrangian, namely when we derived the Euler-Lagrange equations from the variational principle in Chapter II. Equation (36) of that chapter is

$$\frac{\partial L_\varepsilon}{\partial \varepsilon} = -\Lambda_\alpha \frac{\partial \bar{q}_\alpha}{\partial \varepsilon} + \frac{d}{dt}\left[\frac{\partial L}{\partial \dot{q}_\alpha}\frac{\partial \bar{q}_\alpha}{\partial \varepsilon}\right],$$

where we have written

$$\Lambda_\alpha(\ddot{\bar{q}}, \dot{\bar{q}}, \bar{q}, t) = \frac{d}{dt}\frac{\partial L}{\partial \dot{q}_\alpha} - \frac{\partial L}{\partial q_\alpha}.$$

Here as always $\partial L/\partial q_\alpha$, for instance, is the function of $q, \dot{q}, t$ obtained from $L(q, \dot{q}, t)$ by taking the partial derivative with respect to q_α. In the last equation and the one preceding it, however, $q, \dot{q}$ are replaced by $\bar{q}, \dot{\bar{q}}$ in all such functions. These equations are true for all ε and in particular for $\varepsilon = 0$. Now let us assume that $q(t) \equiv \bar{q}(t, 0)$ is a physical motion, which means that the Euler-Lagrange equations are satisfied for the Lagrangian L in the original coordinate system (i.e., at $\varepsilon = 0$). Then for $\varepsilon = 0$ the Λ_α vanish, and we have

$$\frac{\partial L_\varepsilon}{\partial \varepsilon} = \frac{d}{dt}\left[\frac{\partial L}{\partial \dot{q}_\alpha}\frac{\partial \bar{q}_\alpha}{\partial \varepsilon}\right], \qquad \varepsilon = 0. \tag{21}$$

The importance of this equation lies in the time derivative on the right-hand side, for at $\varepsilon = 0$ it is the ordinary formal time derivative (see footnote 2, Chapter II). An obvious way to obtain a constant of the motion presents itself immediately: find an ε-family of transformations such that the expression on the left-hand side is the formal time derivative of some function $G(q, \dot{q}, t)$ (since it is evaluated at $\varepsilon = 0$, the ε-dependence drops out). Then (21) will read

$$\frac{d}{dt}\left[\frac{\partial L}{\partial \dot{q}_\alpha}\frac{\partial \bar{q}_\alpha}{\partial \varepsilon}\bigg|_{\varepsilon=0} - G\right] = 0, \tag{22}$$

so that the expression in square brackets is a constant of the motion. Moreover, all constants of the motion can in principle be found in this way. A more detailed discussion and proof of this statement will be found in Section 5(d).

Let us give an example. Consider a Lagrangian that does not depend explicitly on the time. Since the kinetic energy does not in general depend on the time explicitly, this means that neither does the potential energy. But then the total energy should be conserved. This may he seen from (21) in the following way. Consider the *time translation* transformation

$$q'(t, \varepsilon) = q(t + \varepsilon) = R(\varepsilon)q(t)$$

and its inverse

$$\bar{q}(t, \varepsilon) = S(\varepsilon)q(t) = q(t - \varepsilon).$$

Obviously

$$\frac{\partial \bar{q}_\alpha}{\partial \varepsilon} = -\frac{d\bar{q}_\alpha}{dt} = -\dot{\bar{q}}_\alpha.$$

Since L_ε depends on t and ε in the same way (except for a minus sign), namely through the t-dependence and ε-dependence of $\bar{q}$ and $\dot{\bar{q}}$, it is clear that $\partial L_\varepsilon/\partial\varepsilon = -dL_\varepsilon/dt$. On inserting these results into (21), we arrive at

$$-\frac{dL}{dt} = -\frac{d}{dt}\left[\frac{\partial L}{\partial \dot{q}_\alpha}\dot{q}_\alpha\right]$$

or the result that

$$\frac{\partial L}{\partial \dot{q}_\alpha}\dot{q}_\alpha - L \equiv H(q, \dot{q})$$

is a constant of the motion. It can be shown that under certain further assumptions about the form of the Lagrangian, H is the total energy of the system (see Problem 2). Thus the explicit time independence of L implies conservation of energy.

There is a special case of this procedure which is of particular interest. It is sometimes possible to find an ε-family that leaves the form of the Lagrangian invariant, that is one for which $L_\varepsilon = L_0$ is not a function of ε[1]. Then the left-hand side of (21) will be zero, and consequently the expression in brackets on the right-hand side itself will be a constant of the motion. When this is possible we say that the Lagrangian possesses a *symmetry*, and then *to every symmetry of the Lagrangian corresponds a constant of the motion.* More explicitly, if $\partial L_\varepsilon/\partial\varepsilon$ vanishes at $\varepsilon = 0$, then

$$\left.\frac{\partial L}{\partial \dot{q}_\alpha}\frac{\partial \bar{q}_\alpha}{\partial \varepsilon}\right|_{\varepsilon=0} = K, \tag{23}$$

where K is a constant.

A slightly more general case is the one in which $L_\varepsilon = L + \dot{\Phi}(q, t; \varepsilon)$. As we know, L_ε will then yield exactly the same equations of motion as will L. In this case

$$\left.\frac{\partial L_\varepsilon}{\partial \varepsilon}\right|_{\varepsilon=0} = \left.\frac{d}{dt}\frac{\partial \Phi}{\partial \varepsilon}\right|_{\varepsilon=0},$$

and then

$$\left[\frac{\partial L}{\partial \dot{q}_\alpha}\frac{\partial \bar{q}_\alpha}{\partial \varepsilon} - \frac{\partial \Phi}{\partial \varepsilon}\right]_{\varepsilon=0}$$

is a constant of the motion.

Remarks. 1. The converse assertion is not true: every constant of the motion does not correspond to a symmetry of the Lagrangian. It does, however, correspond to a time-dependent transformation of a more general kind [see Section 5(d)]. *2.* An interesting question, whose complete answer is not known to us [but see Kilmister (1965)], is how two symmetries must be related if the corresponding constants are to be independent. *3.* In the more general case of Eq. (22) we shall say that the Lagrangian is *quasi-symmetric.* Conservation of H for a time-independent Lagrangian is an example of quasisymmetry.

Let us illustrate with an example how symmetry leads to a constant of the motion. Consider the one-particle Lagrangian

$$L = \tfrac{1}{2}m(\dot{x}_1^2 + \dot{x}_2^2 + \dot{x}_3^2) - \tfrac{1}{2}k(x_1^2 + x_2^2) + f(x_3), \tag{24}$$

[1] Actually the Lagrangian need not be invariant for all ε; we require only that $\partial L_\varepsilon/\partial\varepsilon|_{\varepsilon=0} = 0$. But it can be shown that in all cases it is then possible to find an ε-family such that $\partial L_\varepsilon/\partial \varepsilon = 0$ for all ε.

where the x_α are the Cartesian coordinates of the particle. This Lagrangian is invariant under the transformation

$$x_1' = x_1 \cos \varepsilon + x_2 \sin \varepsilon,$$
$$x_2' = -x_1 \sin \varepsilon + x_2 \cos \varepsilon,$$
$$x_3' = x_3,$$

which represents a counterclockwise rotation of the coordinate system about the *3*-axis through an angle ε. The inverse transformation

$$\begin{aligned} \bar{x}_1 &= x_1 \cos \varepsilon - x_2 \sin \varepsilon, \\ \bar{x}_2 &= x_1 \sin \varepsilon + x_2 \cos \varepsilon, \\ \bar{x}_3 &= x_3, \end{aligned} \tag{25}$$

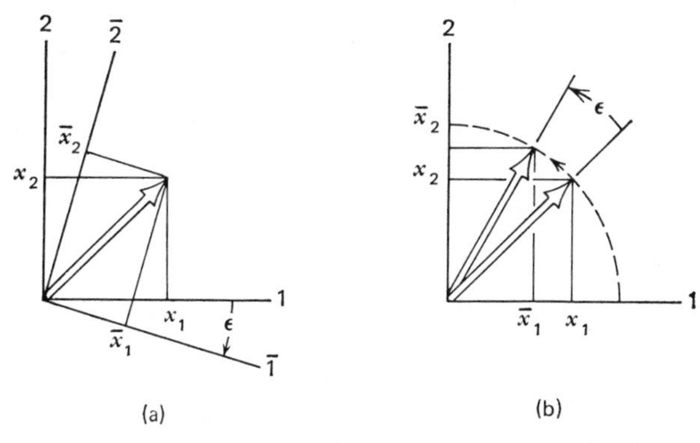

Figure 9. Passive and active views of the same rotation. In (a) the coordinates are rotated clockwise through an angle ε. In (b) the vector is rotated counter-clockwise through the same angle ε. As ε varies the vector sweeps out the dotted circle.

represents a clockwise rotation through an angle ε, as shown in Fig. 9(a). To see the invariance, one must, according to Eqs. (19) and (20), insert the right-hand side of Eqs. (25) into L. Some simple algebra then leads to the result that $L_\varepsilon(x, \dot{x}, t) = L(x, \dot{x}, t)$, so that $\partial L_\varepsilon/\partial \varepsilon = 0$ for all ε. The constant of the motion is given by (23):

$$\left.\frac{\partial L_\varepsilon}{\partial \dot{x}_\alpha} \frac{\partial \bar{x}_\alpha}{\partial \varepsilon}\right|_{\varepsilon=0} = m[\dot{x}_1(-x_1 \sin 0 - x_2 \cos 0) + \dot{x}_2(x_1 \cos 0 - x_2 \sin 0)]$$
$$= x_1 p_2 - x_2 p_1,$$

where $\mathbf{p} = m\dot{\mathbf{x}}$ is the momentum. Thus symmetry under rotation about the *3*-axis implies conservation of angular momentum about the *3*-axis.

III-3 THE ACTIVE VIEW; SOME SPECIAL SYMMETRIES

In order to relate the ideas of Section 2 to cyclic coordinates, as we have been intending, and to obtain a clearer understanding of the physics contained in all this formalism, it is helpful to look at the coordinate transformations from a different point of view. Consider Eq. (19), that is,

$$\bar{q}(t, \varepsilon) = S(\varepsilon)q(t).$$

Up to now we have interpreted it as relating the coordinates of a given configuration-space path in two different coordinate systems. This interpretation is called the *passive view* of the transformation equation. We can just as well say that the transformation equation relates the components of two different paths in a given coordinate system. This is called the *active view*.

In the active view Eq. (20) may be interpreted as saying that the Lagrangian remains the same function: its value along the path q is $L(t) = L(q(t), \dot{q}(t), t)$ and along the different path $\bar{q}$ its value is $L_\varepsilon(t) = L(\bar{q}(t, \varepsilon), \dot{\bar{q}}(t, \varepsilon), t)$. Just as in the passive view, we then arrive at (recall that d/dt is the partial derivative with ε held fixed)

$$\frac{\partial L_\varepsilon(t)}{\partial \varepsilon} = -\Lambda_\alpha \frac{\partial \bar{q}_\alpha}{\partial \varepsilon} + \frac{d}{dt}\left(\frac{\partial L}{\partial \dot{q}_\alpha} \frac{\partial \bar{q}_\alpha}{\partial \varepsilon}\right),$$

so that if $q(t)$ is a physical path, $\Lambda_\alpha|_{\varepsilon=0} = 0$, and hence

$$\left.\frac{\partial L_\varepsilon}{\partial \varepsilon}\right|_{\varepsilon=0} = \frac{d}{dt}\left[\frac{\partial L}{\partial \dot{q}_\alpha} \frac{\partial \bar{q}_\alpha}{\partial \varepsilon}\right]_{\varepsilon=0}.$$

This is obviously identical with Eq. (21). Again, as in the passive view, the Lagrangian is said to be symmetric or *invariant* under $S(\varepsilon)$ if $L(t) = L_\varepsilon(t)$ for all ε (but see footnote 1). In this case we have $\partial L_\varepsilon/\partial \varepsilon = 0$, so that

$$\left.\frac{\partial L}{\partial \dot{q}_\alpha} \frac{\partial \bar{q}_\alpha}{\partial \varepsilon}\right|_{\varepsilon=0}$$

is a constant of the motion. In the passive view a constant of the motion results if in two different coordinate systems related by the transformation $R(\varepsilon)$ the *form* of the Lagrangian is the same; in the active view of the same equations a constant of the motion results if along two paths related by the transformation $S(\varepsilon)$ the *value* (actually the time dependence) of the Lagrangian is the same. As an example, let us again consider the Lagrangian of Eq. (24). Turning to $S(\varepsilon)$ as given by (25), we see that in the active view it subjects the point with coordinates x_α to a counterclockwise rotation through an angle ε, bringing it into the point with coordinates $\bar{x}_\alpha$ as shown in Fig. 9(b). Clearly

L has the same values for $\bar{x}$ and x, so that, as before, the quantity

$$\left.\frac{\partial L}{\partial \dot{x}_\alpha}\frac{\partial \bar{x}_\alpha}{\partial \varepsilon}\right|_{\varepsilon=0} = x_1p_2 - x_2p_1$$

is a constant of the motion.

This example shows the relation between conservation of angular momentum and rotational invariance in a special case. There is, however, a general relation between conservation of linear and angular momentum and invariance of a system under translations and rotations, which we shall now demonstrate.

Consider a system of N particles whose Lagrangian is invariant under the simultaneous translation of each particle position through an arbitrary distance ε in the direction specified by some unit vector $\mathbf{n}$. This transformation is (i is the particle index)

$$\mathbf{x}_i' = \mathbf{x}_i + \varepsilon\mathbf{n} = R(\varepsilon)\mathbf{x}_i,$$

or

$$\bar{\mathbf{x}}_i = \mathbf{x}_i - \varepsilon\mathbf{n} = S(\varepsilon)\mathbf{x}_i.$$

Since L is unchanged by this transformation, we find that

$$\left.\frac{\partial L}{\partial \dot{x}_{\alpha i}}\frac{\partial \bar{x}_{\alpha i}}{\partial \varepsilon}\right|_{\varepsilon=0} = -\sum_{i=1}^{N}\mathbf{p}_i\cdot\mathbf{n}$$

is a conserved quantity. But $\mathbf{P} = \sum \mathbf{p}_i$ is the total linear momentum of the system. Thus we have the general result that if a system is invariant under translation in the direction $\mathbf{n}$, the component $\mathbf{P}\cdot\mathbf{n}$ of the linear momentum in that direction is a constant of the motion.

Next, consider a system of N particles which is invariant under the simultaneous rotation of all the position vectors of the particles through any angle ε about a direction $\mathbf{n}$. This direction can be taken as the 3-axis of a Cartesian coordinate system, so that the equations of the transformation are given by (25). In particular, the components of the ith particle transform according to

$$\bar{x}_{1i} = x_{1i}\cos\varepsilon - x_{2i}\sin\varepsilon,$$

$$\bar{x}_{2i} = x_{1i}\sin\varepsilon + x_{2i}\cos\varepsilon,$$

$$\bar{x}_{3i} = x_{3i}.$$

Then the quantity

$$\left.\frac{\partial L}{\partial \dot{x}_{\alpha i}}\frac{\partial \bar{x}_{\alpha i}}{\partial \varepsilon}\right|_{\varepsilon=0} = (x_{1i}p_{2i} - x_{2i}p_{1i}) = \sum_{i=1}^{N} l_{3i} = L_3$$

is conserved, where l_{3i} is the component of the angular momentum of the ith particle along the 3-axis and L_3 is the component of the total angular momentum along this direction. Thus if a system is invariant under rotations about a direction **n**, the component $\boldsymbol{L} \cdot \mathbf{n}$ of the total angular momentum in the **n** direction is a constant of the motion.

Having seen that conservation of linear and angular momentum is implied by invariance under translation and rotation, we may ask to what extent there are other possible symmetries leading to other possible constants of the motion. Let us consider now only time-independent ε-families of point transformations such that $\partial L/\partial\varepsilon|_{\varepsilon=0} = 0$ and, in particular, one-particle systems. We shall prove the following.

THEOREM. Consider an unconstrained particle moving in a potential. The only possible time-independent ε-families of point transformations under which the Lagrangian is invariant are combinations of rotations and translations.

PROOF. The Lagrangian for the particle is

$$L = \tfrac{1}{2}m(\dot{x}_1^2 + \dot{x}_2^2 + \dot{x}_3^2) - V(x_1, x_2, x_3).$$

In order for it to be invariant under a transformation, the potential and kinetic energy terms must be invariant separately, for there can be no cancellation between these terms: one depends only on the $\dot{x}_\alpha$ and the other only on the x_α. Thus a necessary condition that a transformation leave L invariant is that it leave the kinetic energy T invariant.

We want to find the most general time-independent transformation which leaves T invariant, that is, for which $\partial T_\varepsilon/\partial\varepsilon = 0$ at $\varepsilon = 0$. We have

$$T_\varepsilon(\dot{x}) = T(\dot{\bar{x}}) = \tfrac{1}{2}m\dot{\bar{x}}_\alpha\dot{\bar{x}}_\alpha,$$

and then

$$\frac{\partial T_\varepsilon}{\partial\varepsilon} = \frac{\partial T}{\partial\dot{\bar{x}}_\beta}\frac{\partial\dot{\bar{x}}_\beta}{\partial\varepsilon} = m\dot{\bar{x}}_\beta\frac{\partial\dot{\bar{x}}_\beta}{\partial\varepsilon}.$$

But since the transformations are time independent, we have

$$\dot{\bar{x}}_\beta = \frac{\partial\bar{x}_\beta}{\partial x_\alpha}\dot{x}_\alpha,$$

and we arrive at

$$\frac{\partial T_\varepsilon}{\partial\varepsilon} = m\dot{\bar{x}}_\beta\,\dot{x}_\alpha\frac{\partial^2\bar{x}_\beta}{\partial x_\alpha\,\partial\varepsilon}.$$

Now let us write $\partial\bar{x}_\beta/\partial\varepsilon|_{\varepsilon=0} = z_\beta(x)$, so that at $\varepsilon = 0$ we have

$$\frac{\partial T_\varepsilon}{\partial\varepsilon} = 0 = m\dot{x}_\beta\dot{x}_\alpha\frac{\partial z_\beta}{\partial x_\alpha}. \tag{26}$$

From this equation we will be able to find only the derivatives z_β of the $\bar{x}_\beta$ at $\varepsilon = 0$. But in fact that is all we will ever need, for all the relevant equations are taken at $\varepsilon = 0$. Let us therefore specialize to the simplest functions $\bar{x}_\beta(x;\,\varepsilon)$ that have the correct behavior at $\varepsilon = 0$, namely to

$$\bar{x}_\alpha = x_\alpha + \varepsilon z_\alpha(x), \tag{27}$$

where the z_α are solutions of (26). This may be thought of as either a particular ε-family with the properties we want, one of a large set of such families, or as the first terms in a Taylor's series in ε for a general family with the properties we want.

Since the $\dot{x}_\alpha$ are arbitrary, the expression on the right-hand side of Eq. (26) can be zero only if the coefficient of each distinct factor $\dot{x}_\alpha \dot{x}_\beta$ is zero. This means that

$$\frac{\partial z_\alpha}{\partial x_\beta} = -\frac{\partial z_\beta}{\partial x_\alpha}. \tag{28}$$

Let us take the derivative of this equation with respect to x_γ. Then by changing orders of differentiation and using (28) with different indices we arrive at

$$\frac{\partial^2 z_\alpha}{\partial x_\gamma\,\partial x_\beta} = -\frac{\partial^2 z_\beta}{\partial x_\alpha\,\partial x_\gamma} = \frac{\partial^2 z_\gamma}{\partial x_\beta\,\partial x_\alpha} = -\frac{\partial^2 z_\alpha}{\partial x_\gamma\,\partial x_\beta} = 0.$$

Thus the z_α are linear functions of the x_β, and we have

$$z_\alpha = a_\alpha + b_{\alpha\beta} x_\beta, \tag{29}$$

where, according to (28), $b_{\alpha\beta} = -b_{\beta\alpha}$. This *antisymmetry* of the $b_{\alpha\beta}$ can be expressed by writing[2]

$$b_{\alpha\beta} = \epsilon_{\alpha\beta\gamma} b_\gamma,$$

and then (29) becomes

$$z_\alpha = a_\alpha + \epsilon_{\alpha\beta\gamma} b_\gamma x_\beta. \tag{30}$$

The transformations given by Eq. (27) and (30) are the most general linear ε-families which leave the kinetic energy invariant at $\varepsilon = 0$. To see the nature of these transformations we can study them when only one of the six parameters in Eqs. (30) is nonzero. Thus if only one of the a_α is not zero and $b_\beta = 0$ for all β, the transformation is clearly a translation in the x_α direction. On the other hand if b_3, for instance, is the only nonzero constant, the transformation is

$$\begin{aligned} \bar{x}_1 &= x_1 + \varepsilon b_3 x_2, \\ \bar{x}_2 &= x_2 - \varepsilon b_3 x_1, \\ \bar{x}_3 &= x_3. \end{aligned} \tag{31}$$

As we have mentioned, this can be interpreted as the first two terms of a Taylor expansion of some general transformation. In particular, a comparison with Eq. (25), which gives the transformations for a rotation about the x_3 axis, shows that (31) is the beginning of the expansion for a rotation through the angle $-\varepsilon b_3$ about the x_3 axis.[3] In the same way nonzero values of b_1 and b_2 represent rotations

[2] The b_γ are expressed in terms of the $b_{\alpha\beta}$ by $b_\gamma = \frac{1}{2}\epsilon_{\gamma\alpha\beta} b_{\alpha\beta}$. See Appendix A for the definition of $\epsilon_{\alpha\beta\gamma}$.

[3] The relation between Eqs. (31) and (25) is deeper than just that Eq. (31) is the beginning of the Taylor expansion of (25), for there are many transformations which have (31) for the first two terms of the Taylor expansion. In general for most of these transformations $\partial T_\varepsilon / \partial \varepsilon$ is zero only at $\varepsilon = 0$, but only for (25) is it zero for all ε. (The possibility of obtaining such an extended transformation from one which gives invariance only first order in ε is what was asserted in footnote 1.) We do not, however, prove this statement. Thus what we are proving is really a local or "infinitesimal" version of our theorem, true only around $\varepsilon = 0$.

about the x_1 and x_2 axes. Thus the parameters a_α in Eq. (30) generate translations in the x_α directions, and the b_α generate rotations about the x_α directions. This completes the proof of the theorem.

In order now for the entire Lagrangian to have a symmetry, the potential energy must also be invariant under some transformation which is a special case of Eqs. (27) and (30). Thus a central potential (one depending only on the distance from a fixed point) is invariant under these transformations with $a_\alpha = 0$:

$$\bar{x}_\alpha = x_\alpha + \varepsilon \epsilon_{\alpha\beta\gamma} b_\gamma x_\beta.$$

Actually there are three independent symmetries involved, one for each of three linearly independent choices of the b_α (e.g., we can take all but one of the b_α to be zero, and of course there are three ways to do this). If just one of the b_α is nonzero, the resulting symmetry corresponds to conservation of the component of angular momentum in the α-direction. Consequently in a central potential the vector angular momentum is conserved, for the Lagrangian is invariant under rotation about any axis.

We have seen that invariance under translation yields conservation of momentum and that invariance under rotation yields conservation of angular momentum. We now see the special nature of linear and angular momenta. The form of the kinetic energy term in the Lagrangian imposes the restriction that every symmetry of an unconstrained one-particle system be a combination of rotations and translations, and hence that every constant of the motion which can be obtained from such a symmetry be a combination of the linear and angular momenta.

Let us try to generalize some of these ideas. We started this discussion in an attempt to find coordinate systems in which one or more coordinates are cyclic. What we have achieved is not really very surprising. Consider some time-independent ε-family of transformations that leave the Lagrangian invariant. Under these transformations $S(\varepsilon)$, looked at in the active view, points move so as to sweep out certain curves. For instance in Fig. 9(b) the point with coordinates $\bar{x}_\alpha$ moves, as ε is varied, so as to sweep out the circle indicated by the dotted line. These curves can be taken as part of a new coordinate grid. In other words, new generalized coordinates q_α can be chosen such that for displacement along such a curve all but one of the coordinates remain fixed. Call the varying coordinate q_1. Then the operator $S(\varepsilon)$ changes only q_1. Now since L is symmetric under $S(\varepsilon)$, it remains fixed as q_1 is varied, so that $\partial L/\partial q_1 = 0$, or q_1 is a cyclic coordinate.

In our example of Eq. (24) the circles about the *3*-axis are the curves along which q_1 varies (see Fig. 9). An obvious choice is then cylindrical coordinates.

We write

$$q_1 = \theta = \arc\tan(x_2/x_1),$$
$$q_2 = r = (x_1^2 + x_2^2)^{1/2},$$
$$q_3 = z = x_3;$$

or

$$x_1 = r\cos\theta,$$
$$x_2 = r\sin\theta,$$
$$x_3 = z.$$

Then the Lagrangian becomes

$$L = \tfrac{1}{2}m(\dot{r}^2 + r^2\dot{\theta}^2 + \dot{z}^2) - \tfrac{1}{2}kr^2 + f(z).$$

which is cyclic in θ, so that

$$\frac{\partial L}{\partial \dot{\theta}} = mr^2\dot{\theta} = m(x_1\dot{x}_2 - x_2\dot{x}_1)$$

is a constant of the motion.

There are several natural questions that remain to be answered. Among them are the relation between time-dependent transformations and constants of the motion, the connection between nonzero ε and derivatives at $\varepsilon = 0$, and several others. Some of these will be discussed later, especially in Chapter IX. See also Section 5(d).

III-4 THE LAGRANGIAN OF THE FREE PARTICLE. GALILEI INVARIANCE

The first of the two principles stated in Chapter I says that a free (isolated) particle moves with zero acceleration in every inertial system. We wish to analyze this statement in some more detail, and for this purpose we turn to a brief discussion of transformations between coordinate systems.

Figure 10 shows two Cartesian coordinate systems S and S' with different origins 0 and 0′ and with axes that are rotated with respect to each other. Let $\mathbf{x}$ and $\mathbf{y}$ be the position vectors of some point relative to 0 and 0′ respectively. If the vector from 0 to 0′ is $\mathbf{R}$, the three vectors are related by

$$\mathbf{x} = \mathbf{y} + \mathbf{R}. \tag{32}$$

We shall continue, as we have been, to write vectors in boldface. Thus by $\mathbf{a}$ we mean a physical vector, independent of coordinate system. Given

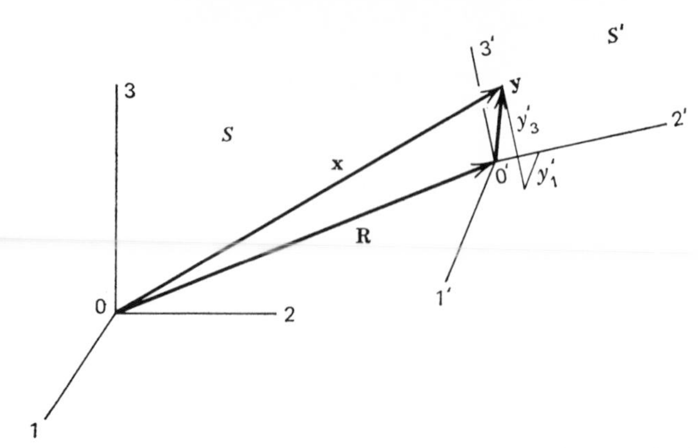

Figure 10. Two Cartesian frames with different origins and rotated with respect to each other.

a coordinate system, however, **a** can be represented by its components in that system. We shall henceforth write $a = (a_1, a_2, a_3)$ and $a' = (a_1', a_2', a_3')$ for the components of **a** in systems S and S', respectively. This is in keeping with our previous notation, where we have written q for the collection of generalized coordinates q_α.

Thus, for instance, for the set of components of **y** in systems S and S' we shall write y and y'. Then y and y' are related by a linear transformation of the form

$$y_\alpha = D_{\alpha\beta} y_\beta' \tag{33}$$

where the $D_{\alpha\beta}$ are a generalization to three dimensions of the two-dimensional rotation of Eq. (25). (We shall discuss the $D_{\alpha\beta}$ more fully in Section 5(c) of Chapter IV and again in Chapter V.) These equations may be abbreviated in the form

$$y = Dy', \tag{34}$$

where we may now think of D as the operator which is defined by (33). In the unprimed coordinate system Eq. (32) can now be written

$$x = Dy' + R. \tag{35}$$

If we know the relative orientation of the axes, we know D, and if we know the relative positions of the origins, we know R. Then, given the coordinates y' of a point particle in S', Eq. (35) tells us its coordinates in S.

With these preliminaries, we pass to a discussion of the dynamics. If S and S' are both inertial, Principle 1 asserts that $\ddot{x} = 0$ implies that $\ddot{y}' = 0$. Indeed, given one inertial system, the transformations to all others (and therefore all other inertial systems themselves) are determined by the condition that $\ddot{x} = 0$ imply $\ddot{y}' = 0$. These transformations are called *Galilei transformations*. We want now to reverse the argument to show that if we know all Galilei transformations, the only equations of motion that will be consistent with them are $\ddot{x} = 0$. In other words, from the requirement that the equations of motion are the same in all inertial systems, that is, in all systems connected by Galilei transformations, we shall derive both the equations of motion of a free particle and the free-particle Lagrangian [to within a term of the form $d\Phi(x, t)/dt$].

The possible Galilei transformations between inertial frames are of four kinds. *1.* The origins of the two systems do not coincide (in particular, at time $t = 0$). *2.* The clocks in the two systems, though going, we shall assume, at the same rate, are set to zero at different times. *3.* The origins of the two systems are moving at a constant relative velocity. *4.* The axes of the two systems, though both rectangular, are not parallel. These transformations fit into the scheme of Eq. (35) with constant $D_{\alpha\beta}$ and with $\mathbf{R} = \mathbf{v}t + \boldsymbol{\xi}$, where $\mathbf{v}$ and $\boldsymbol{\xi}$ are constant vectors. In addition, however, a fourth equation must be added connecting the times as measured in the two systems. Thus the general Galilei transformation is

$$\begin{aligned} x &= Dy' + vt + \xi, \\ t &= t' + \tau, \end{aligned} \tag{36}$$

where τ is a constant.

Now we want to find free-particle equations of motion which have the same form in terms of y', t' as they do in terms of x, t. This will certainly be true if $L(x, \dot{x}, t)$ and $L'(y', \dot{y}', t')$ have the same form when we calculate L' in the usual way from L. Let us try to proceed, then, with this as our requirement. Let $\Gamma x = y'$ be the inverse of the first of Eqs. (36). Then the equality

$$L'(x, \dot{x}, t) \equiv L\left(\Gamma^{-1}x, \frac{d}{dt}[\Gamma^{-1}x], t + \tau\right) = L(x, \dot{x}, t) \tag{37}$$

must hold. When Γ is known, this places certain restrictions on L, and from these restrictions we can obtain the form of the Lagrangian for a free particle.

Let us insert some special Galilei transformations into (37). First let $D = \mathbb{1}$, $v = \xi = 0$ (S and S' coincide but have different clocks). Then $L(x, \dot{x}, t) = L(x, \dot{x}, t + \tau)$, and this shows that L cannot depend on t. Similarly, if we set $D = \mathbb{1}$, $v = 0$, $\tau = 0$ (only the origins are different), we see that L cannot depend on x. Finally, having established that L can depend only on $\dot{x}$, we let $v = \xi = 0$, $\tau = 0$ (coincident origins, but rotated axes), and then $L(\dot{x}) = L(D\dot{x})$ for all rotations D. But this means that L cannot depend on the direction of $\dot{x}$, so it must depend on its magnitude (or the square of its magnitude). Thus

$$L = L(\dot{x}^2).$$

But this is actually too glib, for a Galilei transformation on x induces a transformation on $\dot{x}$. We find by calculation that

$$\dot{x} = \frac{dx}{dt} = D\dot{y}' + v.$$

This means that by applying a Galilei transformation for which $D = \mathbb{1}$, $\xi = 0$, $\tau = 0$ (parallel axes, nonzero relative velocity), $L(\dot{x}^2)$ is changed to $L([\dot{x} + v]^2)$, so that L cannot even depend on $\dot{x}^2$ and must therefore be a constant. Then the equations of motion of a free particle become trivial identities: $0 = 0$. This is an absurd result, and in order to avoid it we shall relax our requirements.

Let us no longer require that the free-particle Lagrangians in the two frames be identical but, instead, that the equations of motion in terms of the primed and unprimed coordinates be identical. Then the form of the Lagrangians in the two frames may actually differ, so long as this difference consists of a term of the form $d\Phi(x, t)/dt$. This weaker requirement is actually more physical, for all it says is that a particle which is moving freely according to an observer on one inertial frame should be moving freely according also to an observer on the other.

We now proceed as follows. First we write down the equations of motion in the primed (or unprimed) frame in terms of the Lagrangian, and then we perform various Galilei transformations, as we did before. The invariance of the equations imposes certain conditions on them, and through them on the Lagrangian. These are the conditions we wish to find.

In Cartesian coordinates the Euler-Lagrange equations for a single particle are

$$\frac{\partial^2 L}{\partial \dot{x}_\alpha\, \partial \dot{x}_\beta}\ddot{x}_\beta + \frac{\partial^2 L}{\partial \dot{x}_\alpha\, \partial x_\beta}\dot{x}_\beta + \frac{\partial^2 L}{\partial \dot{x}_\alpha\, \partial t} - \frac{\partial L}{\partial x_\alpha} = 0.$$

Consider just the first term. Since $\ddot{x}$ does not change under Galilei transformations with $D = \mathbb{1}$, and since it appears only in this term (all the others are functions of x, $\dot{x}$, and t only), its coefficient must be the same in all inertial frames. Then we may use the same argument we used before to show that if L is invariant, it is a constant, this time to show that

$$\frac{\partial^2 L}{\partial \dot{x}_\alpha\, \partial \dot{x}_\beta} = k_{\alpha\beta}, \tag{38}$$

where the $k_{\alpha\beta}$ are constants (and because the order of differentiation is arbitrary, $k_{\alpha\beta} = k_{\beta\alpha}$). By integrating (38) we arrive at

$$L = \tfrac{1}{2}k_{\alpha\beta}\dot{x}_\alpha\dot{x}_\beta + \dot{x}_\beta F_\beta(x, t) + G(x, t), \tag{39}$$

and then the Euler-Lagrange equations become

$$k_{\alpha\beta}\ddot{x}_\beta + \dot{x}_\beta\left[\frac{\partial F_\alpha}{\partial x_\beta} - \frac{\partial F_\beta}{\partial x_\alpha}\right] + \frac{\partial F_\alpha}{\partial t} - \frac{\partial G}{\partial x_\alpha} = 0. \tag{40}$$

The first term is now guaranteed to be Galilei invariant, so the remaining three must also be Galilei invariant. Again by the same argument they must form a constant. Now since this constant must be independent of $\dot{x}$, and since neither the F nor G depend on $\dot{x}$, we have

$$\frac{\partial F_\alpha}{\partial x_\beta} - \frac{\partial F_\beta}{\partial x_\alpha} = 0,$$

or in vector notation $\nabla \times \mathbf{F} = 0$. This means that $\mathbf{F}$ can be written as the gradient of a (time-dependent) scalar $\psi(x, t)$:

$$F_\alpha = \frac{\partial \psi}{\partial x_\alpha}. \tag{41}$$

Finally, the last two terms of (40) form a constant K_α, and then with (41) we have

$$\frac{\partial G}{\partial x_\alpha} = \frac{\partial}{\partial x_\alpha}\frac{\partial \psi}{\partial t} - K_\alpha,$$

which, when solved for G, yields

$$G = \frac{\partial \psi}{\partial t} - K_\beta x_\beta + C(t).$$

Now we combine this with (41) and (39) to arrive at

$$\begin{aligned} L &= \tfrac{1}{2}k_{\alpha\beta}\dot{x}_\alpha\dot{x}_\beta + \dot{x}_\beta\frac{\partial \psi}{\partial x_\beta} + \frac{\partial \psi}{\partial t} - K_\beta x_\beta + C \\ &= \tfrac{1}{2}k_{\alpha\beta}\dot{x}_\alpha\dot{x}_\beta - K_\beta x_\beta + \frac{d}{dt}\left[\psi + \int C\,dt\right]. \end{aligned}$$

The last term here can be dropped, for it does not contribute to the equations of motion; in any case, we expect the free-particle Lagrangian to be arbitrary to within an additive term of the form $d\Phi(x, t)/dt$. Now the Euler-Lagrange equations become

$$k_{\alpha\beta}\ddot{x}_\beta + K_\alpha = 0. \tag{42}$$

In order to determine the $k_{\alpha\beta}$ and the K_α, let us consider transformations for which $v = \xi = 0$ and $\tau = 0$, namely, pure rotations. We replace x_α by $D_{\alpha\beta}x_\beta$ in the Lagrangian, obtaining

$$L' = \tfrac{1}{2}k_{\alpha\beta}D_{\alpha\gamma}D_{\beta\delta}\dot{x}_\gamma\dot{x}_\delta - K_\beta D_{\beta\gamma}x_\gamma.$$

The equations of motion obtained from this Lagrangian are (changing some indices)

$$k_{\gamma\delta} D_{\gamma\alpha} D_{\delta\beta} \ddot{x}_\beta + K_\gamma D_{\gamma\alpha} = 0. \tag{43}$$

Now these equations are to be identical with (42), so that for all rotations we have

$$\begin{aligned} K_\gamma D_{\gamma\alpha} &= K_\alpha, \\ k_{\gamma\delta} D_{\gamma\alpha} D_{\delta\beta} &= k_{\alpha\beta}. \end{aligned} \tag{44}$$

It is shown in Chapter IV (see Problem 32 of that chapter) that this means that

$$\begin{aligned} K_\alpha &= 0, \\ k_{\alpha\beta} &= \delta_{\alpha\beta} k, \end{aligned} \tag{45}$$

where $\delta_{\alpha\beta}$ is the Kronecker δ, and k is some constant. Here we shall not prove these results but shall merely try to make them plausible. First, think of the K_α as being the three components of a vector $\mathbf{K}$. Then the first of Eqs. (44) says that the components of $\mathbf{K}$ in the original system (the right-hand side) are the same as its components in the rotated system (the left-hand side). But the only vector that has the same components in all such systems is the null vector, so $K_\alpha = 0$. This is the first of Eqs. (45). As for the second, we may think of the $k_{\alpha\beta}$ as being the coefficients of a quadratic equation of the form

$$k_{\alpha\beta} x_\alpha x_\beta = a.$$

Then in a rotated coordinate system this equation becomes

$$k_{\gamma\delta}\, D_{\gamma\alpha}\, D_{\delta\beta} x'_\alpha x'_\beta = a,$$

which is a quadratic equation of the same kind, but with new coefficients

$$k'_{\alpha\beta} = k_{\gamma\delta}\, D_{\gamma\alpha}\, D_{\delta\beta}.$$

In terms of these the second of Eqs. (44) says that the quadratic equation has the same coefficients (it is the same equation) in all rotated coordinate systems. But the only quadratic surface with this property, that is, the only one that looks the same in all such coordinate systems, is a sphere, and its coefficients are of the form $k\delta_{\alpha\beta}$, which is the result in the second of Eqs. (45).

Finally, then, the Lagrangian is

$$L = \tfrac{1}{2} k \delta_{\alpha\beta} \dot{x}_\alpha \dot{x}_\beta = \tfrac{1}{2} k \dot{x}^2, \tag{46}$$

where $\dot{x}^2$ is the square of the magnitude of the velocity. The Euler-Lagrange equations are then

$$k\ddot{x}_\alpha = 0, \tag{47}$$

as desired. Thus we have found from the principle of Galilei invariance that the free-particle Lagrangian is a multiple of the kinetic energy (up to a term of the form $d\Phi/dt$). Note that so long as k is finite, its value is irrelevant. And, in fact, we should not expect to discover what k is, or in other words to have the mass enter, on these considerations. The mass is involved only when we begin to consider the interaction of particles, and it first enters the principles of mechanics through the ratio of particle accelerations. Let us then turn to interactions.

In order to fit interactions into this framework, consider two particles. First let them be free, that is, let there be no interaction, so that each particle moves as though there were no other. Then the Lagrangian for this system will clearly be (no sum on A or B)

$$L = \tfrac{1}{2}k_A\dot{x}_A^2 + \tfrac{1}{2}k_B\dot{x}_B^2,$$

where x_A and x_B are the coordinates of the two particles in a given inertial frame, and k_A and k_B are arbitrary constants. The Euler-Lagrange equations are trivially shown to be the free-particle equations of motion.

Assume now that if the two particles interact, the Lagrangian is altered by an additive term, which we may call the interaction Lagrangian L_I, so that

$$L = \tfrac{1}{2}k_A\dot{x}_A^2 + \tfrac{1}{2}k_B\dot{x}_B^2 + L_I(x_A, x_B, \dot{x}_A, \dot{x}_B, t).$$

Assume further that this L is also Galilei invariant (except for a term of the form $d\Phi/dt$), but now under *simultaneous* Galilei transformation of x_A and x_B. This will follow if L_I is a function only of the differences $x = x_A - x_B$ and $\dot{x} = \dot{x}_A - \dot{x}_B$. Since, in addition, L is to be a scalar function,[4] it may depend only on scalars that can be formed of x and $\dot{x}$. The only such scalars are functions of the three independent scalars x^2, $\dot{x}^2$, and $(x, \dot{x}) \equiv x_\alpha\dot{x}_\alpha$. Consequently the general two-particle Lagrangian we shall consider will be of the form

$$L = \tfrac{1}{2}k_A\dot{x}_A^2 + \tfrac{1}{2}k_B\dot{x}_B^2 + L_I(x^2, \dot{x}^2, x_\alpha\dot{x}_\alpha). \tag{48}$$

If k_A and k_B are both positive (more generally, if $k_A \neq -k_B$) we may take $X = (k_Ax_A + k_Bx_B)/(k_A + k_B)$ and x as independent coordinates, and then it follows simply that

$$P = (k_A + k_B)\dot{X}$$

is a constant of the motion, for X is cyclic. We identify P with the total momentum, and then k_A and k_B can he taken as the masses of the two particles.

Let us summarize these results in the form of a set of statements. We shall not call these statements *axioms* or *principles*, for they apply to a relatively

[4] By a *scalar* function we mean merely one that associates a coordinate-system independent number with each point of configuration space [actually of $(\mathbf{x}, \dot{\mathbf{x}}, t)$-space]. This distinguishes it from, for instance, a vector function, which associates vectors, or three coordinate-system dependent numbers, with points in configuration space.

limited though fundamental range of phenomena. Nor shall we try to state them with complete rigor. In part this is because we assume that certain fundamental notions (e.g., time and inertial frame) are understood.[5] *1.* The variational principle. There exists a scalar Lagrangian function L such that the motion of a system is obtained from the condition that $\int L dt$ be an extremum along the motion. *2.* The Lagrangian of a free particle can be found from the conditions that (a) space is isotropic, (b) space and time are homogeneous, and (c) all inertial frames are equivalent. [These three conditions make up Galilei invariance. Condition (a) contributes the D, condition (b) the ξ and τ, and condition (c) the v to the Galilei transformations.] *3.* The Lagrangian of a system of interacting particles is the sum of the free-particle Lagrangians plus an interaction Lagrangian, itself Galilei invariant. Each free-particle term contains the mass of the particular particle.

From these three statements we have been able to derive the equations of motion for a free particle and conservation of momentum for two interacting particles. It is almost obvious that conservation of energy and angular momentum are also implied.

III-5 EXAMPLES, APPLICATIONS, AND EXTENSIONS

(a) Constants of the Motion, Symmetry, and Cyclic Coordinates

The Lagrangian

$$L = \tfrac{1}{2}m(\dot{q}_1^2 + \dot{q}_2^2) - (\alpha q_1 + \beta q_2),$$

where α and β are constants, describes a system with two degrees of freedom. According to the discussion of Section 1 there must therefore exist four independent constants of the motion. By using the Euler-Lagrange equations for the system one easily verifies that the following four independent functions are constants of the motion:

$$F_1 = \tfrac{1}{2}m(\dot{q}_1^2 + \dot{q}_2^2) + (\alpha q_1 + \beta q_2),$$

$$F_2 = m(\beta \dot{q}_1 - \alpha \dot{q}_2),$$

$$F_3 = m\dot{q}_2 + \beta t,$$

$$F_4 = q_1 - \frac{\alpha}{2m} t^2 - \dot{q}_1 t.$$

The first constant F_1 is the total energy of the system. The second F_2 is the component of the linear momentum $\mathbf{p} = m(\dot{q}_1 \mathbf{e}_1 + \dot{q}_2 \mathbf{e}_2)$ in the direction $\mathbf{n} = \beta \mathbf{e}_1 - \alpha \mathbf{e}_2$ which is perpendicular to the constant applied force $-(\alpha \mathbf{e}_1 + \beta \mathbf{e}_2)$.

[5] It is assumed also that space is Euclidean.

That is, $F_2 = \mathbf{p} \cdot \mathbf{n}$. Finally, F_3 and F_4 are time-dependent constants of the motion.

One can obtain F_2 from the fact that under the transformation

$$\begin{aligned} \bar{q}_1 &= q_1 + \varepsilon\beta, \\ \bar{q}_2 &= q_2 - \varepsilon\alpha, \end{aligned} \tag{49}$$

the Lagrangian L is transformed to

$$L_\varepsilon(q, \dot{q}) = L(\bar{q}, \dot{\bar{q}}) = L(q, \dot{q}),$$

so that $\partial L_\varepsilon / \partial \varepsilon = 0$. Then from Eq. (23), we find that

$$F_2 = \frac{\partial L}{\partial \dot{q}_\alpha} \frac{\partial \bar{q}_\alpha}{\partial \varepsilon} = m(\beta \dot{q}_1 - \alpha \dot{q}_2)$$

is a constant of the motion.

This example also illustrates the relation between a symmetry and a cyclic coordinate. The transformation of Eq. (49) leave the quantity

$$Q_1 = \alpha q_1 + \beta q_2$$

unchanged. If we choose Q_1 and

$$Q_2 = \beta q_1 - \alpha q_2$$

as new generalized coordinates, the Lagrangian becomes

$$L = \tfrac{1}{2} m(\alpha^2 + \beta^2)(\dot{Q}_1^2 + \dot{Q}_2^2) - Q_1.$$

It is seen that Q_2 is cyclic. Thus $m\dot{Q}_2 = F_2$ is a constant of the motion.

(b) A Time-Dependent Transformation for the Simple Harmonic Oscillator

The one-dimensional simple harmonic oscillator Lagrangian is

$$L = \tfrac{1}{2} m\dot{q}^2 - \tfrac{1}{2} kq^2.$$

Consider the time-dependent transformation

$$\bar{q} = q + \varepsilon \sin \omega t,$$

where $\omega = \sqrt{k/m}$. Under this transformation the Lagrangian is changed to

$$L_\varepsilon = \tfrac{1}{2} m(\dot{q} + \varepsilon\omega \cos \omega t)^2 - \tfrac{1}{2} k(q + \varepsilon \sin \omega t)^2,$$

so that

$$\left.\frac{\partial L_\varepsilon}{\partial \varepsilon}\right|_{\varepsilon=0} = m\omega\dot{q} \cos \omega t - kq \sin \omega t = \frac{dG}{dt},$$

where $G = m\omega q \cos \omega t$. It then follows from Eq. (22) that

$$F = G - \frac{\partial L}{\partial \dot{q}} \frac{\partial \bar{q}}{\partial \varepsilon} = m\omega \dot{q} \cos \omega t - m\dot{q} \sin \omega t$$

is a constant of the motion. This is easily verified by writing the solution in the form $q(t) = A \sin \omega t + B \cos \omega t$.

(c) The Kepler Problem

Perhaps the most famous problem in mechanics is the Kepler problem: to describe the motion of a particle in three dimensions moving in the potential $V = -k/x$. It can be solved by use of Eq. (51) of Chapter II (see Chapter II, Problem 14). However, an interesting alternative approach is to find the trajectory (or orbit) by showing that the Lagrangian

$$L = \tfrac{1}{2}m\dot{x}^2 + k/x$$

is quasisymmetric under each of the three transformations

$$\bar{x}_\alpha = x_\alpha - \varepsilon m[2\dot{x}_\alpha x_\lambda - x_\alpha \dot{x}_\lambda - \mathbf{x} \cdot \dot{\mathbf{x}} \delta_{\alpha\lambda}], \qquad \alpha, \lambda = 1, 2, 3.$$

(There is a different transformation for each value of λ.)

It is a straightforward calculation to show that under these transformations the Lagrangian becomes

$$\begin{aligned} L_\varepsilon = L &- \varepsilon m^2(2\dot{\mathbf{x}} \cdot \ddot{\mathbf{x}} x_\lambda - \mathbf{x} \cdot \dot{\mathbf{x}} \ddot{x}_\lambda - \mathbf{x} \cdot \ddot{\mathbf{x}} \dot{x}_\lambda) \\ &+ \varepsilon mk(\mathbf{x} \cdot \dot{\mathbf{x}} x_\lambda - x^2 \dot{x}_\lambda)/x^3 + 0(\varepsilon^2), \end{aligned}$$

where $0(\varepsilon^2)$ involves only terms containing ε to the second power. The derivative of this is

$$\left.\frac{\partial L_\varepsilon}{\partial \varepsilon}\right|_{\varepsilon=0} = \frac{dG_\lambda}{dt}$$

$[0(\varepsilon^2)$ does not contribute to the derivative at $\varepsilon = 0]$, where

$$G_\lambda = -m^2(\dot{x}^2 x_\lambda - \dot{\mathbf{x}} \cdot \mathbf{x} \dot{x}_\lambda) - mkx_\lambda/x.$$

Thus from Eq. (22) it follows that at $\varepsilon = 0$ the three quantities

$$\begin{aligned} A_\lambda &= G_\lambda - m\dot{x}_\alpha(\partial \bar{x}_\alpha/\partial \varepsilon) \\ &= m^2\dot{x}^2 x_\lambda - m^2(\dot{\mathbf{x}} \cdot \mathbf{x})\dot{x}_\lambda - mkx_\lambda/x \end{aligned}$$

are constants of the motion. These A_λ are the components of the vector

$$\mathbf{A} = \mathbf{p} \times \mathbf{L} - km\mathbf{x}/x, \tag{50}$$

known as the Runge-Lenz vector.

The fact that $\mathbf{A}$ is a constant of the motion can be used to obtain the orbit. Since the angular momentum $\mathbf{L}$ is also a constant of the motion and since

$\mathbf{x} \cdot \mathbf{L} = 0$ and $\mathbf{L} \cdot \mathbf{A} = 0$, it is clear that the motion is in the plane perpendicular to $\mathbf{L}$ and that $\mathbf{A}$ lies in this plane. If we measure angles in this plane from the constant vector $\mathbf{A}$, we have

$$\mathbf{x} \cdot \mathbf{A} = xA \cos \theta = L^2 - mkx$$

or

$$x = L^2[mk(1 + e \cos \theta)]^{-1}, \tag{51}$$

where $L = |\mathbf{L}|$, and $e = A/mk$. This is the equation of a conic section of eccentricity e with one focus at the origin (the center of force). The magnitude of $\mathbf{A}$ is emk, and $\mathbf{A}$ points along the direction of the major axis of the orbit.

(d) The General Relation Between Constants of the Motion and Transformations

We want to prove that every constant of the motion is related to at least one ε-family of transformations as in Eq. (22). More formally, we prove the following (Noether 1918).

THEOREM. Consider a system with no constraints. Then there exists an ε-family of transformations under which L is quasisymmetric, that is, such that

$$\left.\frac{\partial L_\varepsilon}{\partial \varepsilon}\right|_{\varepsilon=0} = \frac{dG}{dt} \tag{52}$$

for some function $G(q, \dot{q}, t)$, if and only if $F = G - [(\partial L/\partial \dot{q}_\alpha)(\partial \bar{q}_\alpha/\partial \varepsilon)]_{\varepsilon=0}$ is a constant of the motion.

Remark. Equation (52) is assumed to hold formally, or identically, that is, for all functions $q(t)$ (see Footnote 2 of Chapter II). Of course, by Eq. (21), $\partial L_\varepsilon/\partial \varepsilon|_0$ is equal to the total time derivative of $[(\partial L/\partial \dot{q}_\alpha)(\partial \bar{q}_\alpha/\partial \varepsilon)]_0$, but only for functions $q_\alpha(t)$ which are motions, that is, for $q_\alpha(t)$ which are solutions of the Euler-Lagrange equations.

PROOF. To prove the necessity ("only if"), consider an ε-family and assume a G exists such that (52) is true. Then for physical motions Eq. (21) yields

$$\left.\frac{\partial L}{\partial \varepsilon}\right|_{\varepsilon=0} = \frac{dG}{dt} = \frac{d}{dt}\left[\frac{\partial L}{\partial \dot{q}_\alpha}\frac{\partial \bar{q}_\alpha}{\partial \varepsilon}\right]_0,$$

and it follows immediately [remember that $\bar{q}(t, 0) = q(t)$ is a motion] that along a motion

$$\frac{d}{dt}\left[\frac{\partial L}{\partial \dot{q}_\alpha}\frac{\partial \bar{q}_\alpha}{\partial \varepsilon} - G\right]_0 = 0,$$

or that $dF/dt = 0$. Thus F is a constant of the motion.

Now we prove the sufficiency, that is, we prove that for every constant of the motion an ε-family of transformations can be found for which (52) is true. We assume that F is a constant of the motion and we shall find functions

$z_\alpha(q, \dot{q}, t)$ such that Eq. (52) holds for the special type of ε-family of the form

$$\bar{q}_\alpha = q_\alpha + \varepsilon z_\alpha(q, \dot{q}, t). \tag{53}$$

Actually this is not obviously an ε-family, for it depends on $\dot{q}$, but we shall discuss this (and the special ε-dependence) later. Rewrite (52) in the form

$$\left.\frac{\partial L_\varepsilon}{\partial \varepsilon}\right|_0 - \frac{d}{dt}\left[\frac{\partial L}{\partial \dot{q}_\alpha}\frac{\partial \bar{q}_\alpha}{\partial \varepsilon}\right]_0 = \frac{dF}{dt}. \tag{54}$$

We assert that if the z_α are chosen to satisfy the linear equations

$$\frac{\partial^2 L}{\partial \dot{q}_\alpha\, \partial \dot{q}_\beta} z_\alpha = -\frac{\partial F}{\partial \dot{q}_\beta}, \tag{55}$$

where F is the given constant of the motion, then Eq. (54) will hold. To demonstrate this, we will calculate each side of (54) separately and show that they are equal, which will essentially prove the theorem. The left-hand side is easily seen to be (ignoring the order of differentiation)

$$\Lambda = z_\alpha\left[\frac{\partial L}{\partial q_\alpha} - \frac{\partial^2 L}{\partial \dot{q}_\alpha\, \partial \dot{q}_\beta}\ddot{q}_\beta - \frac{\partial^2 L}{\partial \dot{q}_\alpha\, \partial q_\beta}\dot{q}_\beta - \frac{\partial^2 L}{\partial \dot{q}_\alpha\, \partial t}\right],$$

and the right-hand side is

$$\frac{dF}{dt} = \frac{\partial F}{\partial \dot{q}_\beta}\ddot{q}_\beta + \frac{\partial F}{\partial q_\gamma}\dot{q}_\gamma + \frac{\partial F}{\partial t}.$$

If the z_α are chosen in accordance with (55), the second term of Λ is equal to the first term of dF/dt. Thus what we must show is that

$$M(q, \dot{q}, t) = \frac{\partial F}{\partial q_\gamma}\dot{q}_\gamma + \frac{\partial F}{\partial t} - z_\alpha\left[\frac{\partial L}{\partial q_\alpha} - \frac{\partial^2 L}{\partial \dot{q}_\alpha \partial q_\beta}\dot{q}_\beta - \frac{\partial^2 L}{\partial \dot{q}_\alpha\, \partial t}\right] \tag{56}$$

vanishes. To do this, we solve (55) for the z_α, writing the solution in the form[6]

$$z_\alpha = -W_{\alpha\gamma}\frac{\partial F}{\partial \dot{q}_\gamma},$$

where the $W_{\alpha\beta} = W_{\beta\alpha}$ (see Chapter IV, Problem 33 to verify this symmetry) are functions of q, $\dot{q}$, t. With this, M becomes

$$M = \frac{\partial F}{\partial q_\gamma}\dot{q}_\gamma + \frac{\partial F}{\partial t} + \frac{\partial F}{\partial \dot{q}_\gamma}W_{\alpha\gamma}\left[\frac{\partial L}{\partial q_\alpha} - \frac{\partial^2 L}{\partial \dot{q}_\alpha\, \partial q_\beta}\dot{q}_\beta - \frac{\partial^2 L}{\partial \dot{q}_\alpha\, \partial t}\right].$$

Now consider a motion $q(t)$. The Euler-Lagrange equations read

$$\frac{\partial^2 L}{\partial \dot{q}_\alpha\, \partial \dot{q}_\gamma}\ddot{q}_\gamma = \frac{\partial L}{\partial q_\alpha} - \frac{\partial^2 L}{\partial \dot{q}_\alpha\, \partial q_\beta}\dot{q}_\beta - \frac{\partial^2 L}{\partial \dot{q}_\alpha\, \partial t}.$$

[6] The necessary and sufficient condition that unique solutions exist is that the determinant of the $\partial^2 L/\partial \dot{q}_\alpha\, \partial \dot{q}_\beta$ fail to vanish, which we assume to be true (see Chapter II, the Remark at the end of Section 1).

Note the similarity between this and (55); the difference between this equation for the $\ddot{q}_\alpha$ and the equation for the z_α is only on the right-hand side. The solutions are equally similar: replace $\partial F/\partial \dot{q}_\alpha$ by the new right-hand side, obtaining

$$\ddot{q}_\gamma = W_{\alpha\gamma}\left[\frac{\partial L}{\partial q_\alpha} - \frac{\partial^2 L}{\partial \dot{q}_\alpha\, \partial q_\beta}\dot{q}_\beta - \frac{\partial^2 L}{\partial \dot{q}_\alpha\, \partial t}\right].$$

Hence if $q(t)$ is a motion,

$$M(q, \dot{q}, t) = \frac{\partial F}{\partial q_\gamma}\dot{q}_\gamma + \frac{\partial F}{\partial t} + \frac{\partial F}{d\dot{q}_\gamma}\ddot{q}_\gamma = \frac{dF}{dt} = 0,$$

vanishing because F is a constant of the motion. Consequently $M(q, \dot{q}, t) = 0$ for every motion $q(t)$. But this is a contradiction unless M is identically zero, for otherwise it is equivalent to a constraint condition, and we have assumed no constraints. Since this is what we set out to prove at Eq. (56), the proof is now complete.

Now return to Eq. (55). In general the $W_{\alpha\beta}$ and $\partial F/\partial \dot{q}_\beta$ depend on $\dot{q}$ (although $W_{\alpha\beta}$ does not if L is quadratic in $\dot{q}$), and therefore so do the z_α. But then how can Eq. (53) be viewed as an ε-family, and why is the ε-dependence linear? The answers to these questions are related, and have to do with the kind of transformation we used to prove energy conservation. For instance, generalizing that transformation somewhat, we may have something like

$$\bar{q}_\alpha = f_\alpha(q(t + \varepsilon), t, \varepsilon), \tag{57}$$

where

$$f_\alpha(q, t, 0) = q_\alpha(t).$$

Then

$$\left.\frac{\partial \bar{q}_\alpha}{\partial \varepsilon}\right|_{\varepsilon=0} = \left[\frac{\partial f_\alpha}{\partial q_\beta}\dot{q}_\beta + \frac{\partial f_\alpha}{\partial \varepsilon}\right]_{\varepsilon=0}. \tag{58}$$

Now the transformation

$$\bar{q}_\alpha = q_\alpha + \varepsilon\left[\frac{\partial f_\alpha}{\partial q_\beta}\dot{q}_\beta + \frac{\partial f_\alpha}{\partial \varepsilon}\right]_{\varepsilon=0} \tag{59}$$

has the same ε-derivative at $\varepsilon = 0$ as does (57), namely Eq. (58). Since the ε-derivative at $\varepsilon = 0$ is all we need, Eq. (59), which is linear in ε, is as good a transformation for our purposes as is (57); in fact, it represents the first two terms in the Taylor expansion of (57) about $\varepsilon = 0$. When (59) is compared with (53), it is seen that z_α corresponds to the expression in square brackets. Thus the $\dot{q}$ dependence of z_α has to do with the complicated way that (57) depends on ε.

On the other hand, the transformation in (57) is no longer a transformation in configuration space: It does not transform every point q to some other point $\bar{q}$, but rather every path $q(t)$ to another one $\bar{q}(t)$. This is reflected in (59) by the $\dot{q}$ dependence. Transformations mixing the q_α and $\dot{q}_\alpha$ in this way are called transformations in phase space, and are discussed in a slightly altered form in Chapter VI.

We give two simple examples. Consider the one-dimensional free particle, $L = \frac{1}{2}m\dot{q}^2$. We know that the momentum $F = m\dot{q}$ is conserved. To find z we insert this F and L into (55), obtaining

$$mz = -m, \qquad \text{or} \qquad z = -1.$$

The ε-family is then given to terms linear in ε by (53). We arrive at

$$\bar{q} = q - \varepsilon.$$

The Lagrangian is quasisymmetric (in this case also symmetric) under translation. It is generally true, incidentally, that if F is linear in $\dot{q}$, then the transformation given by (53) will be an ε-family in configuration space.

Now consider general one-dimensional motion in a time-independent potential, $L = \frac{1}{2}m\dot{q}^2 - V(q)$. We know that the energy $F = \frac{1}{2}m\dot{q}^2 + V(q)$ is conserved. In this case (55) is

$$mz = -m\dot{q}, \qquad \text{or} \qquad z = -\dot{q}$$

and (53) is then

$$\bar{q} = q - \varepsilon\dot{q}. \tag{60}$$

We leave the rest to the student (Problem 12).

Problems

1. Find $2n$ independent constants of the motion for each of the following systems. (a) A freely falling body. (b) A particle in a central $1/x$ potential (the two-dimensional Kepler problem). (c) Two particles attracting each other with a $1/x^2$ force. (d) The simple harmonic oscillator. For all cases find the motion algebraically in terms of these functions.
2. Consider a Lagrangian $L = T - V$. Show that if T is homogeneous quadratic in the $\dot{q}_\alpha$ and if V is velocity-independent, then $\dot{q}_\alpha\, \partial L/\partial \dot{q}_\alpha - L$ is the total energy.
3. Show that if the determinant of the $\partial^2 L/\partial \dot{q}_\alpha\, \partial \dot{q}_\beta$ is zero, then the system described by the Lagrangian L has less than n degrees of freedom.
4. Consider the Lagrangian for a three-dimensional free particle, namely $L = \frac{1}{2}m\dot{x}^2$. Perform the transformation $\mathbf{x}' = \mathbf{x} + \mathbf{v}t + \frac{1}{2}\mathbf{g}t^2$, where $\mathbf{v}$ and $\mathbf{g}$ are constant vectors. Find the Lagrangian L' in terms of $\mathbf{x}'$, and check explicitly that the motion predicted by $L'(\mathbf{x}')$ is the same as that predicted by $L(\mathbf{x})$. How can this be used to solve the problem of a freely falling body? Can this be extended to other potentials? Is it useful?
5. Consider the one-dimensional Lagrangian $L = 2q^2\dot{q}^2$. Solve for the motion. Find a time-independent coordinate transformation which makes q' cyclic.

6. From symmetry considerations alone find a conserved quantity for the Lagrangian (in three dimensions)

$$L = \tfrac{1}{2}m\dot{x}^2 - V_0 x_1 \sin(2\pi x_3/R) - V_0 x_2 \cos(2\pi x_3/R),$$

where V_0 and R are constants. Does this agree with the theorem of Section 3? What is the symmetry of L? Find generalized coordinates, one of which is cyclic.

7. The three-dimensional isotropic harmonic oscillator Lagrangian is

$$L = \tfrac{1}{2}m\dot{x}^2 - \tfrac{1}{2}kx^2.$$

Show that the six quantities (why six?)

$$F_{\alpha\beta} = \tfrac{1}{2}m\dot{x}_\alpha\dot{x}_\beta + \tfrac{1}{2}kx_\alpha x_\beta$$

are conserved. Find $\dot{q}$-dependent transformations corresponding to these constants [see Eq. (55)] and then others of the form of (57). Check that L is quasisymmetric. Are the $F_{\alpha\beta}$ independent? If not, find a relation between them and find other independent constants of the motion.

8. Obtain the equations of motion of the system described by the Lagrangian

$$L = (\mathbf{x} + \dot{\mathbf{x}})^2 + 2\mathbf{x} \cdot \dot{\mathbf{x}}t.$$

Perform a general Galilei transformation and find the transformed Lagrangian. Obtain equations of motion from the new Lagrangian. Show explicitly that L differs from the free-particle Lagrangian of Eq. (46) by a term of the form $d\Phi/dt$.

9. Consider a universe invariant under space and time translations, under transformations to nonrotated coordinate systems moving at constant velocity, and under rotations about a fixed direction. Take this direction to be the x_3 axis of some coordinate system. Show then that the general form of the free-particle Lagrangian would be

$$L = k(\dot{x}_1^2 + \dot{x}_2^2) + h\dot{x}_3^2 + d\Phi(\mathbf{x}, t)/dt.$$

What quantities are conserved in the real world that are not conserved in this invented universe? (Such a Lagrangian arises in describing the motion of an electron in an anisotropic crystal.)

10. Consider the Lagrangian $L = \dot{q}^2(6q^2 - 4q\dot{q}t + \dot{q}^2t^2)$. Find the motion and compare the result with a free particle in one dimension. What is the relation of this to the theorem near the end of Chapter II, Section 2?

11. Let q_γ be cyclic in $L(q, \dot{q})$, so that the conjugate momentum p_γ is conserved. If L is replaced by $L' = L + d\Phi(q, t)/dt$, and if $\partial\Phi/\partial q_\gamma \neq 0$, q_γ will not be cyclic in L', and the conjugate momentum will therefore not be conserved. Yet we know that L and L' yield the same equations of motion. Explain the paradox: how can the same equations of motion give both conservation and nonconservation?

12. Equation (60) should be related to the transformation $q'(t, \varepsilon) = q(t + \varepsilon)$ which we used to prove energy conservation. Explain the relationship.

CHAPTER IV

Vector Spaces

Vector spaces play an important role in many branches of modern physics, especially in quantum mechanics. They are also very useful in classical mechanics and, since we shall be using vector spaces extensively as we proceed, we shall devote this chapter to the general theory. Readers who are already familiar with this subject may want to skip to the discussion of some important applications in Section 5. Vector spaces are used throughout the rest of the book.

IV-1 DEFINITIONS AND GENERAL PROPERTIES

In elementary mechanics vectors are defined in three dimensions as physical objects having direction and magnitude and adding according to some version of the parallelogram law. If addition is understood properly, one can then show that forces, velocities, displacements, electric fields, and many other things are vectors. But the three-dimensional notion of a vector is an example of a more abstract one, which we shall now describe:

A *linear space* or *vector space* S is a set of elements $\mathbf{x}$, $\mathbf{y}$, . . . , called vectors, with the following properties. First, there is an operation defined on S called addition and denoted by the symbol $+$, which must satisfy four axioms or conditions.[1]

G1. Closure. If $\mathbf{x}, \mathbf{y} \in S$, then there is in S a vector called their sum and written $\mathbf{x} + \mathbf{y} = \mathbf{y} + \mathbf{x}$.

G2. Associativity. If $\mathbf{x}, \mathbf{y}, \mathbf{z} \in S$, then $(\mathbf{x} + \mathbf{y}) + \mathbf{z} = \mathbf{x} + (\mathbf{y} + \mathbf{z})$; that is, the order of adding vectors does not matter.

G3. Existence of a null vector. There exists a vector $0 \in S$ such that $0 + \mathbf{x} = \mathbf{x}$ for all $\mathbf{x} \in S$. This vector 0 is called a *null vector.*[2]

G4. Existence of an inverse. For each $\mathbf{x} \in S$ there is an *inverse* (or *negative*) $\mathbf{y} \in S$ such that $\mathbf{x} + \mathbf{y} = 0$.

[1] The symbol $\in$ used below means "belongs to" or "belong to."

[2] We shall use the symbol 0 for both the null vector and the number zero. This should lead to little or no confusion.

Remarks. 1. It can be shown (see Problem 1) that the inverse of a given vector is unique. We shall write $-\mathbf{x}$ for the inverse of $\mathbf{x}$. 2. These four axioms taken together state that S forms an Abelian group under addition (see Chapter IX).

Every vector space has structure in addition to that deriving from the above four axioms. For those vector spaces with which we shall be dealing, this structure is contained in the statement that S is defined *over the real or complex numbers.* What this means is that every vector in S may be multiplied by any number $\alpha, \beta, \ldots$ (we may use either the real or complex numbers), and this multiplication must satisfy five additional axioms or conditions.

M1. Closure. For every $\mathbf{x} \in S$ and every number α, there is in S a vector called their product and written $\alpha\mathbf{x}$.

M2. Associativity. For $\mathbf{x} \in S$ and α and β numbers, $(\alpha\beta)\mathbf{x} = \alpha(\beta\mathbf{x}) \equiv \alpha\beta\mathbf{x}$.

M3. Distributive with respect to vector addition. For $\mathbf{x}, \mathbf{y} \in S$ and α a number, $\alpha(\mathbf{x} + \mathbf{y}) = \alpha\mathbf{x} + \alpha\mathbf{y}$.

M4. Distributive with respect to addition of numbers. For $\mathbf{x} \in S$ and α and β numbers, $(\alpha + \beta)\mathbf{x} = \alpha\mathbf{x} + \beta\mathbf{x}$.

M5. Multiplication by 1. For every $\mathbf{x} \in S$, $1\mathbf{x} = \mathbf{x}$.

We shall have occasion now and then to talk of parts of vector spaces, or subsets of the set of all vectors in S. When such a subset itself satisfies the vector-space axioms, it is called a *subspace* of S.

One of the most important relations between vectors is that of linear dependence or independence. Let $\mathbf{x}_1, \mathbf{x}_2, \ldots, \mathbf{x}_k$ be a set of k nonnull vectors in S. Then we shall say that the $\mathbf{x}_i$ are *linearly dependent* if there exist k numbers $\alpha_1, \alpha_2, \ldots, \alpha_k$, not all zero, such that

$$\sum_{i=1}^{k} \alpha_i \mathbf{x}_i = 0.$$

If no such numbers exist, or if this equation can be satisfied only if all of the α_i are zero, the $\mathbf{x}_i$ are said to be *linearly independent.*

As a concrete example which is not too familiar, consider the set P_m of all polynomials $f(p)$ with real coefficients and of degree m or less in some variable p. To form a vector space out of this set, we must define addition and multiplication in agreement with the nine axioms. When these operations are taken to be ordinary addition and multiplication, P_m is found to be a vector space over the real numbers, as is easily verified. The null vector is the polynomial which is identically zero (all of whose coefficients are zero), and the negative of $f(p)$ is just $-f(p)$. Two polynomials are linearly independent if (but not only if!) they are of different degrees. If $q \leq m$ then P_q

is a subspace of P_m. Because we shall often use this example to illustrate points of the theory as we proceed, the reader should verify the axioms explicitly and become familiar with the operations in P_m.

A *basis* in S is a set of vectors $\mathbf{x}_1, \ldots, \mathbf{x}_n \in S$ such that (1) it is linearly independent, and (2) every other vector in S can be written as a linear combination of these vectors. That is, for every $\mathbf{y} \in S$ there exists a set of numbers $\eta_1, \ldots, \eta_n$ such that $\mathbf{y} = \sum_{i=1}^{n} \eta_i \mathbf{x}_i$. (For instance a basis in P_m is formed by the monomials $1, p, p^2, \ldots, p^m$.) It follows at once that the η_i are unique; for if we assume that there exists another set of numbers $\eta_1', \ldots, \eta_n'$ such that $\mathbf{y} = \sum_{i=1}^{n} \eta_i' \mathbf{x}_i$ then we must have $\mathbf{y} - \mathbf{y} = 0 = \sum_{i=1}^{n} (\eta_i - \eta_i')\mathbf{x}_i$. But by assumption the $\mathbf{x}_i$ are linearly independent, so that the coefficient of each $\mathbf{x}_i$ is zero. Thus $\eta_i = \eta_i'$. These unique numbers η_i are called the *components* of $\mathbf{y}$ in the basis $\mathbf{x}_i$.

If we assume that S has one basis

$$\mathbf{x}_1, \mathbf{x}_2, \ldots, \mathbf{x}_n \tag{1}$$

containing n vectors,[3] then every other set of n linearly independent vectors in S is also a basis. To prove this, let

$$\mathbf{y}_1, \mathbf{y}_2, \ldots, \mathbf{y}_n \tag{2}$$

be such a linearly independent set. We shall prove the assertion by showing that any vector $\mathbf{z}$ that can be written as a linear combination of the vectors in (1) can also be written as a linear combination of the vectors in (2). Now since (1) is a basis there exist numbers $\alpha_1, \ldots, \alpha_n$ such that $\mathbf{y}_n = \sum_{i=1}^{n} \alpha_i \mathbf{x}_i$. This means that for any k such that $\alpha_k \neq 0$, we can write $\mathbf{x}_k$ as a linear combination of the vectors in the set

$$\mathbf{y}_n, \mathbf{x}_1, \ldots, \mathbf{x}_{k-1}, \mathbf{x}_{k+1}, \ldots, \mathbf{x}_n. \tag{3}$$

Then instead of writing some arbitrary vector $\mathbf{z} \in S$ in terms of the vectors in (1) we may write it in terms of (3) by replacing $\mathbf{x}_k$ by its expression in terms of the vectors in (3). Thus (3) satisfies the second property of a basis. But this means that numbers $\beta_0, \beta_1, \ldots, \beta_{k-1}, \beta_{k+1}, \ldots, \beta_n$ exist such that

$$\mathbf{y}_{n-1} = \beta_0 \mathbf{y}_n + \beta_1 \mathbf{x}_1 + \cdots + \beta_{k-1}\mathbf{x}_{k-1} + \beta_{k+1}\mathbf{x}_{k+1} + \cdots + \beta_n \mathbf{x}_n.$$

[3] We are assuming that there exists such a basis in S with n finite. There are, on the other hand, *infinite-dimensional* spaces in which there are no finite bases. Many of the definitions and theorems that follow are not valid in infinite-dimensional spaces, and we shall sometimes point this out. Some of the Problems involve spaces of infinite dimension in order to exhibit some of the differences.

This means that for any m such that $\beta_m \neq 0$ we can write $\mathbf{x}_m$ as a linear combination of the vectors in the set

$$\mathbf{y}_{n-1}, \mathbf{y}_n, \mathbf{x}_1, \ldots, \mathbf{x}_{k-1}, \mathbf{x}_{k+1}, \ldots, \mathbf{x}_{m-1}, \mathbf{x}_{m+1}, \ldots, \mathbf{x}_n. \tag{4}$$

(We can be sure that there exists at least one $\beta_m \neq 0$ for $m > 0$ because the $\mathbf{y}_i$ are linearly independent.) Then instead of writing $\mathbf{z}$ in terms of vectors in (3) we may write it in terms of (4) by replacing $\mathbf{x}_m$ by its expression in terms of the vectors in (4). Thus (4) satisfies the second property of a basis. We continue in this manner step by step, at each step replacing one of the $\mathbf{x}_i$ from (1) by one of the $\mathbf{y}_i$ from (2). At each step $\mathbf{z}$ can be written in terms of the new set. After n steps we will have eliminated all the $\mathbf{x}_i$ and will have thus shown that any vector $\mathbf{z}$ can be written in terms of the $\mathbf{y}_i$. Since the $\mathbf{y}_i$ are already assumed to be linearly independent, this proves that they form a basis. Every set of n linearly independent vectors is thus a basis.

If the set of n vectors $\mathbf{x}_1, \ldots, \mathbf{x}_n$ is a basis in S, then every set of $n + 1$ (or more) vectors is linearly dependent. To see this, assume the contrary: let $\mathbf{z}_1, \ldots, \mathbf{z}_m$ $(m > n)$ be a set of linearly independent vectors. Every subset of this set must also be linearly independent (check this!) and so from the preceding paragraph the subset $\mathbf{z}_1, \ldots, \mathbf{z}_n$ is a basis. Then $\mathbf{z}_k$ $(n < k \leq m)$ can be expressed as a linearly combination of $\mathbf{z}_1, \ldots, \mathbf{z}_n$ and so the set $\mathbf{z}_1, \ldots, \mathbf{z}_m$ is not linearly independent. It follows from this that there cannot exist a basis with more than n vectors. Furthermore, by reversing the argument we can see that there cannot exist a basis with fewer than n vectors. For if $\mathbf{y}_1, \ldots, \mathbf{y}_m$ is a basis with $m < n$, then the n vectors $\mathbf{x}_1, \ldots, \mathbf{x}_n$ must be linearly dependent and so cannot form a basis, contrary to our original assumption.

The result of this rather lengthy discussion is that every vector space S has associated with it a unique number n (assumed finite) such that every set of n vectors which satisfies either one of the two properties in the definition of a basis also satisfies the other property. This n is called the *dimension* of S. No set of more or less than n vectors can form a basis in S. From now on we shall use the summation convention in an n-dimensional space to indicate sums from 1 to n.

Note that the components of a vector depend on the basis in which they are expressed, although, as we have seen, once the basis is chosen the components are unique. It is left as an exercise (see Problem 5) to show that if $\mathbf{x} + \mathbf{y} = \mathbf{z}$, the ith component of $\mathbf{z}$ is the sum of the ith components of $\mathbf{x}$ and $\mathbf{y}$, and similarly that if $\mathbf{y} = \alpha\mathbf{x}$, the ith component of $\mathbf{y}$ is α times the ith component of $\mathbf{x}$. Given a basis, then, we may identify a vector with the set of its components, which we may display in order, say in a column. Such a column of numbers, called a *column vector*, is said to *represent* the vector. In adding vectors we add the components in their column vectors, entry by entry, to obtain a new column vector; in multiplying a vector by a number, we multiply each component of the column vector by that number. The column vector representing a vector $\mathbf{x}$ will be denoted by italic x. It is convenient to use such column vectors in performing actual calculations, and we shall make use of them later, when we apply the theory to specific problems.

Since it is often convenient to represent vectors in this way, as well as for other reasons (e.g., in one basis the column vector of a given vector may have fewer or simpler entries than in another), it is helpful to be able to change bases at will. Suppose we know the column vector in one basis. How do we find it in another? Let $\mathbf{x}_1, \ldots, \mathbf{x}_n$ and $\mathbf{y}_1, \ldots, \mathbf{y}_n$ be bases in S. Since the $\mathbf{y}_i$ form a basis, each of the $\mathbf{x}_i$ can be expressed in terms of the $\mathbf{y}_i$. Let these components of $\mathbf{x}_i$ be τ_{ji}, so that

$$\mathbf{x}_i = \tau_{ji}\mathbf{y}_j, \tag{5}$$

and let $\mathbf{z}$ be some vector in S with components ζ_i and ζ_i' in the two bases, that is

$$\mathbf{z} = \zeta_i'\mathbf{y}_i = \zeta_j\mathbf{x}_j. \tag{6}$$

In order to find the ζ_i' in terms of the ζ_i we need only insert (5) into the second equality of (6). Then we have (note the changes of indices)

$$\mathbf{z} = \zeta_i'\mathbf{y}_i = \zeta_j\tau_{ij}\mathbf{y}_i.$$

Since the components of $\mathbf{z}$ in a given basis are unique, this means that

$$\zeta_i' = \tau_{ij}\zeta_j. \tag{7}$$

Thus if we know the components of one set of basis vectors in the other basis, we can find the components of any vector in the second basis in terms of its components in the first. The collection of n^2 coefficients τ_{ij} is called the *transformation matrix*, and Eq. (7) is called the transformation equation from the basis $\mathbf{x}_i$ to the basis $\mathbf{y}_i$.

IV-2 LINEAR OPERATORS

We have already come across operators in Chapter III. We shall define operators here in much the same way. An *operator* on S is an association or mapping by which with each vector $\mathbf{x} \in S$ we associate another vector $\mathbf{y} \in S$. We write $\mathbf{y} = \mathbf{A}\mathbf{x}$, and call $\mathbf{A}$ the operator. An operator $\mathbf{A}$ is called *linear* if for each pair $\mathbf{x}$, $\mathbf{y} \in S$ and for each pair of numbers α and β

$$\mathbf{A}(\alpha\mathbf{x} + \beta\mathbf{y}) = \alpha(\mathbf{A}\mathbf{x}) + \beta(\mathbf{A}\mathbf{y}).$$

In what follows we shall speak only of linear operators, and shall therefore usually omit the qualification that they are linear.

An example of a linear operator on P_m is the derivative operator $\mathbf{D}$, defined by $\mathbf{D}f = df/dp$. The operator which maps $f(p)$ into $f(p) + p$ is nonlinear, as is easily verified. It is important to note that an operator must map every vector into another vector. Thus multiplication $\mathbf{M}$ on P_m, defined by $\mathbf{M}f = pf$, is not an admissible operator, for $\mathbf{M}p^m = p^{m+1}$ is not in P_m. On the other hand $\mathbf{MD} \equiv \mathbf{T}$ is an admissible operator.

Remark. It is sometimes useful to define operators that act not on all of S, but only on some part of it. The part of the space on which the operator is defined is then called the *domain* of the operator. Thus, for instance, $\mathbf{M}$ can be defined with domain P_{m-1} in P_m. Note, however, that although the domain of $\mathbf{M}$ is P_{m-1}, the domain of $\mathbf{MD}$ is all of P_m. On the other hand, although the domain of $\mathbf{DM}$ is P_{m-1}, it can be extended in a natural way to all of P_m. This is because if $f \in P_{m-1}$, then $\mathbf{DM}f = p\, df/dp + f$, an operation which is valid for all of P_m.

It is helpful to define addition of operators and multiplication by numbers so that the operators themselves form a vector space in accordance with the general definition. We leave it to the student to show that this result will be achieved if $\alpha\mathbf{A}$ is defined by

$$(\alpha\mathbf{A})\mathbf{x} = \alpha(\mathbf{A}\mathbf{x}) = \mathbf{A}(\alpha\mathbf{x}),$$

and if $\mathbf{A} + \mathbf{B}$ is defined by

$$(\mathbf{A} + \mathbf{B})\mathbf{x} = \mathbf{A}\mathbf{x} + \mathbf{B}\mathbf{x}$$

for all $\mathbf{x} \in S$ and for all operators $\mathbf{A}$ and $\mathbf{B}$. The *null operator*, also written 0 like the null vector and the number zero, is defined as the operator that maps every $\mathbf{x} \in S$ into the null vector.

In addition to this vector-space structure, moreover, we shall define multiplication of operators in the same way as we did in Chapter III. The operator $\mathbf{AB}$ is defined by

$$(\mathbf{AB})\mathbf{x} = \mathbf{A}(\mathbf{B}\mathbf{x}) \equiv \mathbf{AB}\mathbf{x}.$$

It follows immediately that multiplication is associative, that is, that $\mathbf{A}(\mathbf{BC}) = (\mathbf{AB})\mathbf{C} \equiv \mathbf{ABC}$. On the other hand, multiplication of operators is not commutative; that is, $\mathbf{AB}$ may not equal $\mathbf{BA}$.

Consider the two operators $\mathbf{D}$ and $\mathbf{T}$ which we have defined on P_m. Direct calculation yields

$$\mathbf{TD}f = p\frac{d^2f}{dp^2}$$

and

$$\mathbf{DT}f = \frac{df}{dp} + p\frac{d^2f}{dp^2}.$$

Thus **TD** and **DT** are not equal; in fact their difference is defined by $(\mathbf{DT} - \mathbf{TD})f = df/dp = \mathbf{D}f$ for all $f \in P_m$, or

$$\mathbf{DT} - \mathbf{TD} = \mathbf{D}.$$

Of special importance is the *unit* or *identity operator* **1** which maps every $\mathbf{x} \in S$ into itself: $\mathbf{1x} = \mathbf{x}$. The difference between the extension of **DM** and **MD**, operators which are discussed in the last Remark, is just the unit operator.

An operator **A** is said to be *nonsingular* if it maps every basis into a basis.[4] That is, let $\mathbf{x}_1, \ldots, \mathbf{x}_n$ be any basis in S, and let $\mathbf{y}_k = \mathbf{Ax}_k$, $k = 1, \ldots, n$. Then **A** is nonsingular if $\mathbf{y}_1, \ldots, \mathbf{y}_n$ is a basis. Otherwise **A** is said to be *singular*. Nonsingular operators are of importance because for them we can define an *inverse*, another operator which undoes the mapping of the original operator. But before we show this, we shall exhibit some properties of nonsingular operators.

Property 1. If **A** is nonsingular then $\mathbf{y} \neq 0$ implies that $\mathbf{Ay} \neq 0$. *Proof.* Let $\mathbf{y}$ have components η_i in some basis $\mathbf{x}_1, \ldots, \mathbf{x}_n$. Then $\mathbf{Ay} = \mathbf{A}(\eta_i \mathbf{x}_i) = \eta_i(\mathbf{Ax}_i) = \mathbf{z}$. By assumption the $\mathbf{Ax}_i$ form a basis, so that the η_i are components of $\mathbf{z}$ in this new basis. But $\mathbf{y} \neq 0$ implies that $\eta_i \neq 0$, and therefore $\mathbf{z} \neq 0$.

Property 2. Let $\mathbf{z}$ be any vector in S. Then if **A** is nonsingular there exists a unique vector $\mathbf{y}$ such that $\mathbf{Ay} = \mathbf{z}$. *Proof.* Let $\mathbf{x}_1, \ldots, \mathbf{x}_n$ be a basis, so that $\mathbf{Ax}_1, \ldots, \mathbf{Ax}_n$ is a basis. Let $\mathbf{z}$ have components ζ_i in this second basis. Then if $\mathbf{y} = \zeta_i \mathbf{x}_i$, it follows immediately that $\mathbf{Ay} = \mathbf{z}$. To show uniqueness, let $\mathbf{y}' = \eta_i \mathbf{x}_i$ also be mapped into $\mathbf{z}$ by **A**. Then as we showed in proving Property 1, the η_i are the components of $\mathbf{z}$ in the second basis, and these are known to be the ζ_i. Hence $\eta_i = \zeta_i$, and $\mathbf{y} = \mathbf{y}'$.

Property 3. If **A** is nonsingular there exists an operator which we may call $\mathbf{A}^{-1}$ such that $\mathbf{AA}^{-1} = \mathbf{A}^{-1}\mathbf{A} = \mathbf{1}$. (This operator $\mathbf{A}^{-1}$ is the *inverse operator* of which we have already spoken.) *Proof.* Let $\mathbf{x}_1, \ldots, \mathbf{x}_n$ be a basis. Then by assumption $\mathbf{y}_1, \ldots, \mathbf{y}_n$ is a basis, where $\mathbf{y}_k = \mathbf{Ax}_k$. Define $\mathbf{A}^{-1}$ by $\mathbf{A}^{-1}\mathbf{y}_k = \mathbf{x}_k$. Consider an arbitrary vector $\mathbf{z} \in S$ with components ζ_i in the basis of $\mathbf{y}_1, \ldots, \mathbf{y}_n$. Then $\mathbf{A}^{-1}\mathbf{z} = \zeta_i(\mathbf{A}^{-1}\mathbf{y}_i) = \zeta_i \mathbf{x}_i$, and $\mathbf{AA}^{-1}\mathbf{z} = \mathbf{A}(\zeta_i \mathbf{x}_i) = \zeta_i(\mathbf{Ax}_i) = \zeta_i \mathbf{y}_i = \mathbf{z}$. That $\mathbf{A}^{-1}\mathbf{A} = \mathbf{1}$ is proved similarly.

Furthermore, for a space of finite dimension, Properties 1, 2, and 3 each separately imply that **A** is nonsingular. Let us, in fact, prove the converses.

Property 1 (converse). If $\mathbf{z} \neq 0$ implies that $\mathbf{Az} \neq 0$, then **A** is nonsingular. *Proof.* We prove the equivalent statement that if **A** is singular, then there

[4] If S is a space of infinite dimension, an operator **A** is said to be nonsingular (a) if it maps every basis into a basis and (b) if it annihilates no nonzero vector (that is if there is no nonzero $\mathbf{x} \in S$ such that $\mathbf{Ax} = 0$). In this connection, see Problem 7.

exists at least one nonzero $\mathbf{z} \in S$ such that $\mathbf{Az} = 0$ (or then $\mathbf{z} \neq 0$ does not imply that $\mathbf{Az} \neq 0$). Let $\mathbf{x}_1, \ldots, \mathbf{x}_n$ be a basis. Since $\mathbf{A}$ is singular, $\mathbf{y}_1, \ldots, \mathbf{y}_n$ is not a basis, where $\mathbf{y}_i = \mathbf{Ax}_i$. This means that the $\mathbf{y}_i$ are a linearly dependent set, or that there exist α_i, not all zero, such that $\alpha_i \mathbf{y}_i = 0$. Because the α_i are not all zero, $\mathbf{z} = \alpha_i \mathbf{x}_i$ is a nonzero vector, yet $\mathbf{Az} = \alpha_i \mathbf{y}_i = 0$.

Property 2 (*converse*). If for every $\mathbf{z} \in S$ there exists a unique $\mathbf{y} \in S$ such that $\mathbf{Ay} = \mathbf{z}$, then $\mathbf{A}$ is nonsingular. *Proof.* Let $\mathbf{z}$ be an arbitrary vector in S, and let $\mathbf{y} = \eta_i \mathbf{x}_i$ be the unique vector mapped into $\mathbf{z}$ by $\mathbf{A}$, where the $\mathbf{x}_i$ form a basis. This means that $\mathbf{z} = \eta_i(\mathbf{Ax}_i)$, or that every arbitrary vector in S can be written as a linear combination of the $\mathbf{Ax}_i$. Thus the $\mathbf{Ax}_i$ form a basis, or $\mathbf{A}$ is nonsingular.

Property 3 (*converse*). If there exists an operator $\mathbf{A}^{-1}$ such that $\mathbf{AA}^{-1} = \mathbf{A}^{-1}\mathbf{A} = \mathbf{1}$, then $\mathbf{A}$ is nonsingular. *Proof.* We prove the equivalent statement that if $\mathbf{A}$ is singular, then there exists no such operator. As we have seen, Property 1 is equivalent to $\mathbf{A}$ being nonsingular. Thus assume that $\mathbf{A}$ does not possess Property 1, that is, that there exists some nonzero $\mathbf{z} \in S$ such that $\mathbf{Az} = 0$. But $\mathbf{B}0 = 0$ for any operator $\mathbf{B}$ (by linearity), so that $\mathbf{BAz} = \mathbf{B}0 = 0$ no matter what $\mathbf{B}$ may be. Thus there can exist no $\mathbf{A}^{-1}$ such that $\mathbf{A}^{-1}\mathbf{Az} = \mathbf{z}$, or $\mathbf{A}$ has no inverse, as asserted.

Thus we see that each of Properties 1, 2, and 3 is entirely equivalent to the statement that $\mathbf{A}$ is nonsingular and to each other. We shall often make use of this equivalence.

Just as a vector has components in a given basis, there is associated with each linear operator a set of numbers in a given basis. Let $\mathbf{x}_1, \ldots, \mathbf{x}_n$ be a basis, and let $\mathbf{A}$ be an operator. Then for each i the vector $\mathbf{Ax}_i$ can itself be expanded in this basis. Thus we may write

$$\mathbf{Ax}_i = \alpha_{ji}\mathbf{x}_j, \tag{8}$$

defining the *matrix elements* α_{ij} of $\mathbf{A}$. The set of matrix elements is arranged in a square array of the form

$$\begin{Vmatrix} \alpha_{11} & \alpha_{12} & \cdots & \alpha_{1n} \\ \alpha_{21} & \alpha_{22} & \cdots & \alpha_{2n} \\ \cdot & \cdot & & \cdot \\ \cdot & \cdot & & \cdot \\ \cdot & \cdot & & \cdot \\ \alpha_{n1} & \alpha_{n2} & \cdots & \alpha_{nn} \end{Vmatrix}$$

to make up the *matrix of* $\mathbf{A}$ *in the basis* $\mathbf{x}_1, \ldots, \mathbf{x}_n$. This matrix uniquely determines $\mathbf{A}$, for it tells how every element of the basis is mapped by $\mathbf{A}$, and therefore by linearity how every other vector in S is mapped. For instance, let $\mathbf{y}$ be a vector with components η_i (that is $\mathbf{y} = \eta_i \mathbf{x}_i$), and let

$\mathbf{z} = \mathbf{Ay}$ have components ζ_i. To find the ζ_i if we know the η_i and the α_{ij}, we write

$$\mathbf{z} = \mathbf{Ay} = \eta_i \mathbf{Ax}_i = \eta_i \alpha_{ji} \mathbf{x}_j = \zeta_j \mathbf{x}_j.$$

Comparison immediately yields the ζ_j:

$$\zeta_j = \alpha_{ji}\eta_i \qquad (j = 1, \ldots, n). \tag{9}$$

Remark: Note that in Eq. (8) the sum is over the first index, whereas in (9) it is over the second index. The vectors $\mathbf{Ax}_i$ have a different expansion in terms of the $\mathbf{x}_i$ than the components of $\mathbf{Ay}$ have in terms of the components of $\mathbf{y}$.

Given a basis, therefore, (9) tells us how to find the column vector corresponding to $\mathbf{Ay}$ if we know the column vector corresponding to $\mathbf{y}$. In the same sense that a column vector forms a representation of a vector, a matrix forms a representation of an operator. We shall use the notation $\mathbf{A}$ and A for an operator and its matrix, as we use $\mathbf{x}$ and x for a vector and its column vector. As with vectors, the matrix A of $\mathbf{A}$ depends on the basis. As an example, consider the derivative operator $\mathbf{D}$ on P_m in the basis $1, p, p^2, \ldots, p^m$. Since

$$\mathbf{D}p^j = jp^{j-1} = \Delta_{ij}p^i \qquad (j = 0, 1, \ldots, m),$$

the matrix elements of $\mathbf{D}$ are $\Delta_{ij} = j\,\delta_{i,j-1}$. The matrix itself is (note that i and j run from 0 to m, not from 1 to m)

$$D = \begin{Vmatrix} 0 & 1 & 0 & 0 & \cdots & 0 \\ 0 & 0 & 2 & 0 & \cdots & 0 \\ 0 & 0 & 0 & 3 & \cdots & 0 \\ \cdot & \cdot & \cdot & \cdot & & \cdot \\ \cdot & \cdot & \cdot & \cdot & & \cdot \\ \cdot & \cdot & \cdot & \cdot & & \cdot \\ 0 & 0 & 0 & 0 & \cdots & n \\ 0 & 0 & 0 & 0 & \cdots & 0 \end{Vmatrix}.$$

Remark. Equations (9) and (7), although they look very similar, actually say quite different things. Equation (7) gives the components or the column vector of a fixed vector in two different bases, and may be written

$$z' = Tz,$$

where z' and z are the column vectors representing $\mathbf{z}$ in the two bases, and where T is the transformation matrix. Equation (9), on the other hand, gives the column vectors representing two different vectors in a fixed basis.

and may be written

$$z = Ay,$$

where z and y are the column vectors representing **z** and **y** in a fixed basis, and A is the matrix of **A** in that basis. We have come across this distinction before, in discussing the active and passive views in Chapter III.

We leave to the problems the proofs of the following assertions about the matrix elements, or about the matrices, of operators.

Assertion 1. Let **A** and **B** be operators with matrix elements α_{ij} and β_{ij}, respectively, in some basis. Then the operator $\mathbf{C} = \mathbf{A} + \mathbf{B}$ has matrix elements $\gamma_{ij} = \alpha_{ij} + \beta_{ij}$ in this basis. We write $C = A + B$ for the matrices.

Assertion 2. Let **A** be as in Assertion 1. Then the operator $\mathbf{M} = \nu\mathbf{A}$ has matrix elements $\mu_{ij} = \nu\alpha_{ij}$ in the given basis. We write $M = \nu A$ for the matrices.

Assertion 3. Let **A** and **B** be as in Assertion 1. Then the operator $\mathbf{L} = \mathbf{AB}$ has matrix elements $\lambda_{ij} = \alpha_{ik}\beta_{kj}$ in the given basis. We then say that L is the matrix product of A and B and write $L = AB$.

Assertion 4. The matrix elements of the unit operator **1** are the δ_{ij} (the Kronecker δ symbol) in any basis. We shall write $\mathbb{1}$ for the matrix of **1**.

As with vectors, it is often convenient to choose a special basis in which to write out the matrix of a given operator. Let us assume, therefore, that **A** has matrix elements α_{ij} in a basis $\mathbf{x}_1, \ldots, \mathbf{x}_n$, and elements α'_{ij} in a basis $\mathbf{y}_1, \ldots, \mathbf{y}_n$. Further, let the transformation from the $\mathbf{y}_i$ to the $\mathbf{x}_i$ be given by an equation like Eq. (5), namely

$$\mathbf{x}_j = \tau_{ij}\mathbf{y}_i.$$

Since, on the other hand, the $\mathbf{x}_i$ form a basis, each $\mathbf{y}_i$ can also be expressed in terms of the $\mathbf{x}_i$:

$$\mathbf{y}_i = \sigma_{ji}\mathbf{x}_j.$$

Then to find the matrix elements of **A** in the basis $\mathbf{y}_1, \ldots, \mathbf{y}_n$, we calculate

$$\mathbf{A}\mathbf{y}_i = \alpha'_{ji}\mathbf{y}_j = \mathbf{A}\sigma_{ki}\mathbf{x}_k = \sigma_{ki}\alpha_{mk}\mathbf{x}_m = \sigma_{ki}\alpha_{mk}\tau_{jm}\mathbf{y}_j.$$

Direct comparison yields (recall that the components of a vector are unique in a given basis)

$$\alpha'_{ji} = \tau_{jm}\alpha_{mk}\sigma_{ki}. \tag{10}$$

We see in Problem 6 that the τ_{ij} and the σ_{ij} have the property that

$$\tau_{ij}\sigma_{jk} = \delta_{ik}.$$

Since the δ_{ij} are the elements of the unit matrix $\mathbb{1}$, we shall say that the matrix formed of the σ_{ij} is the inverse of that formed of the τ_{ij}, writing T^{-1} and T, respectively, for these two matrices. Then in matrix form Eq. (10) may be written

$$A' = TAT^{-1}, \tag{10'}$$

where A and A' are the matrices of **A** in the two bases. Thus if we know the transformation matrix from one basis to another, as well as its inverse (and we shall soon see how to calculate the one from the other), we can find the matrix of any operator in one of the bases when we know it in the other.

The question now has arisen as to how one goes about calculating the inverse of a matrix; or equivalently, if one knows an operator, how does one find its inverse? This is not in general a simple task, becoming rapidly more complicated as one considers higher dimensions. To perform the calculation one must make use of the *determinant* of the matrix. The determinant of a matrix may be defined as the number det A which satisfies the relation[5] (here $j_1, \ldots, j_n$ are each summed from 1 to n)

$$\epsilon_{i_1\cdots i_n} \det A = \epsilon_{j_1\cdots j_n}\alpha_{i_1j_1}\cdots\alpha_{i_nj_n}. \tag{11}$$

From this definition one can prove (see Problem 10) many of the properties of determinants in a simple and straightforward way. Now for each choice of $i_1, \ldots, i_n$ such that the left-hand side of (11) is not identically zero, the right-hand side consists of $n!$ terms, and each term contains exactly one element α_{ij} from any fixed given row, say the ith. Furthermore all the elements of the fixed ith row will appear in the whole expression, so that (11) may be written in the form det $A = \alpha_{ij}\beta_{j(i)}$ (no sum on i), where β_{ji} is merely the factor multiplying α_{ij} in the expression which is obtained for det A by rewriting (11). Note that β_{ji} is independent of the elements of the ith row, for each of the $n!$ terms in (11) contains one and only one element from the ith row. Suppose now that we have found the β_{ji} for the ith row. What would happen if we tried summing them with the elements of some other row, say the kth? The result would be $\alpha_{kj}\beta_{ji} = 0$, for this is just the determinant obtained from A when the ith row is replaced by the kth (see Problem 10). Combining these results, we obtain

$$\alpha_{kj}\beta_{ji} = \delta_{ji} \det A, \tag{12}$$

which is another expression for the determinant of A. It can be shown, though we shall not show it, that except for sign β_{ji} is the determinant of the $(n - 1)$-dimensional matrix obtained from A when the row and column containing α_{ij} are removed; β_{ji} is called the *cofactor* of α_{ij}.

[5] The quantity $\epsilon_{i_1\cdots i_n}$ is defined to be zero if any two of the subscripts are equal, to be $+1$ if sequence $i_1, \ldots, i_n$ is an even permutation of $1, 2, \ldots, n$, and to be -1 if the sequence is an odd permutation. Thus $\epsilon_{i_1\cdots i_n}$ is just the n-dimensional generalization of $\epsilon_{\alpha\beta\gamma}$ defined in Appendix A.

From (12) we see immediately that if $\det A \neq 0$, the matrix A has an inverse and the elements of this inverse are just

$$\alpha_{ij}^{-1} = \frac{\beta_{ij}}{\det A}. \tag{13}$$

We may go even further. The matrix A has an inverse *only* if $\det A \neq 0$. For assume that A^{-1} exists, and let its determinant be $\det A^{-1}$. Then, as follows from Problem 10,

$$\det (A^{-1}A) = (\det A^{-1})(\det A) = \det \mathbb{1} = 1,$$

which means that neither $\det A$ nor $\det A^{-1}$ can vanish. Obviously the operator $\mathbf{A}$ has an inverse if and only if its matrix A has an inverse, so that we may conclude that an operator is nonsingular if and only if its matrix has nonzero determinant in every coordinate system.

The value of the determinant is moreover independent of basis. This is easily seen by using Eq. (10′). If A' is the representation of an operator $\mathbf{A}$ in one basis, and A is its representation in another, then (again, see Problem 10)

$$\begin{aligned} \det A' &= \det TAT^{-1} = (\det T)(\det A)(\det T^{-1}) \\ &= \det A, \end{aligned}$$

where we use the fact, established above, that $(\det T)(\det T^{-1}) = 1$. Thus we may speak of the determinant of an operator rather than the determinant of a matrix, and we may restate our conclusion in the following form. An operator is nonsingular if and only if its determinant is nonzero.

IV-3 LINEAR ALGEBRAIC EQUATIONS AND EIGENVALUE PROBLEMS

Consider a set of n linear inhomogeneous algebraic equations in the n variables $\xi_1, \ldots, \xi_n$, of the form

$$\alpha_{ij}\xi_j = \gamma_i, \tag{14}$$

where the γ_i and the α_{ij} are given constants. This set of equations can be reinterpreted in the following way. Let S be a linear vector space in n dimensions, and let there be a basis given in S. Let $\mathbf{x}$ be a vector with unknown components ξ_i in this basis, let $\mathbf{c}$ be the vector with components γ_i, and let $\mathbf{A}$ be the operator whose matrix elements in this basis are α_{ij}. Then (14) may be viewed as the representation in this basis of the equation

$$\mathbf{Ax} = \mathbf{c}. \tag{15}$$

But we have already seen that given $\mathbf{c}$, a unique $\mathbf{x}$ will exist if and only if $\mathbf{A}$ is nonsingular (Property 2 of nonsingular operators). Then (Property 3) $\mathbf{A}$ has an inverse, and

$$\mathbf{x} = \mathbf{A}^{-1}\mathbf{c}. \tag{16}$$

When this equation is represented in the original basis in S, the components of $\mathbf{x}$, which are the unknown ξ_i, are given in terms of the components of $\mathbf{c}$ (the γ_i) and the matrix elements of $\mathbf{A}^{-1}$. It is left to the student to show (Problem 11) that when (13) is used to express the matrix elements of $\mathbf{A}^{-1}$, the result is just Cramer's familiar rule for the solution of a set of linear algebraic equations.

Although a unique solution will exist if and only if $\mathbf{A}$ is nonsingular, solutions may exist also if $\mathbf{A}$ is singular. We have seen that if $\mathbf{A}$ is singular, there is at least one nonzero vector $\mathbf{y} \in S$ such that $\mathbf{Ay} = 0$ (Property 1). Then if $\mathbf{x}$ is a solution of (15), so is $\mathbf{x} + \lambda\mathbf{y}$ with any number λ, for

$$\mathbf{A}(\mathbf{x} + \lambda\mathbf{y}) = \mathbf{Ax} + \lambda\mathbf{Ay} = \mathbf{Ax} = \mathbf{c}.$$

Consequently, for singular $\mathbf{A}$ the solution of (14), if it exists at all, is certainly not unique. It is easily shown, moreover, that any two solutions differ by a solution of the homogeneous equation

$$\mathbf{Ay} = 0, \tag{17}$$

which, in the original basis, becomes

$$\alpha_{ij}\eta_j = 0,$$

where the η_i are the components of $\mathbf{y}$ in this basis. Thus, in order to understand the singular case, we must study the homogeneous equation. We know that this equation has nontrivial (i.e., nonnull) solutions if and only if $\mathbf{A}$ is singular (i.e., if and only if $\det A = 0$), and it is clear that if $\mathbf{y}$ is a solution, so is $\lambda\mathbf{y}$. One may ask, however, how many linearly independent solutions exist. The answer depends on the detailed structure of $\mathbf{A}$, and we shall restrict ourselves to some formal statements. The set of all vectors of the form $\mathbf{Az}$, $\mathbf{z} \in S$, is called the range of $\mathbf{A}$. The set of all vectors $\mathbf{y} \in S$ such that $\mathbf{Ay} = 0$ is called the null space of A. It is a simple matter to show that the range and null space are subspaces (in general not disjoint), and it can be shown moreover that their dimensions add up to the dimension of S. Equation (15) will thus have a solution if and only if $\mathbf{c}$ is in the range of $\mathbf{A}$, and the solution $\mathbf{x}$ will be undetermined up to an arbitrary vector in the null space. We emphasize again that these are formal definitions only; to find these subspaces and their dimensions is not always simple, and when solving such homogeneous equations we shall proceed in an ad hoc manner.

Closely related to the question of the homogeneous equations is the following problem. Given an operator $\mathbf{A}$, we wish to find what we shall call the *eigenvectors* of $\mathbf{A}$, namely those vectors $\mathbf{x}$, which are mapped by $\mathbf{A}$ into multiples of themselves. In other words, we wish to solve an equation of the form

$$\mathbf{Ax} = \lambda\mathbf{x} \tag{18}$$

for both λ and $\mathbf{x}$. In order to relate this equation to what we have been discussing, we write it in the form

$$(\mathbf{A} - \lambda\mathbf{1})\mathbf{x} = 0, \tag{19}$$

and it then becomes clear that what we must do is first find the values of λ for which $\mathbf{A} - \lambda\mathbf{1}$ becomes singular, and then solve (19) for $\mathbf{x}$. If (19) is to have a solution, the determinant of $\mathbf{A} - \lambda\mathbf{1}$ must vanish. We write

$$\det(A - \lambda\mathbb{1}) = 0, \tag{20}$$

which is a polynomial of degree n in λ. The roots of this polynomial, that is, the values of λ for which the determinant vanishes, are called the *eigenvalues* of $\mathbf{A}$; they are the only values of λ for which (18) may have a solution. In fact, it is clear that for each value of λ there is at least one vector $\mathbf{x}_\lambda$ satisfying (19) and therefore also (18). We shall say that the eigenvector $\mathbf{x}_\lambda$ *belongs to* the eigenvalue λ. When (20) has multiple roots, there may be more than one $\mathbf{x}_\lambda$ for each value of λ, and in most cases of physical interest, but not in all, the number of linearly independent vectors $\mathbf{x}_\lambda$ belonging to each multiple eigenvalue λ is equal to the multiplicity of the root.

Let us consider two examples. First let $\mathbf{I}$ be the operator on P_1 defined by (here $p^0 = 1$)

$$\mathbf{I}p^0 = p,$$
$$\mathbf{I}p = -p^0.$$

In the basis $\mathbf{x}_1 = p^0$, $\mathbf{x}_2 = p$, therefore, the matrix of $\mathbf{I}$ is

$$\begin{vmatrix} 0 & -1 \\ 1 & 0 \end{vmatrix}.$$

The eigenvalues are given by (20), which in this case becomes

$$\lambda^2 + 1 = 0.$$

Thus the eigenvalues are $\pm i$. To find the eigenvectors, construct the general vector (the general polynomial of first degree) $\alpha + \beta p$, and apply $\mathbf{I}$. If this vector is to be an eigenvector belonging to the eigenvalue $+i$, we obtain

$$\mathbf{I}(\alpha + \beta p) = i(\alpha + \beta p) = -\beta + \alpha p.$$

Since the components of a vector are unique, this means that $\beta = -i\alpha$, but that we are otherwise free to choose α. All such vectors are multiples of the vector $1 - ip \equiv p^0 - ip \equiv \mathbf{x}_1 - i\mathbf{x}_2$. Similarly, all multiples of $\mathbf{x}_1 + i\mathbf{x}_2$ are eigenvectors of $\mathbf{I}$ belonging to the eigenvalue $-i$. These eigenvectors may be represented by the column vectors

$$\begin{vmatrix} 1 \\ -i \end{vmatrix} \quad \text{and} \quad \begin{vmatrix} 1 \\ i \end{vmatrix}.$$

Note that these two eigenvectors are linearly independent (check this), so that they can be taken as a new basis in P_1. It is shown in Problem 13 that in this new basis the matrix of $\mathbf{I}$ is particularly simple in that its only nonzero elements lie along the *main diagonal* (the one going from upper left to lower right of the matrix), and these elements are the eigenvalues of $\mathbf{I}$. This procedure of transforming to the basis in which the basis vectors are the eigenvectors is called *diagonalizing* a matrix. The matrix elements on the main diagonal of a diagonalized matrix are its eigenvalues. Diagonalizing matrices is very important in physics.

Remark. As we have seen in this example, the roots of (20) need not be real, even if we had originally defined S as a vector space over the real numbers. Thus the eigenvalue problem may in fact have no solution over the real numbers, and it may be impossible to diagonalize a given matrix without extending the field to the complex numbers. Thus we may be forced to deal with complex vector spaces even if we start with real ones.

In order to show that a matrix can be diagonalized, one must show that the eigenvectors are linearly independent, for otherwise they cannot form a basis. In general, however, this can only be shown for unequal eigenvalues; that is, that eigenvectors belonging to different eigenvalues are linearly independent. In fact let $\mathbf{Ax} = \lambda\mathbf{x}$ and $\mathbf{Ay} = \mu\mathbf{y}$, where $\lambda \neq \mu$, and therefore either λ or μ is nonzero. Assume (without loss of generality) that $\mu \neq 0$. If we assume that $\mathbf{x}$ and $\mathbf{y}$ are linearly dependent, we shall arrive at a contradiction, for this assumption means that there exists an α such that $\mathbf{y} = \alpha\mathbf{x}$ and therefore that $\mathbf{A}(\mathbf{y} - \alpha\mathbf{x}) = 0 = \mu\mathbf{y} - \lambda\alpha\mathbf{x}$, or $\mathbf{y} = (\lambda\alpha/\mu)\mathbf{x}$. Thus $\lambda/\mu = 1$, which is contrary to assumption. Consequently $\mathbf{x}$ and $\mathbf{y}$ are linearly independent.

The second example we wish to give is a somewhat trivial but pathological one. Consider the derivative operator $\mathbf{D}$ on P_m. In a rather obvious way, no polynomial is mapped into a multiple of itself by $\mathbf{D}$, and indeed the only root of $\det(D - \lambda\mathbb{1}) = 0$ is $\lambda = 0$, its multiplicity being $m + 1$. In spite of its high multiplicity, this eigenvalue has but one eigenvector, namely the solution of $Df = 0$, which is $f = \alpha p^0 = \alpha$, a constant. Thus the matrix D cannot be diagonalized.

IV-4 INNER PRODUCTS AND UNITARY SPACES

The three-dimensional space of elementary mechanics has additional structure which is usefully generalized to n dimensions. In three dimensions this structure is contained in the definition of the dot product, or scalar product. We shall now impose on the general n-dimensional vector space S the analog of this additional structure by defining an analogous operation.

With every pair of vectors, $\mathbf{x}, \mathbf{y} \in S$ we associate a *number*, called their *scalar* or *inner product* and written $(\mathbf{x}, \mathbf{y})$; this association is to have the following properties.

1. Linearity. $(\mathbf{x}, \alpha\mathbf{y} + \beta\mathbf{z}) = \alpha(\mathbf{x}, \mathbf{y}) + \beta(\mathbf{x}, \mathbf{z})$.

2. Hermiticity. $(\mathbf{x}, \mathbf{y}) = (\mathbf{y}, \mathbf{x})^*$, where the star indicates the complex conjugate. (Such an inner product is said to be *Hermitian*. If S is defined over the real numbers, so that the complex conjugate of a number reduces to the number itself, the inner product is said to be *symmetric*.)

3. Positive-definiteness. $(\mathbf{x}, \mathbf{x}) = 0$ if and only if $\mathbf{x} = 0$. Otherwise, $(\mathbf{x}, \mathbf{x}) > 0$.

Remarks. *1.* A vector space with this additional structure is said to be *unitary* (or, in the case of the real numbers, *orthogonal*). Thus, for instance, the three-dimensional space of elementary mechanics is an orthogonal vector space. *2.* In analogy to the concept of the length of a vector in three dimensions we shall write $|\mathbf{x}|^2$ for $(\mathbf{x}, \mathbf{x})$, and shall call its positive square root $|\mathbf{x}|$ the length or *norm* of $\mathbf{x}$. The analogy will soon become obvious. *3.* Note that linearity and Hermiticity imply that $(\alpha\mathbf{x}, \mathbf{y}) = \alpha^*(\mathbf{x}, \mathbf{y})$.

Also as in three dimensions, the inner product allows us to formulate another relationship between vectors, namely, *orthogonality*. Two vectors $\mathbf{x}$ and $\mathbf{y}$ are said to be orthogonal if $(\mathbf{x}, \mathbf{y}) = 0$. To show that this is not an empty definition, we show that given any two linearly independent vectors, they can be combined to form a pair of orthogonal ones. For this purpose, let $\mathbf{x}$ and $\mathbf{y}$ be linearly independent but not necessarily orthogonal and let $(\mathbf{x}, \mathbf{y}) = \alpha$. Consider the vector

$$\mathbf{y}' = \mathbf{y} - \frac{\alpha}{|\mathbf{x}|^2}\mathbf{x}. \tag{21}$$

Then by linearity of the inner product we find that

$$(\mathbf{x}, \mathbf{y}') = (\mathbf{x}, \mathbf{y}) - \frac{\alpha}{|\mathbf{x}|^2}(\mathbf{x}, \mathbf{x}) = \alpha - \alpha = 0.$$

Thus either $\mathbf{y}'$ is orthogonal to $\mathbf{x}$ or $\mathbf{y}' = 0$. But if $\mathbf{y}' = 0$, the original two vectors $\mathbf{x}$ and $\mathbf{y}$ are linearly dependent, contrary to assumption. Thus $\mathbf{y}'$ is orthogonal to $\mathbf{x}$. It is shown in Problem 17 that any two nonzero orthogonal vectors are linearly independent.

A set of vectors is said to be orthogonal if every vector in the set is orthogonal to every other. Given a set of linearly independent vectors, it can always be converted into an orthogonal set of linearly independent vectors each one of which is a linear combination of vectors from the original set. We shall demonstrate this on a basis in S, for orthogonal bases are of particular importance. Thus let $\mathbf{x}_1, \ldots, \mathbf{x}_n$ be a basis. Consider first the vectors $\mathbf{x}_1$ and $\mathbf{x}_2$. We define

$$\mathbf{x}_2' = \mathbf{x}_2 - (\mathbf{x}_1, \mathbf{x}_2) \frac{\mathbf{x}_1}{|\mathbf{x}_1|^2},$$

and then as in Eq. (21) $\mathbf{x}_2'$ is orthogonal to $\mathbf{x}_1$. Furthermore, since $\mathbf{x}_2'$ is a linear combination of $\mathbf{x}_1$ and $\mathbf{x}_2$, it is linearly independent of $\mathbf{x}_i$, $i > 2$. (It is, of course, also linearly independent of $\mathbf{x}_1$.) Now consider $\mathbf{x}_1$ and $\mathbf{x}_3$, and form $\mathbf{y}_3 = \mathbf{x}_3 - (\mathbf{x}_1, \mathbf{x}_3)\,|\mathbf{x}_1|^{-2}\mathbf{x}_1$. As in Eq (21), $\mathbf{y}_3$ is orthogonal to $\mathbf{x}_1$. Furthermore, any amount of $\mathbf{x}_2'$ may be added to $\mathbf{y}_3$ without leading to a vector which is not orthogonal to $\mathbf{x}_1$, for

$$(\mathbf{x}_1, \mathbf{y}_3 + \beta\mathbf{x}_2') = (\mathbf{x}_1, \mathbf{y}_3) + \beta(\mathbf{x}_1, \mathbf{x}_2') = 0.$$

Now β may be chosen so as to make $\mathbf{x}_3' = \mathbf{y}_3 + \beta\mathbf{x}_2'$ orthogonal to $\mathbf{x}'$ [chosen, that is, in accordance with Eq. (21) for the pair of vectors $\mathbf{y}_3$, $\mathbf{x}_2'$]. Then $\mathbf{x}_1$, $\mathbf{x}_2'$, $\mathbf{x}_3'$ will be an orthogonal set of linearly independent vectors, all linearly independent of $\mathbf{x}_i$, $i > 3$. We may proceed in this way, finally replacing the entire basis by a new basis of orthogonal vectors. This procedure is called Gram-Schmidt orthogonalization.

We have succeeded in showing that it is always possible to construct an orthogonal basis. In addition we may specify that each vector of the basis have unit norm (see Problem 18). Let us call such vectors with unit norm *unit vectors*. We then see that there exist what we shall call *orthonormal bases*, consisting of orthogonal unit vectors. Let $\mathbf{e}_1, \ldots, \mathbf{e}_n$ be such an orthonormal basis. Then it is obvious that

$$(\mathbf{e}_i, \mathbf{e}_j) = \delta_{ij}, \tag{22}$$

and the components of a vector $\mathbf{x}$ in such a basis are just

$$\xi_i = (\mathbf{e}_i, \mathbf{x})$$

(see Problem 20). Henceforth because of the simplicity so obtained, we shall use only orthonormal bases unless specified otherwise.

Remark. It is these orthonormal bases which bring out the analogy between norm and length in three dimensions. In fact we have

$$|\mathbf{x}|^2 = (\xi_i\mathbf{e}_i, \xi_j\mathbf{e}_j) = \xi_i^*\xi_j(\mathbf{e}_i, \mathbf{e}_j) = \xi_i^*\xi_j\,\delta_{ij} = \sum |\xi_i|^2.$$

If S is a vector space over the real numbers, in particular, the norm becomes just the Euclidean length of the vector in n dimensions. Note that the usual basis of unit vectors along the *1*-, *2*-, and *3*-axes in three dimensions is indeed orthonormal. Moreover, $(\mathbf{x}, \mathbf{y}) = \xi_i^* \eta_i$.

The matrix elements of an operator $\mathbf{A}$ can also be found in an (orthonormal) basis by using the inner product. The matrix elements of $\mathbf{A}$ in the basis $\mathbf{e}_1, \ldots, \mathbf{e}_n$ are defined by

$$\mathbf{A}\mathbf{e}_i = \alpha_{ji}\mathbf{e}_j.$$

Taking the inner product of this with $\mathbf{e}_k$, we obtain

$$(\mathbf{e}_k, \mathbf{A}\mathbf{e}_i) = \alpha_{ji}(\mathbf{e}_k, \mathbf{e}_j) = \alpha_{ji}\delta_{kj} = \alpha_{ki}. \tag{23}$$

Let $\mathbf{e}_1, \ldots, \mathbf{e}_n$ and $\mathbf{e}_1', \ldots, \mathbf{e}_n'$ be two (orthonormal) bases, and let the transformation matrix U from one to the other have elements ω_{ji}:

$$\mathbf{e}_i' = \omega_{ji}\mathbf{e}_j.$$

Then

$$(\mathbf{e}_k', \mathbf{e}_i') = \delta_{ki} = \omega_{ji}(\mathbf{e}_k', \mathbf{e}_j).$$

On the other hand

$$(\mathbf{e}_i', \mathbf{e}_m) = \omega_{ji}^*(\mathbf{e}_j, \mathbf{e}_m) = \omega_{ji}^*\,\delta_{jm} = \omega_{mi}^*,$$

so that

$$\omega_{ji}\omega_{jk}^* = \delta_{ki}. \tag{24}$$

Let ω_{kj}^{-1} be the elements of U^{-1}. Then (24) means that $\omega_{kj}^{-1} = \omega_{jk}^*$, for the inverse of a matrix is unique. A matrix with this property is said to be *unitary*.

At this point it is helpful to make some definitions. Let $\mathbf{A}$ be an operator on a unitary space S. Then for every pair of vectors $\mathbf{x}, \mathbf{y} \in S$ the value of $(\mathbf{A}\mathbf{x}, \mathbf{y})$ can be calculated. We now assert that for every pair $\mathbf{x}, \mathbf{y}$ there exists a vector $\mathbf{z}$ such that

$$(\mathbf{A}\mathbf{x}, \mathbf{y}) = (\mathbf{x}, \mathbf{z}),$$

and that $\mathbf{A}$ uniquely determines an operator which we shall call its *Hermitian conjugate* $\mathbf{A}^\dagger$, such that $\mathbf{z} = \mathbf{A}^\dagger\mathbf{y}$. We show that $\mathbf{z}$ exists by constructing it. In some coordinate system $\mathbf{e}_1, \ldots, \mathbf{e}_n$ let $\mathbf{y}$ have components η_i, and let $\mathbf{A}$ have elements α_{ij}. Then $\mathbf{z}$ is the vector with components $\zeta_j = \alpha_{ij}^*\eta_i$, as is trivially verified by calculating $(\mathbf{x}, \mathbf{z})$ and $(\mathbf{A}\mathbf{x}, \mathbf{y})$. To show that $\mathbf{A}^\dagger$ is an operator, we must show that $\mathbf{z}$ is unique for fixed $\mathbf{y}$. Assume the contrary: let $(\mathbf{x}, \mathbf{z}_1) = (\mathbf{A}\mathbf{x}, \mathbf{y})$ and $(\mathbf{x}, \mathbf{z}_2) = (\mathbf{A}\mathbf{x}, \mathbf{y})$ for all $\mathbf{x} \in S$. Then $(\mathbf{x}, \mathbf{z}_1) - (\mathbf{x}, \mathbf{z}_2) = (\mathbf{x}, \mathbf{z}_1 - \mathbf{z}_2) = 0$, or $\mathbf{z}_1 = \mathbf{z}_2$ (see Problem 17). This proves the assertion.

Thus we see that the equation

$$(\mathbf{A}\mathbf{x}, \mathbf{y}) = (\mathbf{x}, \mathbf{A}^\dagger\mathbf{y}) \qquad \text{for all } \mathbf{x}, \mathbf{y} \in S, \tag{25}$$

defines the Hermitian conjugate operator $\mathbf{A}^\dagger$. The matrix elements of $\mathbf{A}^\dagger$ are

$$\alpha_{ki}^\dagger = (\mathbf{e}_k, \mathbf{A}^\dagger\mathbf{e}_i) = (\mathbf{A}\mathbf{e}_k, \mathbf{e}_i) = (\mathbf{e}_i, \mathbf{A}\mathbf{e}_k)^* = \alpha_{ik}^* \tag{26}$$

which means that $A^\dagger$, the matrix of $\mathbf{A}^\dagger$, is obtained from A, the matrix of $\mathbf{A}$, by interchanging rows and columns (or flipping across the main diagonal) and taking the complex conjugate. As we say that the operator $\mathbf{A}^\dagger$ is the Hermitian conjugate of $\mathbf{A}$, so we say that the matrix $A^\dagger$ is the Hermitian conjugate of A. In matrix form, for instance, Eq. (24) says

$$UU^\dagger = \mathbb{1}, \tag{27}$$

or a unitary matrix is one whose inverse is its Hermitian conjugate. Finally, an operator or matrix which is itself equal to its Hermitian conjugate is called a *Hermitian operator or matrix*.

Remarks. *1.* For an orthogonal space (i.e., a real space with an inner product), Eq. (25) is said to define not the Hermitian conjugate but the *transpose* $\mathbf{A}^T$ of the operator $\mathbf{A}$. Then the complex conjugate does not appear in the real analog of (26), and we have $\alpha_{ij}^T = \alpha_{ji}$, that is, the transpose matrix is obtained simply by flipping across the main diagonal. An operator (or matrix) whose inverse is its transpose is said to be *orthogonal* (the real analog of unitary), and one which is equal to its transpose is said to be *symmetric* (the real analog of Hermitian). *2.* A *unitary operator* $\mathbf{U}$ (like a unitary matrix) is one whose inverse is its Hermitian conjugate: $\mathbf{U}^\dagger = \mathbf{U}^{-1}$. An operator is unitary if and only if it preserves inner products, that is if and only if for all $\mathbf{x}, \mathbf{y} \in S$

$$(\mathbf{Ux}, \mathbf{Uy}) = (\mathbf{U}^\dagger\mathbf{Ux}, \mathbf{y}) = (\mathbf{x}, \mathbf{y}).$$

This is left to the reader to prove. *3.* In an orthogonal space $(\mathbf{Rx}, \mathbf{Rx}) = (\mathbf{x}, \mathbf{x})$ for all $\mathbf{x}$ implies that $\mathbf{R}$ is orthogonal, for if we write $\mathbf{x} = \mathbf{y} + \mathbf{z}$, this tells us that

$$\begin{aligned}(\mathbf{Ry}, \mathbf{Ry}) + 2(\mathbf{Ry}, \mathbf{Rz}) + (\mathbf{Rz}, \mathbf{Rz})\\ = (\mathbf{y}, \mathbf{y}) + 2(\mathbf{y}, \mathbf{z}) + (\mathbf{z}, \mathbf{z}).\end{aligned}$$

Since $(\mathbf{Ry}, \mathbf{Ry}) = (\mathbf{y}, \mathbf{y})$ and similarly for $\mathbf{z}$, we arrive at

$$(\mathbf{Ry}, \mathbf{Rz}) = (\mathbf{y}, \mathbf{z})$$

for all $\mathbf{y}$, $\mathbf{z}$, or $\mathbf{R}$ is orthogonal, as in Remark 2.

The matrix of a Hermitian operator is itself Hermitian in every (orthonormal) basis. This is clear because (26) is calculated in an arbitrary basis, but it can also be seen directly from (10) and (27). Let A be the matrix of a Hermitian operator $\mathbf{A}$ in one basis, and let A' be its matrix in another. If U is the transformation matrix from the first basis to the second, then $A' = UAU^{-1}$. But according to (27) $U^{-1} = U^\dagger$, so that $A' = UAU^\dagger$. That A' is Hermitian now follows from Problem 22. There is, in particular, one very simple form for the matrix of a Hermitian operator. This may be stated in the form of the following theorem.

THEOREM. Given a Hermitian operator $\mathbf{A}$, there exists an orthonormal basis in which its matrix A is diagonal with real matrix elements (see Problem 13). (The diagonal elements are, of course, the eigenvalues of $\mathbf{A}$.)

PROOF. Note that according to (26) the elements on the main diagonal of any Hermitian matrix are real. Thus if A is diagonal, all of its matrix elements are real, and if $\mathbf{A}$ has any complex eigenvalues, it can certainly not be diagonalized. We shall first show, therefore, that all of its eigenvalues are real. Let $\mathbf{Ax} = \lambda\mathbf{x}$. Since $\mathbf{A}$ is Hermitian, we have

$$\begin{aligned}(\mathbf{x}, \mathbf{Ax}) &= (\mathbf{Ax}, \mathbf{x}) = (\mathbf{x}, \lambda\mathbf{x}) = (\lambda\mathbf{x}, \mathbf{x}) \\ &= \lambda\,|\mathbf{x}|^2 = \lambda^*\,|\mathbf{x}|^2.\end{aligned}$$

Thus λ is real and there is some chance that $\mathbf{A}$ can be diagonalized.

We shall first prove the theorem on the assumption that all the eigenvalues of A have multiplicity one; that is, all n roots of the characteristic equation $\det(\mathbf{A} - \lambda\mathbf{1}) = 0$ are distinct. Then there are n eigenvalues, and n linearly independent eigenvectors forming a basis in which $\mathbf{A}$ is diagonal (see Problem 13). All we need show is that this basis is orthogonal. Thus let $\mathbf{x}_\lambda$ be an eigenvector (and hence a basis vector) belonging to the eigenvalue λ, and let $\mathbf{x}_\mu$ be an eigenvector belonging to the eigenvalue μ. Then because $\mathbf{A}$ is Hermitian and λ and μ are real we have

$$\begin{aligned}(\mathbf{x}_\lambda, \mathbf{Ax}_\mu) &= (\mathbf{x}_\lambda, \mu\mathbf{x}_\mu) = \mu(\mathbf{x}_\lambda, \mathbf{x}_\mu) \\ &= (\mathbf{Ax}_\lambda, \mathbf{x}_\mu) = (\lambda\mathbf{x}_\lambda, \mathbf{x}_\mu) = \lambda(\mathbf{x}_\lambda, \mathbf{x}_\mu).\end{aligned}$$

If $\lambda \neq \mu$, this implies that $(\mathbf{x}_\lambda, \mathbf{x}_\mu) = 0$, as asserted. The theorem is now proved for n distinct eigenvalues.

If the multiplicities of some roots of the characteristic equation are greater than one, the theorem is not so easy to prove. It requires first showing that if λ is a root of multiplicity m, then there are m linearly independent eigenvectors belonging to λ. We shall not prove this, but refer the reader to Halmos (1958). Once this is accepted, however, the rest is relatively simple. All we need do is proceed by Gram-Schmidt orthogonalization in the subspace of vectors *spanned by* (i.e., that can be written as linear combinations of) the linearly independent eigenvectors belonging to λ. The result will then clearly be an orthonormal basis in which A is orthogonal. The details are left to the reader.

We make one final remark before proceeding to some examples. It is clear from the above and from the rules for multiplication of matrices that if there exists a single basis in which two Hermitian operators $\mathbf{A}$ and $\mathbf{B}$ both have diagonal matrices, then $\mathbf{A}$ and $\mathbf{B}$ commute. One may go further. Two Hermitian operators commute if *and only if* they can be simultaneously diagonalized. We shall prove this again on the assumption that $\mathbf{A}$ and $\mathbf{B}$ have n distinct eigenvalues each. Let $\mathbf{x}$ be an eigenvector of $\mathbf{B}$ belonging to the eigenvalue β. Then $\mathbf{Ax}$ is also an eigenvector belonging to β, for

$$\mathbf{B}(\mathbf{Ax}) = \mathbf{ABx} = \mathbf{A}\beta\mathbf{x} = \beta(\mathbf{Ax}).$$

Since each eigenvalue has only one eigenvector (up to multiplication by a number) it follows that $\mathbf{Ax}$ is a multiple of $\mathbf{x}$, or

$$\mathbf{Ax} = \alpha\mathbf{x}.$$

Thus if $\mathbf{A}$ and $\mathbf{B}$ commute, the eigenvectors of one are the eigenvectors of the other. Since the basis in which their matrices are diagonal is made up of these eigenvectors, the matrices are diagonalized by the same basis.

IV-5 EXAMPLES, APPLICATIONS, AND EXTENSIONS

(a) The Legendre Polynomials

Let us define an inner product on P_m by

$$(f, g) = \int_{-1}^{1} f(p)g(p)\, dp, \tag{28}$$

where $f(p)$ and $g(p)$ are polynomials of degree m or less. This converts P_m into an orthogonal space. The reader may check that the inner product defined in this way satisfies the three properties at the beginning of Section 4.

We wish to find an orthonormal basis $\mathbf{e}_0, \mathbf{e}_1, \mathbf{e}_2, \ldots, \mathbf{e}_n$, where $\mathbf{e}_j$ is a polynomial of degree j. To construct this basis we start with $\mathbf{e}_0$. Since $\mathbf{e}_0$ is a polynomial of degree zero, we may write $\mathbf{e}_0 = \alpha$. Then if the norm is to be 1, we have

$$|\mathbf{e}_0|^2 = \int_{-1}^{1} \alpha^2\, dp = 2\alpha^2 = 1,$$

or

$$\alpha = \frac{1}{\sqrt{2}}.$$

Now we construct $\mathbf{e}_1 = \beta + \gamma p$ from the condition that it be orthogonal to $\mathbf{e}_0$:

$$(\mathbf{e}_0, \mathbf{e}_1) = \int_{-1}^{1} \frac{1}{\sqrt{2}}(\beta + \gamma p)\, dp = \frac{2\beta}{\sqrt{2}} = 0.$$

Thus $\beta = 0$. As for γ, we find it from the condition that

$$|\mathbf{e}_1|^2 = \gamma^2 \int_{-1}^{1} p^2\, dP = \tfrac{2}{3}\gamma = 1.$$

Thus $\gamma = \sqrt{3/2}$, and $\mathbf{e}_1 = \sqrt{3/2}p$.

Let us calculate just one further basis vector. Let $\mathbf{e}_2 = \lambda + \mu p + \nu p^2$. From the condition that $\mathbf{e}_2$ be orthogonal to $\mathbf{e}_0$ we find that $\nu = -3\lambda$. From the condition that $\mathbf{e}_2$ be orthogonal to $\mathbf{e}_1$ we find that $\mu = 0$. From the condition that $\mathbf{e}_2$ have unit norm we find that $\lambda = \pm(1/2)\sqrt{5/2}$. We choose the negative sign and obtain $\mathbf{e}_2 = (1/2)\sqrt{5/2}\,(3p^2 - 1)$.

Collecting our results from these and similar calculations, we arrive at

$$
\begin{aligned}
e_0 &= \frac{1}{\sqrt{2}}, \\
e_1 &= \sqrt{\frac{3}{2}}\,p, \\
e_2 &= \frac{1}{2}\sqrt{\frac{5}{2}}\,(3p^2 - 1), \\
e_3 &= \frac{1}{2}\sqrt{\frac{7}{2}}\,(5p^3 - 3p), \\
e_4 &= \frac{1}{8}\sqrt{\frac{9}{2}}\,(35p^4 - 30p^2 + 3).
\end{aligned}
$$

The calculation can be continued to obtain an orthonormal basis in P_m for any desired m.

The reader may recognize these polynomials as the first five (normalized) Legendre polynomials. Thus the Legendre polynomials may be characterized in the following way. If P_m is made an orthogonal space by the norm of Eq. (28) and an orthonormal basis is desired in which the basis vectors are polynomials of successively increasing degree, this basis will consist of the Legendre polynomials of degree zero to m. If the norm were defined in some other way, different polynomials would be obtained (see Problem 24).

(b) Small Vibrations

There is one many-body problem in mechanics which can be solved exactly and which is a very good approximation for many other problems. This is the so-called problem of *small vibrations*.

Consider a system with n degrees of freedom, possessing a Lagrangian of the form

$$
L = \tfrac{1}{2}\tau_{\alpha\beta}(q)\dot{q}_\alpha\dot{q}_\beta - V(q), \qquad \alpha = 1, \ldots, n, \tag{29}
$$

where $\tau_{\alpha\beta} = \tau_{\beta\alpha}$.[6] (Note that we have returned to Greek indices for a mechanical system.) We identify the first term, as usual, with the kinetic energy T

[6] Even if $\tau_{\alpha\beta} \neq \tau_{\beta\alpha}$, the $\tau_{\alpha\beta}$ can be replaced by other functions which have this symmetry. Write

$$
\tau_{\alpha\beta} = \tfrac{1}{2}(\tau_{\alpha\beta} + \tau_{\beta\alpha}) + \tfrac{1}{2}(\tau_{\alpha\beta} - \tau_{\beta\alpha}),
$$

and set $\frac{1}{2}(\tau_{\alpha\beta} + \tau_{\beta\alpha}) = \nu_{\alpha\beta} = \nu_{\beta\alpha}$. Then a simple calculation shows that $\tau_{\alpha\beta}\dot{q}_\alpha\dot{q}_\beta = \nu_{\alpha\beta}\dot{q}_\alpha\dot{q}_\beta$, so that the symmetric $\nu_{\alpha\beta}$ can replace the asymmetric $\tau_{\alpha\beta}$ in (29).

and the second with the potential energy. Now assume that the potential energy has a minimum point $q = Q$, and let us look qualitatively at the motion of the system if it is displaced slightly from this minimum and allowed to start from rest. Since the Lagrangian is time indepedent, the energy $T + V$ is conserved. The system, as in the similar one-dimensional problem, will start moving toward lower potential energy, that is, toward the minimum point Q. As the potential energy decreases, the kinetic energy increases, and the system, with its accumulated kinetic energy, will pass through (perhaps by) Q and start losing kinetic and gaining potential energy. In this way it will oscillate near the *equilibrium point* Q; it will never move very far away from Q because its maximum potential energy cannot surpass the initial value of the total energy.

For small enough displacements this motion can be analyzed in detail. We start by expanding V in a Taylor's series about Q. Note first that V (and similarly L) is undetermined up to a constant. If this constant is chosen so that $V(Q) = 0$, the Taylor's series becomes

$$V(q) = \tfrac{1}{2}\kappa_{\alpha\beta}\eta_\alpha\eta_\beta + \tfrac{1}{6}\sigma_{\alpha\beta\gamma}\eta_\alpha\eta_\beta\eta_\gamma + \cdots, \tag{30}$$

where $\eta_\alpha = q_\alpha - Q_\alpha$,

$$\kappa_{\alpha\beta} = \kappa_{\beta\alpha} = \left.\frac{\partial^2 V}{\partial q_\alpha\, \partial q_\beta}\right|_{q=Q}, \tag{31}$$

and

$$\sigma_{\alpha\beta\gamma} = \left.\frac{\partial^3 V}{\partial q_\alpha\, \partial q_\beta\, \partial q_\gamma}\right|_{q=Q}.$$

Note that the first derivatives of V vanish because Q is a minimum point. Now if Q is indeed a minimum, $\kappa_{\alpha\beta}\eta_\alpha\eta_\beta \geq 0$ for all η_α. Furthermore we assume that if there is any displacement η^0 such that $\kappa_{\alpha\beta}\eta_\alpha^0\eta_\beta^0 = 0$, then $V(Q + \lambda\eta^0) = 0$ for all λ; that is, that then all the other terms in the Taylor's series vanish for this displacement. This means that for displacements in this direction in configuration space V remains constant. Thus Q need not be an absolute minimum, but at least it is not a maximum with respect to variations in some direction (not a saddle point). Our assumption also implies that if V varies at all for a given displacement, it varies quadratically in the first approximation. (For a one-dimensional example of an anomolous case in which this is not true, see Problem 26.)

Just as we expand V in a Taylor's series about Q, so we shall expand T. Implicit is the assumption that this is possible, that is, that T is sufficiently smooth at Q (but see Problem 27). Since $\dot{q} = \dot{\eta}$, we have

$$T(q, \dot{q}) = \tfrac{1}{2}\mu_{\alpha\beta}\dot{\eta}_\alpha\dot{\eta}_\beta + \tfrac{1}{2}\rho_{\alpha\beta\gamma}\dot{\eta}_\alpha\dot{\eta}_\beta\eta_\gamma + \cdots, \tag{32}$$

where

$$\mu_{\alpha\beta} = \mu_{\beta\alpha} = \tau_{\alpha\beta}(Q) \tag{33}$$

and

$$\rho_{\alpha\beta\gamma} = \left.\frac{\partial \tau_{\alpha\beta}}{\partial q_\gamma}\right|_{q=Q}.$$

Recall that the kinetic energy is always positive (unless the velocity is zero). This implies that $\frac{1}{2}\mu_{\alpha\beta}\dot{\eta}_\alpha\dot{\eta}_\beta > 0$ for any nonzero $\dot{\eta}$, so that in particular the first term in this Taylor's series never vanishes.

Now let us restrict our discussions to displacements η from equilibrium so small that in Eq. (30) all terms can be neglected compared to the first, except for the special displacements for which V remains zero, and in Eq. (32) all terms can be neglected compared to the first. Thus we proceed in what is, in general, an approximation. It is clear, however, that if the potential is truly quadratic and if the kinetic energy is q-independent, our treatment will be exact.

With these restrictions on the displacement, (29) becomes, approximately,

$$L = \tfrac{1}{2}\mu_{\alpha\beta}\dot{\eta}_\alpha\dot{\eta}_\beta - \tfrac{1}{2}\kappa_{\alpha\beta}\eta_\alpha\eta_\beta. \tag{34}$$

It is a simple matter to obtain the equations of motion from this Lagrangian:

$$\mu_{\alpha\beta}\ddot{\eta}_\beta + \kappa_{\alpha\beta}\eta_\beta = 0. \tag{35}$$

Now if this were a system with one degree of freedom, the equation would be (we may drop the indices since there is only one component)

$$\mu\ddot{\eta} + \kappa\eta = 0,$$

which is the equation of the simple harmonic oscillator. The well-known solution is, in complex form,

$$\eta = \alpha e^{i\omega t} + \beta e^{-i\omega t},$$

where $\omega = (\kappa/\mu)^{1/2}$, and α and β are complex constants to be determined from the initial conditions. Since η is real we must have $\alpha = \beta^*$.

We have already seen that in the n-dimensional case the system oscillates about Q and we now see that in the one-dimentional case this oscillation is simple harmonic. Perhaps we can find a solution to the n-dimensional problem which also has some simple harmonic properties.

In order to proceed with such an attempt, we write Eq. (35) in vector-space terms. Let $\mathbf{y}$ be a time-dependent vector in an n-dimensional complex vector space S, and let $\mathbf{y}$ have components η_α in a certain basis. Let $\mathbf{M}$ and $\mathbf{K}$ be operators on S whose matrix elements in this basis are $\mu_{\alpha\beta}$ and $\kappa_{\alpha\beta}$, respectively. Then (35) becomes

$$\mathbf{M}\ddot{\mathbf{y}} + \mathbf{K}\mathbf{y} = 0, \tag{36}$$

where $\ddot{\mathbf{y}}$ is the vector whose components are the $\ddot{\eta}_\alpha$. We shall attempt to find solutions of the form

$$\mathbf{y} = \mathbf{a}e^{i\omega t} + \mathbf{a}^*e^{-i\omega t}, \tag{37}$$

where $\mathbf{a}$ and $\mathbf{a}^*$ are constant vectors. To ensure that $\mathbf{y}$ is real the components of $\mathbf{a}^*$ must be the complex conjugate of the components of $\mathbf{a}$.

Inserting (37) into (36), we obtain

$$e^{i\omega t}(\mathbf{K} - \lambda\mathbf{M})\mathbf{a} + e^{-i\omega t}(\mathbf{K} - \lambda\mathbf{M})\mathbf{a}^* = 0,$$

where $\lambda = \omega^2$. Since this must hold for all times, the coefficients of $e^{i\omega t}$ and $e^{-i\omega t}$ must vanish separately, and we arrive at

$$(\mathbf{K} - \lambda\mathbf{M})\mathbf{a} = 0, \tag{38}$$

and similarly for $\mathbf{a}^*$. This is a generalization of the eigenvalue problem we treated in Section 3: we must solve for both λ and $\mathbf{a}$. As in the other eigenvalue problem, a vector $\mathbf{a}$ will exist only for those values of λ which are roots of the characteristic equation

$$\det(\mathbf{K} - \lambda\mathbf{M}) = 0.$$

In order to proceed we introduce a scalar product into S. Let $\mathbf{x}$ have components ξ_α and $\mathbf{y}$ have components η_α in the basis we have been using [i.e., in the basis in which (36) appears in the form of (35)]. Then we define

$$(\mathbf{x}, \mathbf{y}) = \xi_\alpha^* \mu_{\alpha\beta} \eta_\beta.$$

This is not the simplest scalar product one could introduce in S, but it is the most useful. To show that it is indeed a scalar product, we must show that it has the three properties listed at the beginning of Section 4. Linearity, the first of these properties, is self-evident. Hermiticity may be shown as follows:

$$\begin{aligned}(\mathbf{x}, \mathbf{y}) &= \xi_\alpha^* \mu_{\alpha\beta} \eta_\beta = \eta_\beta \mu_{\alpha\beta} \xi_\alpha^* \\ &= (\eta_\alpha^* \mu_{\alpha\beta} \xi_\beta)^* = (\mathbf{y}, \mathbf{x})^*,\end{aligned}$$

where we have used the fact that the $\mu_{\alpha\beta} = \mu_{\beta\alpha}$ are real. To show that the scalar product is positive definite we make use of the discussion following Eq. (33), which shows that if $\mathbf{f}$ is a real nonzero vector, that is, a vector with real components, then $(\mathbf{f}, \mathbf{f}) > 0$. Now let $\mathbf{x} = \mathbf{f} + i\mathbf{g}$, where $\mathbf{f}$ and $\mathbf{g}$ are real vectors. Then by hermiticity and linearity we have $(\mathbf{x}, \mathbf{x}) = (\mathbf{f}, \mathbf{f}) + (\mathbf{g}, \mathbf{g})$. Since each term on the right is positive, so is $(\mathbf{x}, \mathbf{x})$, unless $\mathbf{x} = 0$, which shows that the scalar product is indeed positive definite.

We now rewrite (38) in the form

$$\mathbf{M}^{-1}\mathbf{K}\mathbf{a} = \lambda\mathbf{a}. \tag{39}$$

[Note that $\mathbf{M}^{-1}$ exists, for if $\mathbf{M}$ annihilates some vector $\mathbf{x}$ with components ξ_α, then $\mu_{\alpha\beta}\xi_\beta$ is zero and so is $\xi_\alpha^*\mu_{\alpha\beta}\xi_\beta = (\mathbf{x}, \mathbf{x})$, which means that $\mathbf{x} = 0$.]

We now assert that $\mathbf{M}^{-1}\mathbf{K}$ is a Hermitian operator with nonnegative eigenvalues, and that therefore according to the theorem of Section 4 there exists an orthonormal basis which diagonalizes $\mathbf{M}^{-1}\mathbf{K}$. Since each of the basis vectors in such an orthonormal basis is an eigenvector of $\mathbf{M}^{-1}\mathbf{K}$, this means that there are n linearly independent solutions of (39) and hence of (38). To prove the assertion, we note that a simple calculation yields

$$(\mathbf{x}, \mathbf{M}^{-1}\mathbf{K}\mathbf{y}) = \xi_\alpha^* \kappa_{\alpha\beta} \eta_\beta. \tag{40}$$

That $\mathbf{M}^{-1}\mathbf{K}$ is Hermitian follows immediately from the symmetry and reality of the $\kappa_{\alpha\beta}$. From Eq. (40) and the discussion following Eq. (31) it follows that $(\mathbf{f}, \mathbf{M}^{-1}\mathbf{K}\mathbf{f}) \geq 0$ for nonzero real vectors $\mathbf{f}$, and then it can be shown, in the same way as we showed that $(\mathbf{x}, \mathbf{x})$ is positive definite, that $(\mathbf{x}, \mathbf{M}^{-1}\mathbf{K}\mathbf{x}) \geq 0$ for all $\mathbf{x}$. Let $\mathbf{a}$ be a solution of (39). Taking scalar products of both sides of (39) with $\mathbf{a}$, we have

$$(\mathbf{a}, \mathbf{M}^{-1}\mathbf{K}\mathbf{a}) = \lambda \, |\mathbf{a}|^2.$$

Since $|\mathbf{a}|^2 > 0$ and $(\mathbf{a}, \mathbf{M}^{-1}\mathbf{K}\mathbf{a}) \geq 0$, it follows that

$$\lambda \geq 0.$$

Thus Eq. (38) has n linearly independent eigenvectors all belonging to nonnegative eigenvalues.

This nonnegative property of the eigenvalues has physical meaning. If λ is positive, ω is real, and then according to Eq. (37) $\mathbf{y}$ oscillates. If λ could take on negative values, ω could be imaginary, and then according to (37) $\mathbf{y}$ could grow or decay exponentially. Note that $\lambda = 0$ means that $\mathbf{a}$ is annihilated by $\mathbf{K}$. If there exist vectors annihilated by $\mathbf{K}$, Eq. (36) becomes

$$\mathbf{M}\ddot{\mathbf{y}} = 0, \qquad \mathbf{K}\mathbf{y} = 0, \tag{41}$$

for such vectors. The solution is

$$\mathbf{y} = \mathbf{y}_0 + \mathbf{v}t, \tag{42}$$

where $\mathbf{y}_0$ and $\mathbf{v}$ are constant vectors annihilated by $\mathbf{K}$. This might have been expected, for according to the discussion following Eq. (31), $\mathbf{K}\mathbf{y} = 0$ means that the potential V remains unaltered for displacements corresponding to the vector $\mathbf{y}$, and therefore that there is no generalized force in this direction. Thus one would expect the motion to be at uniform velocity, as indicated by Eq. (42).

We have now established that there exists an orthonormal basis of vectors $\mathbf{e}_\alpha$ diagonalizing (i.e., each of which is an eigenvector of) $\mathbf{M}^{-1}\mathbf{K}$. The $\mathbf{e}_\alpha$ are the normalized solutions of the eigenvalue problem, that is, they are the $\mathbf{a}$'s which satisfy (39), but normalized. For our purposes normalization will be unnecessary, so we may choose an orthogonal basis of unnormalized vectors $\mathbf{a}_\alpha$. Each of the $\mathbf{a}_\alpha$ (or its normalized form $\mathbf{e}_\alpha$) belongs to an eigenvalue λ_α

and yields a solution of (36) which has the form of (37). For instance, if we write

$$\mathbf{y}_\alpha = (\rho e^{i\omega_\alpha t} + \rho^* e^{-i\omega_\alpha t})\mathbf{a}_\alpha \qquad \text{(no sum on } \alpha\text{)},$$

where ρ is an arbitrary number and $\omega_\alpha = \sqrt{\lambda_\alpha}$, then

$$\mathbf{M}\ddot{\mathbf{y}}_\alpha + \mathbf{K}\mathbf{y}_\alpha = (\rho e^{i\omega_\alpha t} + \rho^* e^{-i\omega_\alpha t})[-\lambda_{(\alpha)}\mathbf{M} + \mathbf{K}]\mathbf{a}_\alpha = 0 \qquad \text{(no sum)},$$

and $\mathbf{y}_\alpha$ is a solution of (36). The general solution is a linear combination of such solutions.

The general vector $\mathbf{y} \in S$ is of the form $\mathbf{y} = \eta'_\alpha \mathbf{a}_\alpha$, for the $\mathbf{a}_\alpha$ form a basis. We have established above that if $\mathbf{y}$ is to be a solution of (36) its (primed) components in this basis must be

$$\eta'_\alpha = \rho_{(\alpha)} e^{i\omega_\alpha t} + \rho^*_{(\alpha)} e^{-i\omega_\alpha t}.$$

The values of the ρ_α must be found from the initial conditions. It is seen that each of the primed components of a solution vibrates in simple harmonic motion.

Often in a physical problem one is interested not so much in the detailed motion as in what are the simple harmonic solutions and what are their frequencies. These solutions are called the *normal modes* of vibration of the system, and the frequencies are called the *normal frequencies*. In order to find the normal modes in terms of the original generalized coordinate system in configuration space (the original basis in S), we need to find the η'_α in terms of the η_α, or the transformation matrix from the original basis to the $\mathbf{a}_\alpha$. But this comes automatically with the solution of the eigenvalue problem.

Before demonstrating the technique on an example, we summarize the results in matrix notation.

A. Given L, find the $\mu_{\alpha\beta}$ and the $\kappa_{\alpha\beta}$.

B. Obtain the roots of the characteristic equation

$$\det(K - \lambda M) = 0.$$

These are the squares of the normal frequencies.

C. For each root find the components of the eigenvector from the matrix equation

$$(K - \lambda_{(\alpha)} M)a_\alpha = 0.$$

If the root in question has multiplicity m, find m linearly independent solutions. Note that it is not in general necessary to normalize the eigenvectors. These eigenvectors are the normal modes.

D. The general solution is (sum on α)

$$y = a_\alpha(\rho_\alpha e^{i\omega_\alpha t} + \rho^*_\alpha e^{-i\omega_\alpha t}). \tag{43}$$

The ρ_α must be found from the initial conditions. The ω_α are the positive square roots of the λ_α. When expressed in this way the solution will automatically be given not in the basis $\mathbf{a}_1, \ldots, \mathbf{a}_n$, but in the original basis in S (the original generalized coordinates), for the $\mathbf{a}_\alpha$ are represented in this basis by step C of this summary.

As an example of this procedure we consider a sort of linear classical water molecule. Figure 11 illustrates a system of three particles connected by

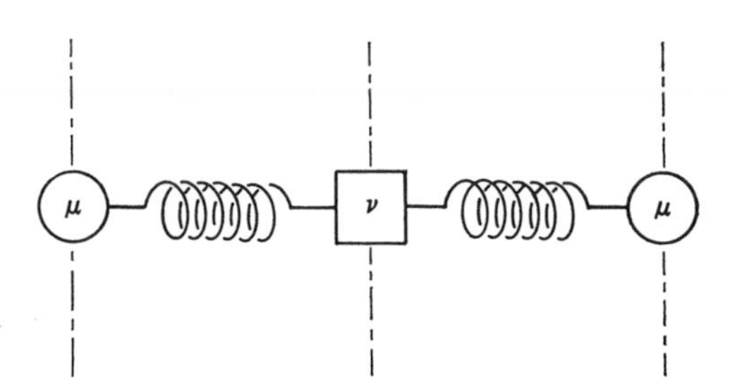

Figure 11. Three particles connected by springs. Each particle is shown at its equilibrium position.

springs. Let the outer two particles have equal masses μ, and let the particle in the center have mass ν. Assume that each spring constant is equal to k. The system is constrained to move in one dimension. Let the coordinate η_3 of mass ν take on the value zero at some arbitrary point, and let the coordinates η_1 and η_2 of the other two masses take on the value zero at their equilibrium positions when $\eta_3 = 0$. Then

$$T = \tfrac{1}{2}\mu(\dot{\eta}_1^2 + \dot{\eta}_2^2) + \tfrac{1}{2}\nu\dot{\eta}_3^2,$$

$$V = \tfrac{1}{2}k[(\eta_1 - \eta_3)^2 + (\eta_2 - \eta_3)^2].$$

Note that in this case the treatment is exact. Note also that if the system is moved as a whole, that is if $\eta_1 = \eta_2 = \eta_3$, the potential energy remains zero. This is an example of a potential energy which is not positive *definite*; as has been seen, we may expect motion at constant velocity in this direction.

From the equations for T and V we obtain

$$M = \begin{vmatrix} \mu & 0 & 0 \\ 0 & \mu & 0 \\ 0 & 0 & \nu \end{vmatrix}, \qquad K = k\begin{vmatrix} 1 & 0 & -1 \\ 0 & 1 & -1 \\ -1 & -1 & 2 \end{vmatrix}.$$

The characteristic equation is

$$\lambda(k - \lambda\mu)[(2\mu + \nu)k - \lambda\mu\nu] = 0,$$

and the three eigenvalues are

$$\lambda_1 = \frac{k}{\mu},$$

$$\lambda_2 = \frac{k}{\mu}\left(1 + \frac{2\mu}{\nu}\right),$$

$$\lambda_3 = 0.$$

To find the eigenvectors we form the matrices $K - \lambda_\beta M$ and solve the equation $(K - \lambda_{(\beta)} M)a_\beta = 0$. For instance with $\beta = 1$ we obtain (let a_1 have components α_γ)

$$(M - \lambda K)a_1 = \begin{vmatrix} 0 & 0 & -k \\ 0 & 0 & -k \\ -k & -k & k\left(2 - \frac{\nu}{\mu}\right) \end{vmatrix} \begin{vmatrix} \alpha_1 \\ \alpha_2 \\ \alpha_3 \end{vmatrix}$$

$$= \begin{vmatrix} -k\alpha_3 \\ -k\alpha_3 \\ -k\left(\alpha_1 + \alpha_2 + \left[\frac{\nu}{\mu} - 2\right]\alpha_3\right) \end{vmatrix} = \begin{vmatrix} 0 \\ 0 \\ 0 \end{vmatrix}$$

This yields $\alpha_3 = 0$, $\alpha_2 = -\alpha_1$. Without worrying about normalization, we choose $\alpha_1 = 1$, and then

$$a_1 = \begin{vmatrix} 1 \\ -1 \\ 0 \end{vmatrix}, \qquad \omega_1 = \sqrt{\frac{k}{\mu}}.$$

In a similar way we obtain (again, not normalized)

$$a_2 = \begin{vmatrix} 1 \\ 1 \\ -2\frac{\mu}{\nu} \end{vmatrix}, \qquad \omega_2 = \sqrt{\frac{k}{\mu}\left(1 + \frac{2\mu}{\nu}\right)},$$

and

$$a_3 = \begin{vmatrix} 1 \\ 1 \\ 1 \end{vmatrix}, \qquad \omega_3 = 0$$

(this last was expected).

In the first normal mode, η_1 and η_2 are displaced equally in opposite directions, as is seen from the first two components of a_1, while the center mass remains stationary ($\eta_3 = 0$). In the second normal mode, the two end masses move in the same direction, while the center mass moves to keep the center of mass of the entire system fixed.

Let us choose the initial conditions to be

$$\eta_1(0) = \sigma, \qquad \eta_2(0) = 0, \qquad \eta_3(0) = -\sigma\frac{\mu}{\nu},$$
$$\dot{\eta}_1(0) = \dot{\eta}_2(0) = \dot{\eta}_3(0) = 0.$$

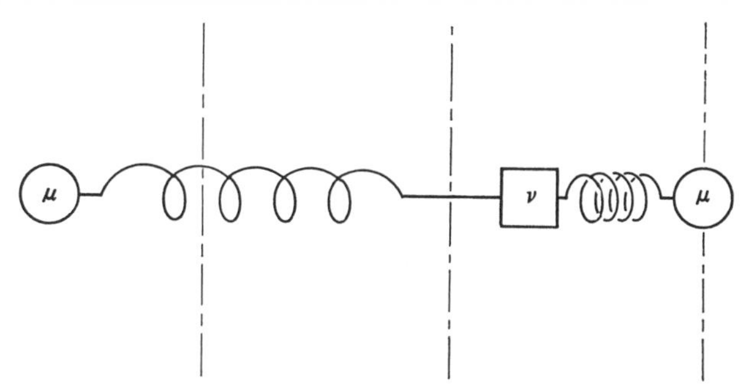

Figure 12. Three particles connected by springs. Each particle is shown in its initial position.

The system starts at rest with the left-hand mass and the center mass displaced, but with the center of mass at zero (see Fig. 12). Then at $t = 0$, Eq. (43) yields

$$y(0) = \sigma\begin{vmatrix} 1 \\ 0 \\ -\mu/\nu \end{vmatrix} = (\rho_1 + \rho_1^*)\begin{vmatrix} 1 \\ -1 \\ 0 \end{vmatrix} + (\rho_2 + \rho_2^*)\begin{vmatrix} 1 \\ 1 \\ -2\mu/\nu \end{vmatrix} + (\rho_3 + \rho_3^*)\begin{vmatrix} 1 \\ 1 \\ 1 \end{vmatrix}.$$

From this equation we obtain

$$\rho_1 + \rho_1^* = \rho_2 + \rho_2^* = \tfrac{1}{2}\sigma,$$
$$\rho_3 + \rho_3^* = 0.$$

The initial condition on the velocity $\dot{y}(0) = 0$ yields $\rho_1 = \rho_1^* = \rho_2 = \rho_2^* = \frac{1}{4}\sigma$. Inserting this into (43) we obtain

$$y(t) = \begin{vmatrix} \eta_1 \\ \eta_2 \\ \eta_3 \end{vmatrix} = \begin{vmatrix} \frac{1}{2}\sigma\cos\omega_1 t + \frac{1}{2}\sigma\cos\omega_2 t \\ -\frac{1}{2}\sigma\cos\omega_1 t + \frac{1}{2}\sigma\cos\omega_2 t \\ -\sigma\frac{\mu}{\nu}\cos\omega_2 t \end{vmatrix}.$$

It is seen that although mass ν will perform simple harmonic motion, the two end masses will move with a combination of the two frequencies ω_1 and ω_2. If ω_1 is sufficiently close to ω_2 (if μ/ν is sufficiently small), the well known phenomenon of beats will be observed.

(c) Rotations

At the end of Chapter 3, in deriving the free-particle Lagrangian from Galilei invariance, we made use of rotation operators. We now want to describe these operators carefully. Since in the next chapter they will be used extensively in discussing rigid-body motion, our introduction of them will lean toward that application.

Let us turn to ordinary three-dimensional configuration space S, and consider transformations which map every $\mathbf{x} \in S$ into some vector $\mathbf{f}(\mathbf{x}) \in S$ (we do not assume linearity) and which have the following two properties.

Property 1. The difference between $\mathbf{f}(\mathbf{x})$ and $\mathbf{f}(\mathbf{y})$ depends continuously only on the difference between $\mathbf{x}$ and $\mathbf{y}$; mathematically,

$$\mathbf{f}(\mathbf{x}) - \mathbf{f}(\mathbf{y}) = \mathbf{g}(\mathbf{x} - \mathbf{y}) \tag{44}$$

Property 2. The distance between any two vectors remains invariant. This means that $\mathbf{g}(\mathbf{x} - \mathbf{y})$ has the same length (norm) as $\mathbf{x} - \mathbf{y}$. (Actually, it can be shown that Property 2 alone is sufficient for our purpose.)

These properties, particularly Property 2, are related to rigid bodies. When such a body moves, the vectors connecting points in the body change. At any time (i.e., at any orientation of the body) each vector, originally in one position, is now in a new one. This may be thought of as a time-dependent mapping of the vectors in the body, and this mapping has the two properties we have listed. For instance, in a rigid body the distance between any two points remains constant (Property 2).

Let us first turn to Property 1. Consider three vectors $\mathbf{x}$, $\mathbf{y}$, $\mathbf{z} \in S$. Then (44) implies that

$$\mathbf{f}(\mathbf{x}) - \mathbf{f}(\mathbf{y}) = \mathbf{g}(\mathbf{x} - \mathbf{y})$$

$$\mathbf{f}(\mathbf{y}) - \mathbf{f}(\mathbf{z}) = \mathbf{g}(\mathbf{y} - \mathbf{z}).$$

Adding these equations and using (44) again, we arrive at

$$\mathbf{f}(\mathbf{x}) - \mathbf{f}(\mathbf{z}) = \mathbf{g}(\mathbf{x} - \mathbf{y}) + \mathbf{g}(\mathbf{y} - \mathbf{z}) = \mathbf{g}(\mathbf{x} - \mathbf{z}).$$

This equation holds for any $\mathbf{y}$. Let us, therefore, set $\mathbf{y} = 0$, and write $\mathbf{z} = -\mathbf{w}$. Then we have

$$\mathbf{g}(\mathbf{x}) + \mathbf{g}(\mathbf{w}) = \mathbf{g}(\mathbf{x} + \mathbf{w}).$$

The problem now is to find a continuous function $\mathbf{g}$ which satisfies this equation for all $\mathbf{x}$, $\mathbf{w} \in S$. It is a well-known result (we shall not prove it here) that the only continuous solutions to this problem are the linear ones. In other words, the only solutions are of the form

$$\mathbf{g}(\mathbf{x}) = \mathbf{D}\mathbf{x},$$

where $\mathbf{D}$ is a linear operator. Now we return to (44), setting $\mathbf{y} = 0$. Let $\mathbf{f}(0) = \mathbf{a}$. Then we arrive at an expression for $\mathbf{f}(\mathbf{x})$, namely,

$$\mathbf{f}(\mathbf{x}) = \mathbf{a} + \mathbf{D}\mathbf{x}. \tag{45}$$

Remarks. 1. A transformation such as (45) is called an *inhomogeneous* linear transformation. The inhomogeneous term $\mathbf{a}$ may be eliminated if there is one point in the space that remains fixed under the mapping. Then if that point is taken as the origin of a Cartesian coordinate system, we have $\mathbf{f}(0) = 0$, so that $\mathbf{a} = 0$. *2.* We made no use of Property 2 in arriving at (45). In other words every continuous transformation possessing Property 1 is an inhomogeneous linear transformation. *3.* According to Eq. (45), every displacement of a rigid body is a translation of a point in the body by an amount $\mathbf{a}$ plus a linear transformation $\mathbf{D}$ which keeps that point fixed. We shall see that this linear transformation is a rotation.

We now go on to Property 2. Let us calculate the distance between $\mathbf{f}(\mathbf{x})$ and $\mathbf{f}(\mathbf{y})$. We have

$$\begin{aligned} |\mathbf{f}(\mathbf{x}) - \mathbf{f}(\mathbf{y})|^2 &= |\mathbf{D}\mathbf{x} - \mathbf{D}\mathbf{y}|^2 \\ &= |\mathbf{D}(\mathbf{x} - \mathbf{y})|^2. \end{aligned}$$

Let $\mathbf{x} - \mathbf{y} = \mathbf{z}$. Then Property 2 states that

$$|\mathbf{D}\mathbf{z}|^2 \equiv (\mathbf{D}\mathbf{z}, \mathbf{D}\mathbf{z}) = (\mathbf{z}, \mathbf{z}),$$

for all $\mathbf{z} \in S$. In other words, for all $\mathbf{z} \in S$ (recall that S is a real vector space and that therefore the inner or scalar product is symmetric and that the analog of the Hermitian conjugate is the transpose)[7]

$$(\mathbf{D}^T\mathbf{D}\mathbf{z}, \mathbf{z}) = (\mathbf{z}, \mathbf{z}) = \mathbf{z} \cdot \mathbf{z} = \mathbf{z}_\alpha \mathbf{z}_\alpha.$$

This implies that $\mathbf{D}$ is orthogonal [see Remark 3 after Eq. (27)]:

$$\mathbf{D}\mathbf{D}^T = \mathbf{D}^T\mathbf{D} = \mathbf{1}; \qquad \mathbf{D}^T = \mathbf{D}^{-1}. \tag{46}$$

We may thus say that the inhomogeneous linear transformation of (45) is a translation by $\mathbf{a}$ plus an orthogonal transformation $\mathbf{D}$. We propose now to study the properties of these orthogonal operators.

[7] A notational remark. When dealing with ordinary 3-space, we shall use the conventional notation we have established in the earlier chapters: a vector $\mathbf{x}$ has components $x = (x_1, x_2, x_3)$, and Greek letters will be used for the subscripts. An operator $\mathbf{A}$ has matrix A with elements $A_{\alpha\beta}$.

It is instructive to write out (46) in terms of the matrix elements $D_{\alpha\beta}$ of $\mathbf{D}$ in an orthogonal basis. In terms of the matrix elements, (46) becomes

$$D_{\gamma\alpha}D_{\gamma\beta} = \delta_{\alpha\beta}. \tag{47}$$

Remark. This equation may be thought of as follows. Let each row $(D_{\alpha 1}, D_{\alpha 2}, D_{\alpha 3})$ represent the components of a vector $\mathbf{D}_\alpha$. Then (47) tells us that the three row vectors, $\mathbf{D}_1$, $\mathbf{D}_2$, $\mathbf{D}_3$ form an orthonormal set. It is left to the reader to show that the same is true for the columns.

Equation (47) tells us precisely what conditions the $D_{\alpha\beta}$ must satisfy if D is to be an orthogonal matrix. Since α and β range from 1 to 3, Eq. (47) includes nine equations, but not all are independent. In fact if α and β are interchanged, the equation is not changed. Thus there are six equations, three with $\alpha \neq \beta$ and three with $\alpha = \beta$. This imposes six conditions on the nine $D_{\alpha\beta}$ and therefore only three are independent: once three are chosen, the other six can be found from (47). More generally, three more conditions independent of (47), together with (47), will uniquely specify the $D_{\alpha\beta}$ and therefore also D. These three conditions need not be the values of three of the $D_{\alpha\beta}$, and in practice they hardly ever will be, but in any case, it takes three parameters to specify a particular orthogonal matrix D. Our task will be to pick these three parameters conveniently.

To proceed, therefore, we shall restrict our discussion to orthogonal matrices without translation. That is, we set $\mathbf{a} \equiv \mathbf{f}(0) = 0$ so that the origin of our system is held fixed. Now let us slowly rotate S about the origin, starting with the identity mapping. As we rotate, S is mapped at each stage onto itself, and to each such mapping corresponds a certain orthogonal operator on S. We shall call this a *continuous sequence of rotation operators* starting from the unit operator. It is intuitively obvious that the matrix elements vary continuously through this operation (more accurately, that if the sequence of rotations is labeled by a continuous parameter, then the matrix elements are continuous functions of this parameter). We shall say that all orthogonal operators that can be reached by such a continuous sequence from the identity form the set of *proper rotations* or simply *rotations*, and we now show that the set of all orthogonal operators is larger than the set of rotations, that is, that not all orthogonals can be reached continuously from the identity.

Let us take the determinant of both sides of (46). We have

$$(\det D^T)(\det D) = (\det D)^2 = 1,$$

where we have used the fact that $\det D = \det D^T$. Thus

$$\det D = \pm 1. \tag{48}$$

Now the determinant of a matrix is a simple continuous function of its matrix elements, so that as we go through a continuous sequence of rotation operators the determinant will vary continuously. Since it starts at 1 (the determinant of the unit matrix is 1), it can never get to -1 by any continuous process. Thus all proper rotations have determinant one. There are, in addition, other orthogonal matrices. For instance inversion of S (every point in S is mapped into the point equidistant from the origin in the opposite direction) changes every vector in S to its inverse. The corresponding orthogonal matrix is

$$-\mathbb{1} = \begin{vmatrix} -1 & 0 & 0 \\ 0 & -1 & 0 \\ 0 & 0 & -1 \end{vmatrix},$$

whose determinant is -1. In fact it can be shown (see Problem 31) that every orthogonal is either a proper rotation or the product of such a rotation and $-\mathbb{1}$.

Let us now turn to the rotations themselves, for they are obviously the ones that are of most interest in application to rigid bodies. Let us find their eigenvalues and eigenvectors. It is shown in Problem 23 that all the eigenvalues of a unitary operator must have absolute value one. If we consider only real vectors and real eigenvalues (i.e., orthogonal operators) this means that the eigenvalues must be 1 or -1, except that we must always remember that it may be impossible to diagonalize a given rotation operator over the real numbers. On the other hand if some particular D can be diagonalized over the reals, there is an orthonormal basis (see Problem 23) in which it becomes

$$D = \begin{vmatrix} \lambda_1 & 0 & 0 \\ 0 & \lambda_2 & 0 \\ 0 & 0 & \lambda_3 \end{vmatrix}.$$

The determinant of $\mathbf{D}$ is then obviously (recall that the determinant is independent of basis)

$$\lambda_1 \lambda_2 \lambda_3 = 1,$$

for we are dealing with the rotations. Since each of the λ_α is ± 1, the eigenvalue -1 must occur an even number of times: either twice or not at all. We thus have three cases.

Case 1. The eigenvalue -1 occurs twice. Then there is a basis in which D is diagonal and in which $\lambda_1 = 1$, $\lambda_2 = \lambda_3 = -1$. We shall see that this is a special case (limit point) of Case 3.

Case 2. The eigenvalues are all $+1$. Then D is the unit matrix in one orthonormal basis, and hence also the unit matrix in all bases, or **D** is the unit operator. We shall see that this is also a special case of Case 3.

Case 3. **D** cannot be diagonalized over the reals. This must be because the equation

$$\det(D - \lambda \mathbb{1}) = 0$$

has complex roots. It is known that these always occur in complex conjugate pairs. Thus the roots are λ_1, λ_2, and λ_2^*, and the determinant, as before, is

$$\lambda_1 \, |\lambda_2|^2 = 1.$$

But we know that, like all roots of a unitary operator, λ_2 has absolute value one. Thus $\lambda_1 = 1$: there is just one real eigenvalue and it is equal to one. Every eigenvalue has at least one eigenvector. Let the normalized eigenvector belonging to the eigenvalue 1 be $\mathbf{e}_1$, the first vector in an orthonormal basis. The first column of D then has a $+1$ at the top (in the upper left-hand corner of D). Since the column is, according to the Remark after Eq. (47), a unit vector, the other two elements in the first column must be zero. Then the other two columns must represent unit vectors orthogonal to the (1, 0, 0) vector represented by the first column, so that they are $(0, \cos\theta, -\sin\theta)$ and $(0, \sin\theta, \cos\theta)$ for some value of θ. Consequently we have

$$D = \begin{vmatrix} 1 & 0 & 0 \\ 0 & \cos\theta & \sin\theta \\ 0 & -\sin\theta & \cos\theta \end{vmatrix}. \tag{49}$$

Thus in Case 3, given a proper rotation **D**, there exists an orthonormal basis in which D takes on the form of (49). It is seen that Case 1 occurs when $\theta = \pi$ and Case 2 when $\theta = 0$.

To see how **D** operates on any vector, we operate on the basis vectors of the orthonormal basis of (49). By representing these basis vectors as column vectors, we see immediately that

$$\mathbf{De}_1 = \mathbf{e}_1, \quad \mathbf{De}_2 = \mathbf{e}_2 \cos\theta - \mathbf{e}_3 \sin\theta, \quad \mathbf{De}_3 = \mathbf{e}_2 \sin\theta + \mathbf{e}_3 \cos\theta. \tag{50}$$

As is seen in Fig. 13, this is a rotation through an angle θ about the *1*-axis: $\mathbf{e}_1$ is unaltered, and $\mathbf{e}_2$ and $\mathbf{e}_3$ are rotated. Thus every rotation is a rotation about an axis. There is one direction, lying along the eigenvector of **D**, that remains fixed. This is Euler's theorem on the motions of a rigid body. What it says, for instance, is that in the continuous sequence of rotations we were discussing earlier, at each stage there is one line in S which is in its original position.

We see that every rotation can be specified by giving the axis which remains stationary and the angle of rotation about that axis. We have found three convenient numbers to specify a rotation, as we wanted to: two numbers to specify the axis (say, two direction cosines) and one to specify the angle. We have seen also how to find the axis, given **D**: solve the eigenvalue problem.

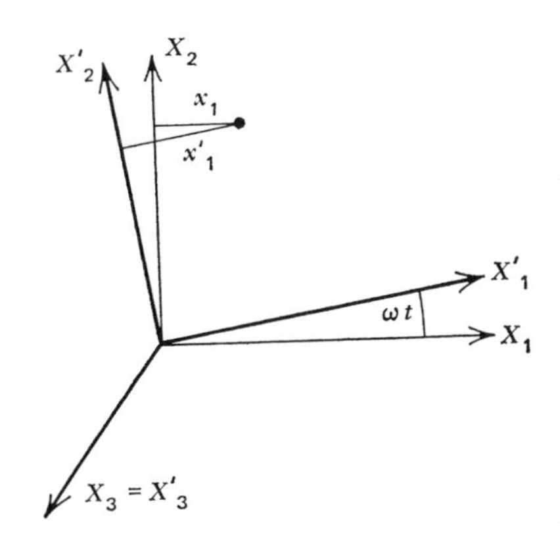

Figure 13. A rotation about the 1-axis.

How then does one find the angle? It follows from (49) that in that special coordinate system

$$\cos\theta = \tfrac{1}{2}\,\mathrm{tr}\,D - \tfrac{1}{2}, \tag{51}$$

where tr A, the *trace* of a matrix A, is defined [see Section 5(d)] in terms of its matrix elements $\alpha_{\beta\gamma}$ by

$$\mathrm{tr}\,A = \alpha_{\beta\beta}.$$

This is the sum of the elements on the main diagonal. The trace of a matrix is independent of the coordinate system in which it is calculated, as we shall now show, so that this prescription does not depend on representing **D** in the special system. That the trace is indeed coordinate independent is seen when one recalls that a matrix A is changed to $T^{-1}AT$ by a coordinate transformation T. But

$$\mathrm{tr}\,(T^{-1}AT) = \overset{-1}{\tau}_{\alpha\beta}\alpha_{\beta\gamma}\tau_{\gamma\alpha} = \alpha_{\beta\beta}$$
$$= \mathrm{tr}\,A,$$

which establishes the result.

Remark. Suppose we use Eq. (50) to define a new coordinate system (basis) by $\mathbf{e}'_\alpha = \mathbf{D}\mathbf{e}_\alpha$. Since $\mathbf{D}$ is an orthogonal operator, the $\mathbf{e}'_\alpha$ are an orthonormal set, and thus define a new Cartesian coordinate system. Now let some vector $\mathbf{x}$ have components x'_α in terms of the $\mathbf{e}'_\alpha$ and x_α in terms of the $\mathbf{e}_\alpha$. Then as we have seen in going from Eq. (5) to (7), the x'_α are given in terms of the x_α by the transpose of D. Since, moreover, D is orthogonal, its transpose is its inverse. We then have the following result. In going from a Cartesian coordinate system to another Cartesian system obtained by rotating the first, the components of a vector transform according to the *inverse* of the rotation matrix.

We shall not go further into a discussion of rotations at this time but shall continue to develop their properties where they will be needed.

(d) Some Properties of the Trace

In Section 5(c) we defined the *trace* of a matrix as

$$\operatorname{tr} A = \alpha_{ii},$$

and we showed that the trace is independent of basis, so that we may call it the trace of the operator. We wish to show now that the trace can be used to make the space of all linear operators in n dimensions a unitary space.

We have already seen that the space of all operators in n dimensions is a vector space under addition of operators and multiplication by numbers. Let $\mathbf{A}$ and $\mathbf{B}$ be two operators, and define their *inner product* $(\mathbf{A}, \mathbf{B})$ by

$$(\mathbf{A}, \mathbf{B}) = \operatorname{tr}(\mathbf{A}^\dagger \mathbf{B}) = \alpha^\dagger_{ij}\beta_{ji} = \alpha^*_{ji}\beta_{ji}.$$

We must show that this definition possesses all the properties of an inner product. Let us check them in order.

1. *Linearity.*

$$\begin{aligned}(\mathbf{A}, \lambda\mathbf{B} + \mu\mathbf{C}) &= \alpha^\dagger_{ij}(\lambda\beta_{ji} + \mu\gamma_{ji}) \\ &= \lambda\alpha^\dagger_{ij}\beta_{ji} + \mu\alpha^\dagger_{ij}\gamma_{ji} = \lambda(\mathbf{A}, \mathbf{B}) + \mu(\mathbf{A}, \mathbf{C}).\end{aligned}$$

2. *Hermiticity.*

$$\begin{aligned}(\mathbf{A}, \mathbf{B}) &= \alpha^*_{ji}\beta_{ji} = (\alpha_{ji}\beta^*_{ji})^* = (\beta^*_{ji}\alpha_{ji})^* \\ &= (\mathbf{B}, \mathbf{A})^*.\end{aligned}$$

3. *Positive Definiteness.* We have

$$|\mathbf{A}|^2 = (\mathbf{A}, \mathbf{A}) = \alpha^*_{ij}\alpha_{ij} = \sum_{i,j} |\alpha_{ij}|^2.$$

This is the sum of positive terms, and is thus positive unless all of the terms are zero. But if all terms are zero, all of the matrix elements of **A** are zero and $\mathbf{A} = 0$. Thus $|\mathbf{A}|^2$ is positive definite.

We note the following properties of the trace, which will be useful in the next chapter. First,

$$\text{tr}\,(ABC) = \alpha_{ij}\beta_{jk}\gamma_{ki} = \beta_{jk}\gamma_{ki}\alpha_{ij} = \text{tr}\,(BCA) = \text{tr}\,(CAB). \tag{53}$$

In particular, $\text{tr}\,(AB) = \text{tr}\,(BA)$. Second, $\text{tr}\,A = \text{tr}\,A^T$, for A^T is obtained from A by flipping across the main diagonal, but leaving the main diagonal invariant. Since the trace is the sum of the elements on the main diagonal, this does not change its value.

Early in the next chapter we shall have occasion to take the derivative of the trace with respect to the matrix elements of one of the matrices appearing in it. Thus let A be a varying matrix, and let us calculate

$$\frac{\partial}{\partial \alpha_{ij}}\,\text{tr}\,AB = \frac{\partial}{\partial \alpha_{ij}}\,\alpha_{lm}\beta_{ml} = \delta_{il}\delta_{mj}\beta_{ml} = \beta_{ji}. \tag{54}$$

Note that this itself may be considered the element of a matrix. We thus may write the whole matrix at once:

$$\frac{\partial}{\partial A}\,\text{tr}\,AB = B^T. \tag{55}$$

Similarly since $\alpha_{ij}^T = \alpha_{ji}$ we find that

$$\frac{\partial}{\partial A^T}\,\text{tr}\,AB = B. \tag{56}$$

(e) The Variational Principle with Constraints

In Chapter II we showed how the Euler-Lagrange equations could be derived from a variational principle. There the generalized coordinates were assumed to have already been chosen so as to eliminate any holonomic constraints. Now we want to show that the equations of motion can be obtained from a variational principle even when there are explicit constraints connecting the coordinates. The discussion has been postponed to this point because vector-space ideas will make it much easier.

We want to discuss not only holonomic constraints but also general velocity-dependent constraints, which, if they cannot be integrated as described at Eq. (3) of Chapter II, are called *nonholonomic*. We shall show that the equations can be obtained from a variational principle in both cases, but that the variations used are somewhat different depending on the type of constraint (Saletan 1970).

Consider then a system with Lagrangian $L(q, \dot{q}, t)$, in n generalized coordinates q_α, $\alpha = 1, \ldots, n$, constrained by $K < n$ constraint equations

$$f_I(q, \dot{q}, t) = 0, \qquad I = 1, \ldots, K, \tag{57}$$

where $q = \{q_1, \ldots, q_n\}$ as usual. The variational problem is almost the same as the one of Chapter II, Section 3, except that the allowed comparison paths $q(t, \varepsilon)$ are now restricted. The restriction is provided by Eqs. (57), but exactly how is still a question.

The simplest type of restriction leads to what is called the variational problem of Lagrange. It is the following: To find the single-valued function or path $q(t)$ which (i) passes through certain fixed points $Q_1 = q(t_1)$ and $Q_2 = q(t_2)$ at fixed times t_1 and $t_2 > t_1$, (ii) satisfies the constraints (57), and (iii) makes the definite integral

$$S = \int_{t_1}^{t_2} L(q, \dot{q}, t)\, dt \tag{58}$$

an extremum *when compared with other paths satisfying* (i) *and* (ii).

To solve this problem we assume that a solution $q(t)$ exists, and we imbed it in a family of paths $q(t, \varepsilon)$ which is differentiable in the parameter ε and such that each path satisfies conditions (i) and (ii). Let $q(t, 0) = q(t)$. When the ε-family is inserted into (58), S becomes a function of ε, and our problem reduces to finding the condition on $q(t)$ such that $S'(0) = 0$ for every ε-family, where $S' = dS/d\varepsilon$.

The derivative is easily calculated. After integrating by parts and using condition (i) in the form $\partial Q_1/\partial\varepsilon = \partial Q_2/\partial\varepsilon = 0$, just as we did in obtaining (37) in Chapter II, we find that $q(t) = q(t, 0)$ satisfies the condition[8] (the second term in Chapter II, Eq. (37) is zero here as there)

$$\int_{t_1}^{t_2} \Lambda_\alpha(t, 0)\eta_\alpha(t, 0)\, dt = 0, \tag{59}$$

where

$$\Lambda_\alpha(t, \varepsilon) = \frac{d}{dt}\frac{\partial L}{\partial \dot{q}_\alpha} - \frac{\partial L}{\partial q_\alpha} \tag{60}$$

(here, as usual, we write d/dt for the t derivative for fixed ε), and

$$\eta_\alpha(t, \varepsilon) = \partial q_\alpha/\partial\varepsilon. \tag{61}$$

It is helpful to view the left-hand side of (59) as an inner product. It is easy to show that all the sets of n real square-integrable functions $g(t) = \{g_1(t), \ldots, g_n(t)\}$, defined on the interval $t_1 < t < t_2$, form an orthogonal vector space F with the inner product

$$(g, h) = \int_{t_1}^{t_2} g_\alpha(t)h_\alpha(t)\, dt. \tag{62}$$

[8] Sums over α run from $\alpha = 1$ to $\alpha = n$, and sums over I run from $I = 1$ to $I = K$.

With this definition of inner product on F, Eq. (59) can be written

$$(\Lambda, \eta) = 0, \tag{63}$$

where $\Lambda(t) = \{\Lambda_1(t, 0), \ldots, \Lambda_n(t, 0)\}$, and $\eta = \{\eta_1(t, 0), \ldots, \eta_n(t, 0)\}$.

What Eq. (63) tells us is that Λ is orthogonal in F to any vector η that can be obtained by (61) from an ε-family, and therefore Λ lies in the subspace which is the *orthogonal complement* of the subspace composed of all such vectors η. If there are no constraints, for instance, that is, if $K = 0$, then the η vectors are arbitrary and span all of F. Hence Λ is orthogonal to all vectors in F, or is the null vector:

$$\Lambda_\alpha = \frac{d}{dt}\frac{\partial L}{\partial \dot{q}_\alpha} - \frac{\partial L}{\partial q_\alpha} = 0. \tag{64}$$

These are the usual Euler-Lagrange equations satisfied by $q(t)$, which we derived in Chapter II.

In general, however, η is not an arbitrary vector, but is restricted through (61) by condition (ii). Since all of the comparison paths satisfy the constraints, the f_I are independent of ε. This implies (but is not equivalent to[9]) the conditions

$$\frac{\partial f_I}{\partial \varepsilon} = \frac{\partial f_I}{\partial q_\alpha}\frac{\partial q_\alpha}{\partial \varepsilon} + \frac{\partial f_I}{\partial \dot{q}_\alpha}\frac{\partial \dot{q}_\alpha}{\partial \varepsilon} = 0. \tag{65}$$

To put this into a form similar to (63), we use the following elementary fact:

Let $h_I(t)$ be K continuous bounded functions of t on the interval $t_1 < t < t_2$. Then $h_I(t) = 0$ if and only if for every set of K sufficiently well behaved functions $\mu_I(t)$ on the same interval we have[8]

$$\int_{t_1}^{t_2} \mu_I(t) h_I(t)\, dt = 0.$$

With this, we can replace (65) by

$$0 = \int_{t_1}^{t_2} \mu_I \frac{\partial f_I}{\partial \varepsilon}\, dt = \int_{t_1}^{t_2} \left[\mu_I \frac{\partial f_I}{\partial q_\alpha} - \frac{d}{dt}\left(\mu_I \frac{\partial f_I}{\partial \dot{q}_\alpha}\right)\right] \frac{\partial q_\alpha}{\partial \varepsilon}\, dt, \tag{66}$$

with arbitrary functions $\mu_I(t)$, where we have again integrated once by parts and used condition (i) as in obtaining (59). At $\varepsilon = 0$, which is where we use it, Eq. (66) reads

$$\left(\sum_I \chi_I, \eta\right) = 0, \tag{67}$$

[9] Equation (65) implies only that the f_I are independent of ε, whereas equation (57) implies that the f_α are identically zero when the q_α are replaced by the $q_\alpha(t, \varepsilon)$. This is important because it explains why we shall still need to use equation (57) to solve the problem.

where the χ_I are vectors in F whose components are given by (no sum on I)

$$\chi_{I\alpha} = \mu_{(I)} \frac{\partial f_I}{\partial q_\alpha} - \frac{d}{dt}\left(\mu_{(I)} \frac{\partial f_I}{\partial \dot{q}_\alpha}\right) \tag{68}$$

with arbitrary $\mu_I(t)$.

This is the restriction on η: it is arbitrary except that it is orthogonal to the subspace F_K of F composed of vectors of the form $\sum_I \chi_I$. Thus Eq. (63) tells us that Λ lies in the orthogonal complement of the orthogonal complement of F_K, or in F_K itself. Consequently, there exist K functions $\lambda_I(t)$ (these are called *Lagrange multipliers*) such that

$$\Lambda_\alpha = \frac{d}{dt}\left(\frac{\partial L}{\partial \dot{q}_\alpha}\right) - \frac{\partial L}{\partial q_\alpha} = \lambda_I \frac{\partial f_I}{\partial q_\alpha} - \frac{d}{dt}\left(\lambda_I \frac{\partial f_I}{\partial \dot{q}_\alpha}\right). \tag{69}$$

This together with Eq. (57) represents the solution of the variational problem of Lagrange.

Does (69) give the physical motion? In the holonomic case it does, for then the f_I are independent of the $\dot{q}_\alpha$ and (69) reduces to

$$\Lambda_\alpha = \lambda_I \frac{\partial f_I}{\partial q_\alpha}. \tag{70}$$

These n second-order differential equations plus the K constraint conditions give us $n + K$ equations for the n function $q_\alpha(t)$ and the K functions $\lambda_I(t)$. Equations (70) give rise to $2n$ constants of integration, of which $2K$ are determined by setting $t = 0$ in the equations $f_I = 0$ and in their time derivatives $df_I/dt = 0$.[10] (One may say that each holonomic constraint eliminates one degree of freedom.) The other $2(n - K)$ constants are at our disposal: they are determined by the initial conditions. In other words, the $2n$ initial q_α and $\dot{q}_\alpha$ are restricted by $2K$ constraint conditions.

We see then that Eqs. (70) are possible equations for holonomic systems. Whether they yield in fact the physical solution is a question that cannot be answered on the basis of the variational principle, but depends on a detailed analysis of the applied forces and the forces of constraint. It turns out that Eqs. (70) and (57) are in fact the correct equations for smooth constraints, and therefore the equations of motion of a holonomic system can be obtained from the variational problem of Lagrange.

[10] The equations $f_I = 0$ and $df_I/dt = 0$ must give $2K$ *independent* conditions in the holonomic case, or the entire procedure will not work. A necessary and sufficient condition for this is that the nonsquare matrix of the $\partial f_I/\partial q_\alpha$ contain a K-by-K square matrix whose determinant is nonzero when $f_I = 0$. For example, if a particle is constrained to a plane we may write $f = \mathbf{a} \cdot \mathbf{x}$, where $\mathbf{a}$ is a fixed vector, for the $K = 1$ constraint function. We may not write $f = (\mathbf{a} \cdot \mathbf{x})^2$. See footnote 1 of Chapter II.

Equations (69) cannot be used for nonholonomic systems, however, when the f_I depend on the $\dot{q}_\alpha$. This is because these equations involve the $\dot{\lambda}_I$ and so will give rise to $2n + K$ constants of integration, K of which are determined by setting $t = 0$ in the equations $f_I = 0$. (One may say that each nonholonomic constraint eliminates one half of a degree of freedom.) This leaves $2n$ (namely $2n - K$ of the initial q_α and $\dot{q}_\alpha$ plus the K initial λ_I) at our disposal. But this is nonphysical, for there is no way to determine the $\lambda_I(0)$. Physically we are allowed only to adjust the $2n - K$ initial conditions on the $q_\alpha(0)$ and $\dot{q}_\alpha(0)$. Note that in this nonholonimic case the equations $df_I/dt = 0$ at $t = 0$ are not available for adjusting constants of integration, for this would involve setting initial values of the $\ddot{q}_\alpha$.

Since the variational problem of Lagrange does not give physically meaningful equations of motion in the nonholonomic case, we look for another variational procedure that does, one in which the ε-families of comparison paths are chosen in a different way.

Let us first return to the holonomic case, in which the two kinds of ε-families will be seen to coincide. The constraints imply certain conditions on the generalized accelerations, obtained by differentiating the constraint equations twice. If we write

$$\frac{df_I}{dt} = g_I(q, \dot{q}, t) = \frac{\partial f_I}{\partial q_\alpha} \dot{q}_\alpha + \frac{\partial f_I}{\partial t}, \tag{71}$$

then

$$\frac{d^2 f_I}{dt^2} = \frac{\partial g_I}{\partial \dot{q}_\alpha} \ddot{q}_\alpha + \frac{\partial g_I}{\partial q_\alpha} \dot{q}_\alpha + \frac{\partial g_I}{\partial t} = 0, \tag{72}$$

which are conditions restricting the accelerations. Equation (71) shows that $\partial g_I/\partial \dot{q}_\alpha = \partial f_I/\partial q_\alpha$, so that (72) may be written

$$\frac{\partial f_I}{\partial q_\alpha} \ddot{q}_\alpha + \frac{\partial g_I}{\partial q_\alpha} \dot{q}_\alpha + \frac{\partial g_I}{\partial t} = 0. \tag{73}$$

Now let us make a sort of minimal assumption about the forces of constraint: they affect the accelerations *only* as implied by Eqs. (72) or (73). To understand this assumption, consider first the case $K = 1$. Then Eq. (73) determines only the projection of $\ddot{q} = \{\ddot{q}_1, \ldots, \ddot{q}_n\}$ onto $\partial f/\partial q = \{\partial f/\partial q_1, \ldots, \partial f/\partial q_n\}$. If the q_α are Cartesian, these projections are the components of acceleration perpendicular to the $f = 0$ surface, and thus only these perpendicular components are affected. Therefore the forces of constraint are also perpendicular to this surface. For $K > 1$ a similar result holds. Equations (73) determine only the projection of $\ddot{q}$ into the K-dimensional space spanned by the $\partial f_I/\partial q_\alpha$, and thereby specify the possible directions of the constraint forces (that this space is K dimensional is implied by footnote 10).

These considerations are related to the ε-families of comparison paths in the following way. Equation (65) becomes

$$\frac{\partial f_I}{\partial q_\alpha}\frac{\partial q_\alpha}{\partial \varepsilon} = 0 \tag{74}$$

in the holonomic case. We have interpreted this to mean that *each path in the ε-family satisfies (locally) the constraint equations.* Let us call this interpretation (A). But another interpretation is also possible, namely that *each ε-family is "perpendicular" to, that is, has zero projection (locally) into the space spanned by the $\partial f_I/\partial q_\alpha$.* We shall call this interpretation (B). Interpretations (A) and (B), since they refer to one and the same set of equations, will not affect the results in the holonomic case. We shall see that in the nonholonomic case, on the other hand, the two interpretations lead to different equations and hence to different results

Before moving on to the nonholonomic case, we remark that interpretation (B) is physically quite reasonable. Consider again $K = 1$ and Cartesian q_α. Then interpretation (B) says, according to (74), that for each fixed value of t the curve $q(t, \varepsilon)$ is locally (i.e., at $\varepsilon = 0$) perpendicular to the accelerations produced by the constraints. In this sense we say that the ε-families are locally perpendicular to these accelerations. Since these accelerations are parallel to the forces of constraint, the ε-families are locally perpendicular to the forces of constraint. Straightforward calculation with $K > 1$ and non-Cartesian q_α will show that this result is quite general. In fact, interpretation (B) is equivalent to *D'Alembert's principle: virtual displacements* (from one path in an ε-family to an "adjacent" one at the same t) do no work (are perpendicular to the forces of constraint).

In the nonholonomic case the first time derivatives of the constraint equations, rather than the second, yield the restrictions on the accelerations. Instead of (73) we have

$$\frac{df_I}{dt} = \frac{\partial f_I}{\partial \dot{q}_\alpha}\ddot{q}_\alpha + \frac{\partial f_I}{\partial q_\alpha}\dot{q}_\alpha + \frac{\partial f_I}{\partial t} = 0.$$

Thus $\partial f_I/\partial \dot{q}_\alpha$ replaces $\partial f_I/\partial q_\alpha$. If we use interpretation (B) of Eq. (74), we obtain

$$\frac{\partial f_I}{\partial \dot{q}_\alpha}\frac{\partial q_\alpha}{\partial \varepsilon} = 0. \tag{75}$$

If we use interpretation (A) we obtain (65). But we have already seen that (65) leads to unacceptable results, and thus we try interpretation (B) and hence Eq. (75).

The rest is simple. Just as in going from (65) to (69) we start by putting (75) in the form of an integral and then conclude after some calculation that there exist K function $\lambda_I(t)$ such that

$$\Lambda_\alpha = \frac{d}{dt}\frac{\partial L}{\partial \dot{q}_\alpha} - \frac{\partial L}{\partial q_\alpha} = \lambda_I \frac{\partial f_I}{\partial \dot{q}_\alpha}. \tag{76}$$

This is the usually accepted result (Rund 1966). Because only λ_I and not $\dot{\lambda}_I$ appear in (76), the solutions are capable of complete physical interpretation. As in the holonomic case, whether these yield in fact the physical solution is a question that cannot be answered on the basis of the variational principle.

Remark. For the problem of Lagrange, and hence in the holonomic case, one can write $\mathscr{L} = L - \lambda_I f_I$, and then the equations of motion become simply

$$\frac{d}{dt}\left(\frac{\partial \mathscr{L}}{\partial \dot{q}_\alpha}\right) - \frac{\partial \mathscr{L}}{\partial q_\alpha} = 0. \tag{77}$$

This is important and will be used in Chapter V. However, there exists no such $\mathscr{L}$ for the nonholonomic case. Nevertheless, both cases can be treated in a unified way, namely according to Eq. (76), if in the holonomic case one takes not Eqs. (57) as the constraint equations, but their time derivatives $\dot{f}_I = 0$. Then the difference between the holonomic and nonholonomic case is that in the former the integrals of the constraint equations exist (and are used in the final solution); in the latter they do not.

As a concrete example of a problem with velocity-dependent constraints, consider a disc of radius R rolling on a perfectly rough horizontal plane, and let the disc be constrained to remain always vertical (Fig. 14). We may choose

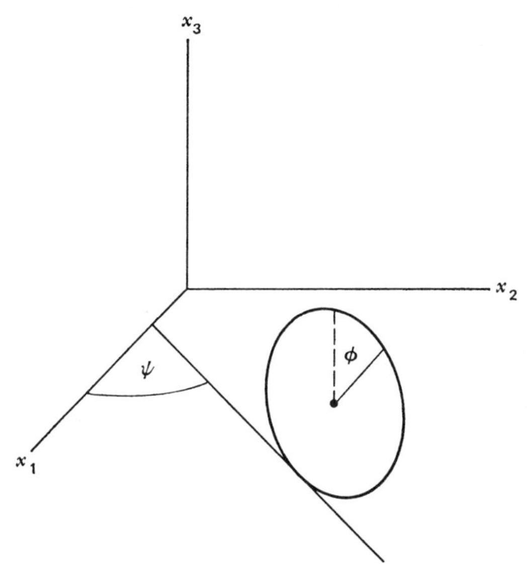

Figure 14. A rolling disc constrained to remain vertical.

the x_1 and x_2 coordinates of its center of mass as two of the generalized coordinates, and the other two as the angle ψ the plane of the disc makes with the x_1x_3 plane and the angle ϕ some line on the disc makes with the vertical. Then the condition that the disc is rolling without slipping may be stated in the form of the two velocity-dependent constraint equations

$$\begin{aligned} f_1' &= R\dot{\phi}\cos\psi - \dot{x}_1 = 0, \\ f_2' &= R\dot{\phi}\sin\psi - \dot{x}_2 = 0. \end{aligned} \tag{78}$$

These constraints involve the generalized velocities in a nonintegrable way, and so Eq. (76) must be used to solve the problem.

Instead of starting with the constraints stated in the form of (78), however, we shall state them in an equivalent, but nonlinear way. The reason is that we should like to illustrate the velocity-dependent case in a somewhat more general form. Let us therefore take

$$\begin{aligned} f_1 &= \dot{x}_1^2 + \dot{x}_2^2 - R^2\dot{\phi}^2 = 0, \\ f_2 &= \dot{x}_1\sin\psi - \dot{x}_2\cos\psi = 0, \end{aligned} \tag{79}$$

to be the constraint equations. It is a simple matter to show that (78) and (79) are equivalent.

The Lagrangian, involving only kinetic energy (there are no forces other than those of constraint), is

$$L = \tfrac{1}{2}I_0\dot{\phi}^2 + \tfrac{1}{2}I_1\dot{\psi}^2 + \tfrac{1}{2}m(\dot{x}_1^2 + \dot{x}_2^2),$$

where I_0 is the moment of inertia of the disc about its symmetry axis, and I_1 is its moment of inertia about a diameter. According to Eq. (76) the equations of motion are

$$I_0\ddot{\phi} = -2\lambda_1R^2\dot{\phi}, \tag{80a}$$

$$m\ddot{x}_1 = 2\lambda_1\dot{x}_1 + \lambda_2\sin\psi, \tag{80b}$$

$$m\ddot{x}_2 = 2\lambda_1\dot{x}_2 - \lambda_2\cos\psi, \tag{80c}$$

$$I_1\ddot{\psi} = 0. \tag{80d}$$

On multiplying (80b) by $\dot{x}_1$ and (80c) by $\dot{x}_2$ and adding, we obtain, using (79),

$$\frac{m}{2}\frac{d}{dt}(\dot{x}_1^2 + \dot{x}_2^2) = \frac{m}{2}R^2\frac{d}{dt}\dot{\phi}^2 = mR^2\dot{\phi}\ddot{\phi} = 2\lambda_1R^2\dot{\phi}^2.$$

Then by (80a)

$$2\lambda_1R^2(1 + mR^2/I_0)\dot{\phi}^2 = 0,$$

or

$$\lambda_1 = 0.$$

Now λ_2 can be found from Eqs. (80*b*) and (80*c*). We have

$$\begin{aligned}\lambda_2 &= m(\ddot{x}_1 \sin \psi - \ddot{x}_2 \cos \psi)\\ &= m \frac{d}{dt}(\dot{x}_1 \sin \psi - \dot{x}_2 \cos \psi) - m\dot{\psi}(\dot{x}_1 \cos \psi + \dot{x}_2 \sin \psi).\end{aligned}$$

The first of these terms vanishes by (79). The second is a constant, for $\ddot{\psi} = 0$ and

$$\begin{aligned}&\frac{d}{dt}(\dot{x}_1 \cos \psi + \dot{x}_2 \sin \psi)\\ &\qquad = -\dot{\psi}(\dot{x}_1 \sin \psi - \dot{x}_2 \cos \psi) + \ddot{x}_1 \cos \psi + \ddot{x}_2 \sin \psi = 0.\end{aligned}$$

Thus $\lambda_2 = m\mu$ is a constant.

The equations of motion are now easily solved. The solution is

$$\begin{aligned}\phi &= \omega t + \phi_0,\\ \psi &= \Omega t + \psi_0,\\ x_1 &= a + ut - \frac{\mu}{\Omega^2} \sin (\Omega t + \psi_0),\\ x_2 &= b + vt + \frac{\mu}{\Omega^2} \cos (\Omega t + \psi_0),\end{aligned} \tag{81}$$

where a, b, u, v, ω, ϕ_0, Ω, ψ_0 are the constants of integration. With this solution the second constraint equation implies that

$$u \sin (\Omega t + \psi_0) - v \cos (\Omega t + \psi_0) = 0.$$

Then either (i) ψ is a constant (and $\Omega = 0$) or (ii) $u = v = 0$. Consider first Case (i). Then (81) becomes

$$\begin{aligned}\phi &= \omega t + \phi_0,\\ \psi &= \psi_0,\\ x_1 &= a + ut,\\ x_2 &= b + vt\end{aligned} \tag{82}$$

(we have absorbed all constant terms into a and b in the last two of these equations). The first constraint equation implies that $u^2 + v^2 = R^2\omega^2$, so that only six independent constants result. The disc rolls in a straight line.

Now consider Case (ii). After the first constraint equation is inserted, (81) becomes

$$\begin{aligned}\phi &= \omega t + \phi_0,\\ \psi &= \Omega t + \psi_0,\\ x_1 &= a - \frac{R\omega}{\Omega}\sin(\Omega t + \psi_0),\\ x_2 &= b + \frac{R\omega}{\Omega}\cos(\Omega t + \psi_0).\end{aligned} \tag{83}$$

The disc moves with constant speed $R\omega$ in a circle of radius $R\omega/\Omega$ centered at (a, b). It is seen that (82) is a special case of (83): $\Omega = 0$, the circle has infinite radius and its center is infinitely far away. Another special case is $\omega = 0$: the disc spins in place.

Problems

1. Prove the following basic properties of a vector space by using axioms G1–G4 and M1–M5. Prove the properties in the order stated, using any one you have proved to obtain any later one. It is interesting that M5 is not needed for (a)–(d).
   (a) *Uniqueness of the null vector.* If for some $\mathbf{x} \in S$ there exists a $\mathbf{y} \in S$ such that $\mathbf{x} + \mathbf{y} = \mathbf{x}$, then $\mathbf{y}$ is the null vector.
   (b) *Multiplication by zero.* When a vector is multiplied by the number zero, the result is the null vector.
   (c) *Uniqueness of the inverse.* If $\mathbf{x} + \mathbf{y} = 0$ and $\mathbf{x} + \mathbf{z} = 0$, then $\mathbf{z} = \mathbf{y} = -\mathbf{x}$.
   (d) *Multiplication by* -1. The inverse of $\alpha\mathbf{x}$ is $(-\alpha)\mathbf{x}$. [To go further and show that the inverse of $\mathbf{x}$ is $(-1)\mathbf{x}$, one must use M5.]
   (e) If $\mathbf{x}$ is not the null vector but $\alpha\mathbf{x}$ is, then $\alpha = 0$.
2. Let R be a subset of vectors from S. Prove that all that must be shown in order to prove that R is a subspace is that it is closed under vector addition and multiplication by numbers.
3. Consider the vector space S constructed in the following way from two other vector spaces S_1 and S_2. Let S_1 consist of vectors $\mathbf{x}_1, \mathbf{y}_1, \ldots$, and let S_2 consist of vectors $\mathbf{x}_2, \mathbf{y}_2, \ldots$. Then S, called the *direct sum* of S_1 and S_2 and written $S = S_1 \oplus S_2$, is defined as the space of ordered pairs $(\mathbf{x}_1, \mathbf{x}_2) \equiv \mathbf{x}$, $(\mathbf{y}_1, \mathbf{y}_2) \equiv \mathbf{y}, \ldots$. Addition is defined by $\mathbf{x} + \mathbf{y} = (\mathbf{x}_1 + \mathbf{y}_1, \mathbf{x}_2 + \mathbf{y}_2)$, and multiplication by $\alpha\mathbf{x} = (\alpha\mathbf{x}_1, \alpha\mathbf{x}_2)$. Prove that S so defined is a vector space. In what sense can S_1 be considered a subspace of S?
4. Show that $1 - p, p - p^2, \ldots, p^{m-1} - p^m, p^m$ forms a basis in P_m. Show that if p^m, the last vector in this set, is replaced by $p^m + 1$, the resulting set is still a basis, but that if p^m is replaced by $p^m - 1$, it is not a basis. (In general if fewer than n vectors of a basis are given, there are many ways to complete the set.) What is the dimension of P_m?

5. Let $\mathbf{w}_1, \ldots, \mathbf{w}_n$ be a basis in S, and let $\mathbf{x}, \mathbf{y} \in S$. Let the components of $\mathbf{x}$ be ξ_i, and those of $\mathbf{y}$ be η_i. Show that the components of $\mathbf{z} = \mathbf{x} + \mathbf{y}$ are $\zeta_i = \xi_i + \eta_i$. Show that the components of $\mathbf{v} = \alpha\mathbf{x}$ are $\psi_i = \alpha\xi_i$.
6. Show that transformation matrices can be multiplied according to the rules of matrix multiplication. That is, show that if $\mathbf{x}_i = \alpha_{ji}\mathbf{y}_j$ and $\mathbf{y}_i = \beta_{ji}\mathbf{z}_j$, then $\mathbf{x}_j = \gamma_{ij}\mathbf{z}_i$, where the $\mathbf{x}_i$, the $\mathbf{y}_i$ and the $\mathbf{z}_i$ each form a basis and the γ_{ij} are the elements of the product matrix of the α_{ij} and the β_{ij} (in what order?). If the $\mathbf{z}_i$ are the $\mathbf{x}_i$, that is if the second equation is the inverse of the first, show that the matrix of the β_{ij} is the inverse of the matrix of the α_{ij}.
7. Let P_∞ be the space of all (formal) power series, with real coefficients, in some variable p. For every $\mathbf{f}(p) = \sum_{i=0}^{\infty} \alpha_i p^i$ in P_∞ we define the operator $\mathbf{A}$ by

$$\mathbf{Af} = \sum_{i=0}^{\infty} \alpha_{i+1} p^i,$$

and the operator $\mathbf{B}$ by

$$\mathbf{Bf} = \sum_{i=1}^{\infty} \alpha_{i-1} p^i.$$

Show that each of these operators satisfies one of the two conditions in footnote 4, but not both. Try to find an inverse for $\mathbf{A}$ and one for $\mathbf{B}$, and establish that none exist (remember than an inverse must commute with the operator: $\mathbf{AA}^{-1} = \mathbf{A}^{-1}\mathbf{A} = \mathbf{1}$). Try to construct analogous operators on P_m, $m < \infty$, whose domains are all of P_m, and show that these analogs satisfy either none or both of these conditions.
8. Prove the four assertions concerning the matrix elements of operators which follow the Remark before Eq. (10).
9. We have found the matrix of the derivative operator $\mathbf{D}$ on P_m. Show from the matrix alone that this operator has no inverse. An integration operator can be defined on P_∞, but not on P_m (why?). Try to construct an operator analogous to the integration operator, but on P_m, and show why it fails to be the inverse of the derivative operator. Is the integration operator on P_∞ the inverse of the derivative operator on P_∞? Write down the matrix of $\mathbf{D}$ in the first basis of Problem 4.
10. Use the definition Eq. (11) of the determinant of a matrix to prove the following properties: (i) If two rows (or two columns) of A are equal, $\det A = 0$. (ii) If A is changed to a new matrix B by adding one row (or column) of A to another one, then $\det A = \det B$. (iii) If A is changed to a new matrix B by multiplying one row (or column) by a number α, then $\det B = \alpha \det A$. (iv) If the rows (or columns) of a matrix A are linearly dependent,[11] then $\det A = 0$. (v) $\det(AB) = (\det A)(\det B)$. (vi) If A is

[11] Two rows or columns of a matrix are said to be linearly dependent if the vectors represented by their elements in any basis are linearly dependent.

changed to a new matrix A^T by interchanging rows and columns, then $\det A^T = \det A$.

11. Show that the solution to Eq. (15) as given by (16) is indeed the solution obtained by Cramer's rule.
12. Consider the set of equations

$$\begin{aligned} \xi_1 \qquad - \quad \xi_3 &= \gamma_1, \\ 2\xi_2 + 4\xi_3 &= \gamma_2, \\ -\xi_1 - 2\xi_2 - 3\xi_3 &= \gamma_3. \end{aligned}$$

Find the conditions on the γ_i for a solution to exist. Write down the general solution. Find the null space and the range of the matrix of the coefficients, and see that their dimensions add up properly. Note, incidentally, whether the null space and the range are disjoint.

13. Consider the operator **I** on complex P_1 discussed in the text following Eq. (20). Since the eigenvectors belonging to $+i$ and $-i$ (call them $\mathbf{x}_+$ and $\mathbf{x}_-$) are linearly independent, they can form a basis. Find the matrix of **I** in the basis $\mathbf{x}_+$, $\mathbf{x}_-$. Find the column vectors representing $\mathbf{x}_+$, $\mathbf{x}_-$, $1 = p^0$, and p in this basis. Obtain the transformation matrix from the original basis to the new one. Check Eq. (10) explicitly. *Note:* This procedure of transforming to the basis in which the eigenvectors are the basis vectors is called *diagonalizing* the matrix. It is very important in physics.
14. Show that if two matrices are diagonalized by the same basis (see Problem 13), they commute.
15. Find the eigenvalues of, and diagonalize the matrix

$$\begin{vmatrix} \alpha & \beta \\ \beta & \gamma \end{vmatrix}.$$

16. Find the eigenvalues of, and diagonalize the matrix

$$\begin{vmatrix} \alpha & 0 \\ \beta & \alpha \end{vmatrix}.$$

Discuss any difficulties.

17. Let **x** and **y** be nonzero vectors such that $(\mathbf{x}, \mathbf{y}) = 0$. Show that **x** and **y** are linearly independent. Show that $(\mathbf{x}, \mathbf{z}) = 0$ for all $\mathbf{x} \in S$ implies that $\mathbf{z} = 0$.
18. Show that for every nonzero vector **x** there exists a unique unit vector **e** such that $\mathbf{x} = \mu\mathbf{e}$, where $\mu = |\mathbf{x}|$.
19. Let **x** and **y** be written as in Problem 18:

$$\mathbf{x} = \nu\mathbf{e}_x, \qquad \mathbf{y} = \mu\mathbf{e}_y.$$

Show that if **x** and **y** are linearly dependent, then $\mathbf{e}_x = \lambda\mathbf{e}_y$, where $|\lambda|^2 = 1$.

20. Show that the components of a vector **x** in an orthonormal basis are just $\xi_i = (\mathbf{e}_i, \mathbf{x})$.
21. Consider a two-dimensional real vector space with inner product given by

$$(\mathbf{x}, \mathbf{y}) = 4\xi_1\eta_1 + \tfrac{1}{2}\xi_2\eta_2 + \xi_1\eta_2 + \xi_2\eta_1,$$

where $\mathbf{x} = \xi_i \mathbf{z}_i$, $\mathbf{y} = \eta_i \mathbf{z}_i$, and $\mathbf{z}_1$, $\mathbf{z}_2$ form a basis. Check that this is an admissible inner product. Find an orthonormal basis one of whose vectors is a multiple of $\mathbf{z}_1$.

22. Show that the Hermitian conjugate of **AB** is $\mathbf{B}^\dagger \mathbf{A}^\dagger$ (and similarly for their matrices).
23. Show that the eigenvalues of a unitary operator all have absolute value one and that the eigenvectors belonging to distinct eigenvalues are orthogonal.
24. (a) Instead of Eq. (28), use the expression $(\mathbf{f}, \mathbf{g}) = \int_{-\infty}^{\infty} e^{-p^2} f(p) g(p)\, dp$ to define a scalar product on P_m. Verify that this can indeed be taken as a scalar product. Calculate the first few (say the first five) vectors of an orthonormal basis with this norm, requiring that the jth vector of the basis be a polynomial of degree j or less. These will be the (normalized) Hermite polynomials. (b). Write down the derivative matrix **D** on P_∞ in the Legendre-polynomial basis and in the Hermite-polynomial basis. (Of course, you can not write down the entire ∞-by-∞ matrix, but you can write down the upper left-hand j-by-j corner.) Compare Problem 9.
25. Consider a vector $\mathbf{x}$ in an n-dimensional unitary space S. We wish to "approximate" this vector by a set of $m < n$ vectors $\mathbf{e}_1, \ldots, \mathbf{e}_m$ of an orthonormal basis in the following way. The vector $\mathbf{x}(m) = \sum_{i=1}^{m} \xi_i \mathbf{e}_i$ is said to be the best approximation to $\mathbf{x}$ when the norm of the difference $\mathbf{x} - \mathbf{x}(m)$ is a minimum. Show that if the ξ_i are chosen to give the best approximations they will be the first m components of $\mathbf{x}$. This is the procedure used to find the coefficients of a function when expanding it in a series of orthonormal functions. Try it, for instance, with the Legendre polynomials in P_∞.
26. Figure 15 shows a system consisting of a mass m constrained to move in a horizontal line under the influence of a massless spring of spring-constant

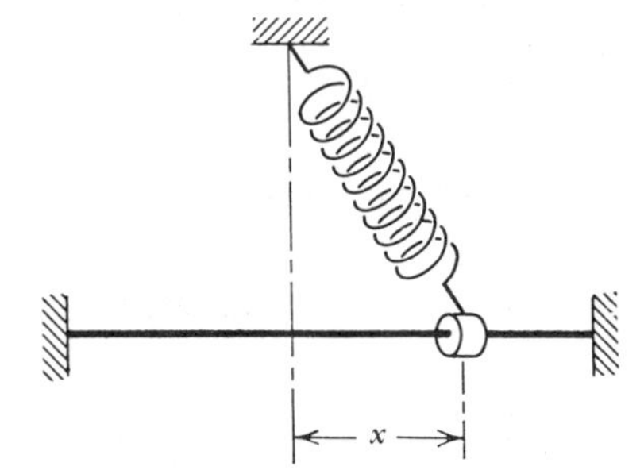

Figure 15. A mass constrained to move on a horizontal line.

k. Find the equilibrium point(s) and describe the motion for small vibrations in the following three cases. (a) When the mass is at $x = 0$ the spring is stretched. (b) When $x = 0$ the spring is at its equilibrium length. (c) When $x = 0$ the spring is compressed.

27. Try solving for the motion of an oscillator whose Lagrangian is $L = \frac{1}{2}m\dot{q}^2 - \frac{1}{2}kq^6$ by making the substitution $r = q^3$. Why can't the resulting Lagrangian be treated by the standard procedure?
28. Two simple pendula, each of length l and mass m, are suspended a distance d apart (see Fig. 16). Their bobs are connected by a spring of unstretched length d and spring constant k. Find the normal modes and normal frequencies for *small* oscillations of this system. Find the general solution for the motion of one of the pendula and show that this motion exhibits the phenomenon of beats if $k/m \ll g/l$.

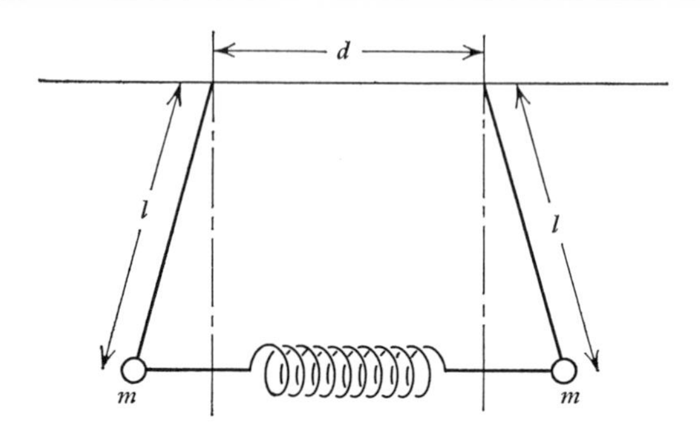

Figure 16. Two pendula connected by a spring.

29. Consider the linear water molecule discussed in Section 5(b) and suppose it is in the normal mode a_3, moving parallel to its axis with constant velocity v. It makes a right-angle collision with a rigid wall, the first hydrogen atom bouncing back with velocity initially $-v$. Investigate the subsequent motion of the molecule and determine how much energy is in each mode after the molecule moves away from the wall.
30. Two equal masses m are hung from the ceiling by springs of equal spring constants k (see Fig. 17). The masses are connected by a massless rigid rod of length d equal to the distance between the points of suspension of the springs. Find suitable coordinates with which to describe this system

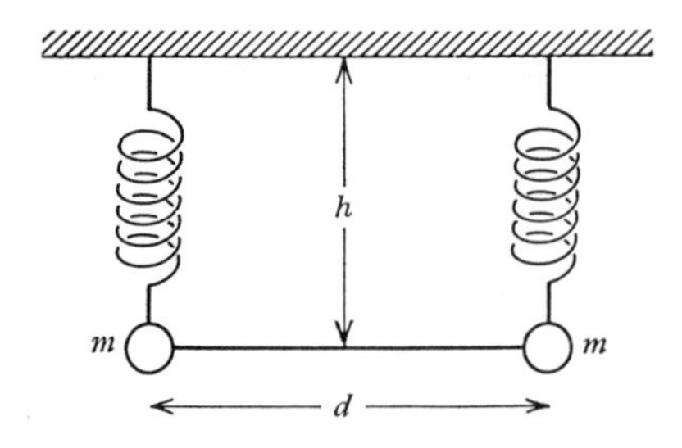

Figure 17. Two spring-suspended pendula connected by a rigid rod.

with five degrees of freedom and find the normal frequencies and normal modes.

31. Let **D** be an orthogonal operator on 3-space. Show that either **D** is a proper rotation or **D** can be written in the form **D** = (−**1**)**R**, where **R** is a proper rotation.
32. Let **K** be an operator on 3-space. Show that if the matrix elements of **K** are to be the same in all Cartesian coordinate systems, **K** must be a multiple of the unit matrix. (That is, if K commutes with all rotation matrices, then $K = \lambda \mathbb{1}$.) We had to use this fact toward the end of our derivation of the free-particle Lagrangian from Galilei invariance in Chapter III. There we gave an argument based on the fact that the quadratic surface $K_{\alpha\beta} x_\alpha x_\beta$ must be a sphere.
33. Show that every matrix M in n dimensions can be written uniquely in the form
$$M = \mu \mathbb{1} + A + S,$$
where $A = -A^T$, and $S = S^T$ and $\operatorname{tr} S = 0$. In some new coordinate system let M become $M' = TMT^{-1}$, where T is the transformation matrix. Then M' can also be written in the form
$$M' = \mu' \mathbb{1} + A' + S',$$
where $A' = -A'^T$, $S' = S'^T$ and $\operatorname{tr} S' = 0$. Show that $\mu' = \mu$, and that A' is obtained from A (how?) and S' from S.
34. (a) A rotation of 90° about the 1-axis is followed by a rotation of 90° about the 2-axis. Find the axis and angle of the resulting rotation. (b) Now try them in the opposite order and find the axis and angle of the resulting rotation. (c) Suppose two similar rigid bodies, initially oriented in the same way, are rotated, one as in (a) and the other as in (b). What rotation will then bring the first into the orientation of the second. Find also the axis and angle of this rotation.
35. Consider a system with a time-independent Lagrangian L subject to non-holonomic (velocity-dependent) constraints. Let H be defined by $H = \dot{q}_\alpha\, \partial L / \partial \dot{q}_\alpha - L$ (this is usually the energy). Prove that $dH/dt = 0$ if each constraint equation (57) is homogeneous in the $\dot{q}_\alpha$. Show that this result can be extended to holonomic constraints if not $f_I = 0$, but $\dot{f}_I = 0$ are taken as the constraint equations, as in the Remark at Eq. (77). (This should agree with our result from Chapter II which says that energy is conserved if the constraints are time independent.)
36. Solve the rolling-disc problem using the constraints in the linear form (78). (It is instructive to try to solve this rolling disc problem with quadratic constraints as though it were the problem of Lagrange. One can see just where additional constants are needed and what type of solutions would be obtained.)

37. Consider a particle in a plane, its speed constrained to be proportional to its distance from the 1-axis. Otherwise the particle is free. Solve for the motion. Is energy conserved?
38. Consider a particle in a plane, the 1-component of its velocity constrained to be proportional to its distance from the 1-axis. Otherwise the particle is free. Solve for the motion. What is conserved?

Note: More nonholonimic constraint problems will be found at the end of Chapter V.

CHAPTER V

Rigid Body Motion

In this chapter we shall discuss the motion of rigid bodies. So far we have been discussing the motion only of point particles, that is, of objects whose location can be described completely by specifying the position of a point. Our results also apply immediately to the motion of the center of mass of an extended system like a rigid body, so long as the external forces acting on the system depend only on the position of its center of mass. A rigid body, on the other hand, is an object whose location is specified by both the position of some point in the body and the orientation of the body about this point.

Any mechanical object may be treated as a collection of point particles, often in the limit as their number approaches infinity. Thus the discussion of the previous chapters can be extended to any mechanical system, although this will in general involve treating a large number N of interacting particles. Usually such a many-body problem cannot be solved exactly, but there exist special cases that can be (e.g., as we saw in Chapter IV, Section 5(b), a system of N particles interacting through the potential $\frac{1}{2}V_{\alpha\beta}\eta_\alpha\eta_\beta$). From this point of view, the motion of a rigid body is another solvable many-body problem.

V-1 EULER'S EQUATIONS FOR THE MOTION OF A RIGID BODY

A rigid body is defined as a distribution of mass such that the distance between any two points in the distribution remains constant. If we were to think of the body as made up of a large number of discrete point masses, this condition of constant separation would provide a system of constraints on their motion. We shall, however, proceed in a more direct manner. We shall use the condition of rigidity in conjunction with what we have learned about rotations in Chapter IV, Section 5(c), to set up generalized coordinates suitable for writing down the Lagrangian for the body, and then we shall proceed by the Lagrangian formalism.

Let an arbitrary point of the body be taken as the origin of a coordinate system $\mathscr{B}$ fixed in the body and moving with it. This coordinate system, or any other system fixed in the body, will be called a *body coordinate system*. In general this is not an inertial system, and so care must be taken in using it to describe the motion of anything. Furthermore in $\mathscr{B}$ the coordinates x of any point $\mathbf{x}$ of the body remain constant: $\dot{x} = 0$. Of course such constant coordinates cannot tell us anything about the motion of the body and can therefore not be used for generalized coordinates. In order to find suitable generalized coordinates, consider $\mathbf{r}$, the position vector of the origin of $\mathscr{B}$, relative to some inertial system $\mathscr{I}$, as shown in Fig. 18. Then if some point in the body has position vector $\mathbf{x}$ relative to the origin of $\mathscr{B}$, this point has position vector

$$\mathbf{y} = \mathbf{r} + \mathbf{x} \tag{1}$$

relative to the origin of $\mathscr{I}$.

We introduce a third coordinate system $\mathscr{S}$ (see Fig. 18), the *space coordinate system*, which has the same origin as $\mathscr{B}$, but whose axes always remain parallel to those of $\mathscr{I}$. The system $\mathscr{S}$ will not itself be an inertial system if its origin is accelerating with respect to $\mathscr{I}$. Now let $\mathbf{z}$ be any vector. Its components z' in $\mathscr{I}$ are the same as its components in $\mathscr{S}$. We can write Eq. (1) in terms of the (primed) components of the vectors in $\mathscr{S}$:

$$y' = r' + x'.$$

As was discussed in Chapter III, Section 4, [see Eq. (35) of that chapter], x' and x are related by

$$x' = Dx,$$

where D is a rotation matrix, so that

$$y' = r' + Dx. \tag{2}$$

A similar equation can be written for every point in the body. What is

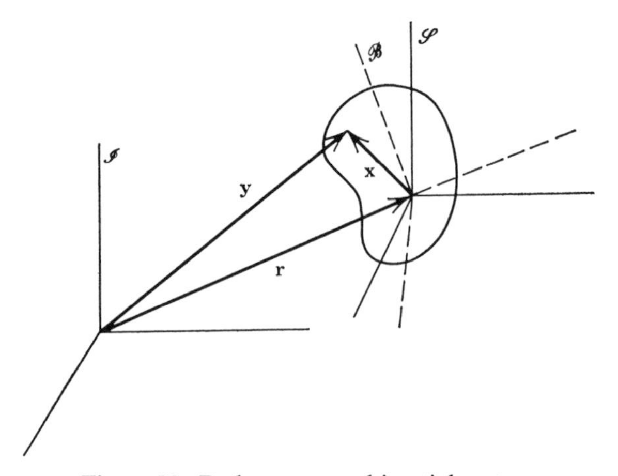

Figure 18. Body space, and inertial systems.

important, however, is that because the body is rigid, the same rotation matrix D will appear in the corresponding equation for each point [see Chapter IV, Section 5(c)].

Now we can define generalized coordinates to specify the position of any point in the body. Given a point in the body (which is conveniently done by specifying its coordinates x in $\mathscr{B}$, though any other method is just as good), its position relative to the inertial system $\mathscr{I}$ is completely determined by the position $\mathbf{r}$ of the origin of $\mathscr{B}$ and by the matrix elements of D. Recall that the elements of a rotation matrix D are not all independent, but that in fact they can be specified by three independent parameters. Thus the position of any point in the body and hence of the body itself is completely determined by six numbers: the three components of $\mathbf{r}$ and the three parameters entering into D. These, then, are the six independent coordinates.

To write the Lagrangian, however, we shall take the three components of $\mathbf{r}$ and all *nine* elements of D as our generalized coordinates and add as a constraint the condition that D be a rotation matrix, namely

$$D^T D = D D^T = \mathbb{1}. \tag{3}$$

If we take the common origin of $\mathscr{B}$ and $\mathscr{S}$ at the center of mass of the body, the kinetic energy is (see Chapter I, Problem 4)

$$T = \tfrac{1}{2} M \dot{r}'^2 + T_{\text{rot}}, \tag{4}$$

where

$$T_{\text{rot}} = \tfrac{1}{2} \int \rho(x')(\dot{x}', \dot{x}')\, d^3x'. \tag{5}$$

Here M is the total mass of the body and the first term in Eq. (4) is the kinetic energy of the center of mass of the body relative to the inertial system. Note that $\dot{r}' = dr'/dt$ is the rate of change of the components of $\mathbf{r}$ in the inertial system $\mathscr{I}$ (or equivalently in $\mathscr{S}$). The second term in Eq. (4) is the kinetic energy of rotation of the body about its center of mass, and $\rho(x')$ is the mass density function. The integral in Eq. (5) is over the entire rigid body, and the velocity $\dot{x}' = dx'/dt$ in the integrand is the rate of change of the components of $\mathbf{x}$ in $\mathscr{I}$ (or equivalently in $\mathscr{S}$), not in $\mathscr{B}$. Thus the first term in (4), is written in terms of our generalized coordinates, but T_{rot} as given by (5) is not.

To write (5) in terms of D we change variables in the integral from x' to x. This is simple because $x' = Dx$, and thus

$$\dot{x}' = \dot{D}x + D\dot{x} = \dot{D}x, \tag{6}$$

since $\dot{x} = 0$. Also $\rho(x') = \hat{\rho}(x)$, where $\hat{\rho}$ is the mass density function expressed in terms of the components of $\mathbf{x}$ in $\mathscr{B}$. Since the Jacobian of the transformation from the Cartesian variables x' to the Cartesian variables x is one, we have

$$T_{\text{rot}} = \tfrac{1}{2}\int \hat{\rho}(x)(\dot{D}x,\, \dot{D}x)\, d^3x$$

$$= \tfrac{1}{2}\int \hat{\rho}(x)\dot{D}_{\alpha\beta}x_\beta \dot{D}_{\alpha\gamma}x_\gamma\, d^3x$$

$$= \tfrac{1}{2}\dot{D}_{\alpha\beta}K_{\beta\gamma}\dot{D}_{\alpha\gamma}$$

$$= \tfrac{1}{2}\operatorname{tr}(\dot{D}K\dot{D}^T), \tag{7}$$

where K is the symmetric matrix with elements

$$K_{\alpha\beta} = K_{\beta\alpha} = \int \hat{\rho}(x)x_\alpha x_\beta\, d^3x. \tag{8}$$

(The indices in these equations and throughout this chapter run from 1 to 3.) The matrix K as defined here is time independent since it is expressed in the body coordinate system. Furthermore, since the integral which defines K is over the entire body, K is not a function of x. (The elements of K are just the *second moments* of the mass distribution.) The kinetic energy, of course, will vary in time, for it depends on $\dot{D}$, and Eqs. (4) and (7) then express the total kinetic energy in terms of the generalized coordinates we have chosen.

In order to proceed we must assume that the potential energy V depends in an additive way on the position of the center of mass and the orientation of the body, that is that it can be written in the form

$$V = U(r') + V(D). \tag{9}$$

This is a very common case. For instance, for a body in a uniform gravitational field, the gravitational potential energy is independent of D; if, in addition, the body is subjected to torques that are independent of the position of the center of mass but dependent on the orientation, the total potential energy is just of this form. [See also the Remark following Eq. (10).] There are cases, however, in which this is not true, and for which the subsequent formalism is not valid. For instance, the potential energy of a body in a nonuniform gravitational field is not of this type.

Then combining Eqs. (4), (7), and (9), we obtain the Lagrangian

$$L = \tfrac{1}{2}M\dot{r}'^2 + \tfrac{1}{2}\operatorname{tr}(\dot{D}K\dot{D}^T) - U(r') - V(D).$$

In addition, we have the constraint equations given by Eq. (3). Each constraint equation is of the form

$$D_{\beta\alpha}D_{\beta\gamma} - \delta_{\alpha\gamma} = 0.$$

We have seen (see the Remark at Eq. (77) of Chapter IV) that the way to take account of such holonomic constraints is to multiply each by a Lagrange multiplier $\lambda_{\gamma\alpha}$ and add all the expressions so obtained to the Lagrangian. The sum of such expressions is

$$\lambda_{\gamma\alpha}(D_{\beta\alpha}D_{\beta\gamma} - \delta_{\alpha\gamma}) = \mathrm{tr}\,(\Lambda D^T D - \Lambda)$$

where Λ is the matrix with elements $\lambda_{\gamma\alpha}$. As a result of adding this to L we have

$$\mathscr{L} = \tfrac{1}{2}M\dot{r}'^2 + \tfrac{1}{2}\,\mathrm{tr}\,(\dot{D}K\dot{D}^T) - U(r') - V(D) + \mathrm{tr}\,(\Lambda[D^T D - \mathbb{1}]). \quad (10)$$

Remark. Consider a rigid body with one point P held fixed. If we pick $\mathscr{B}$ and $\mathscr{S}$ with common origins at P, then $\mathscr{S}$ is an inertial system and we can take $\mathscr{I} = \mathscr{S}$. Then the kinetic energy is simply T_{rot}, the kinetic energy of rotation about P. Everything we have done goes through formally, except that there are no terms depending on r'. Note that in this case the potential energy is always a function of D only. In this way we would obtain

$$\mathscr{L} = \tfrac{1}{2}\,\mathrm{tr}\,(\dot{D}K\dot{D}^T) - V(D) + \mathrm{tr}\,(\Lambda[D^T D - \mathbb{1}]),$$

where K is again given by Eq. (8), but in a coordinate system $\mathscr{B}$ centered at the fixed point P. The reader should satisfy himself that there are only three degrees of freedom in this new problem.

We now proceed to obtain the Euler-Lagrange equations in the usual way. It is a trivial matter to obtain the equations for the three components of r'. These will be the usual center-of-mass equations

$$M\ddot{r}' = -\nabla' U = F'. \quad (11)$$

(Remember that this is an equation for the three components of $\mathbf{r}$ and that F' represents the components of the force $\mathbf{F}$ in the inertial system.)

The equations for the $D_{\alpha\beta}$ are obtained in the usual way also. We may use the results of Chapter IV, Section 5(d) to write down the equations immediately in matrix form. For instance, according to those results,

$$\partial\mathscr{L}/\partial\dot{D} = \tfrac{1}{2}[\dot{D}K + (K\dot{D}^T)^T] = \dot{D}K,$$

where we have used the fact that $K = K^T$ (K is symmetric). Proceeding in this way, we obtain the Euler-Lagrange equations in matrix form:

$$\ddot{D}K + \partial V/\partial D = D(\Lambda + \Lambda^T). \quad (12)$$

Now Eqs. (3), (11), and (12) represent the equations of motion which will give r', D, and Λ as functions of the time. We are not, however, interested in Λ, and it is therefore convenient to eliminate it at this stage. To do this we multiply (12) on the left by D^T and use Eq. (3), obtaining

$$D^T\ddot{D}K + D^T(\partial V/\partial D) = \Lambda + \Lambda^T. \quad (13)$$

The right-hand side of this equation is obviously symmetric, which means that if we take its transpose, namely,

$$K\ddot{D}^T D + (\partial V/\partial D^T)D = \Lambda^T + \Lambda,$$

the right-hand side remains the same. (We leave it to the reader to show that the transpose of $\partial V/\partial D$ is $\partial V/\partial D^T$.) Therefore subtraction of (13) from its transpose eliminates Λ, and we arrive at an equation for D alone:

$$D^T\ddot{D}K - K\ddot{D}^T D = (\partial V/\partial D^T)D - D^T(\partial V/\partial D). \tag{14}$$

Equations (3), (11), and (14) are the equations of motion for r' and D.

Remark. Both the left-hand side and the right-hand side of (14) are antisymmetric matrices. (A matrix M is antisymmetric if $M^T = -M$.) Equation (14) is the dynamical equation of motion, and it tells us that (twice) the antisymmetric part of $D^T\ddot{D}K$ is equal to (twice) the antisymmetric part of $(\partial V/\partial D^T)D$. To find out how the entire matrices are related we must make use of what we know about the symmetry properties of both D and K. This is why we must still use Eq. (3). In fact we shall have also to use the fact that K is symmetric. Another way to say the same thing is that it is the condition that the left-hand side of (13) be symmetric which determines the dynamical equations of motion (see Problem 1).

We can make further use of the fact that D is a rotation matrix by differentiating Eq. (3) with respect to the time. This yields

$$D^T\dot{D} + \dot{D}^T D = \Omega + \Omega^T = 0,$$

where we define the antisymmetric matrix Ω by

$$\Omega = D^T\dot{D}, \qquad \text{or} \qquad \dot{D} = D\Omega. \tag{15}$$

The time derivative of (15) yields

$$D^T\ddot{D} = D^T(\dot{D}\Omega + D\dot{\Omega}) = \Omega^2 + \dot{\Omega},$$

whose transpose is

$$\ddot{D}^T D = (\Omega^2)^T + \dot{\Omega}^T = \Omega^2 - \dot{\Omega},$$

and these may be inserted into (14) to give

$$K\dot{\Omega} + \dot{\Omega}K + \Omega^2 K - K\Omega^2 = N, \tag{16}$$

where we have defined the new antisymmetric matrix N by

$$N = \left(\frac{\partial V}{\partial D^T}\right)D - D^T\left(\frac{\partial V}{\partial D}\right) = -N^T. \tag{17}$$

We shall return to N later. Equation (16) is the matrix form of what are known as *Euler's equations for the motion of a rigid body.*

As before, both sides of Eq. (16) are antisymmetric matrices. Now, an anitisymmetric matrix in three dimensions has only three independent elements, for those above the main diagonal are just the negatives of those below, and those on the main diagonal are all zero. We came across a similar situation in Chapter III [see Eq. (29) of that chapter], where it was shown that if an antisymmetric matrix B has elements $B_{\alpha\beta}$, they can be written in the form

$$B_{\alpha\beta} = -\epsilon_{\alpha\beta\gamma}b_\gamma, \qquad \text{or} \qquad b_\gamma = -\tfrac{1}{2}\epsilon_{\gamma\alpha\beta}B_{\alpha\beta}. \tag{18}$$

The minus sign is included here for convenience. The three numbers (b_1, b_2, b_3) determine the antisymmetric matrix B uniquely and vice versa. If we treat these three numbers as a column vector b,[1] there is a one-to-one correspondence between column vectors and antisymmetric matrices. By writing out the matrix explicitly in terms of the b_γ in accordance with (18), this correspondence may be represented as $b \leftrightarrow B$, or

$$\begin{vmatrix} b_1 \\ b_2 \\ b_3 \end{vmatrix} \leftrightarrow \begin{vmatrix} 0 & -b_3 & b_2 \\ b_3 & 0 & -b_1 \\ -b_2 & b_1 & 0 \end{vmatrix}.$$

The equations of motion in the form of Eq. (16) are not intractable, but it is easier to go from the antisymmetric matrices that appear on both sides of it to the corresponding column vectors. Thus we define the column vectors n and ω corresponding to the antisymmetric matrices N and Ω:

$$N_{\alpha\beta} = -\epsilon_{\alpha\beta\gamma}n_\gamma, \qquad \text{or} \qquad n_\gamma = -\tfrac{1}{2}\epsilon_{\gamma\alpha\beta}N_{\alpha\beta}; \tag{19}$$

$$\Omega_{\alpha\beta} = -\epsilon_{\alpha\beta\gamma}\omega_\gamma, \qquad \text{or} \qquad \omega_\gamma = -\tfrac{1}{2}\epsilon_{\gamma\alpha\beta}\Omega_{\alpha\beta}. \tag{20}$$

We shall see later that n and ω are the components in $\mathscr{B}$ of the torque and angular velocity vectors, respectively.

Now we apply $-\frac{1}{2}\epsilon_{\alpha\beta\gamma}$ to both sides of (16) and first calculate explicitly the first term on the left-hand side. We have

$$\begin{aligned} -\tfrac{1}{2}\epsilon_{\alpha\beta\gamma}(K\dot{\Omega})_{\beta\gamma} &= -\tfrac{1}{2}\epsilon_{\alpha\beta\gamma}K_{\beta\rho}\dot{\Omega}_{\rho\gamma} \\ &= \tfrac{1}{2}\epsilon_{\alpha\beta\gamma}\epsilon_{\rho\gamma\sigma}\dot{\omega}_\sigma K_{\beta\rho} \\ &= \tfrac{1}{2}(\delta_{\alpha\sigma}\,\delta_{\beta\rho} - \delta_{\alpha\rho}\,\delta_{\beta\sigma})\dot{\omega}_\sigma K_{\beta\rho} \\ &= \tfrac{1}{2}(K_{\rho\rho}\,\delta_{\alpha\sigma} - K_{\alpha\sigma})\dot{\omega}_\sigma \\ &= \tfrac{1}{2}I_{\alpha\sigma}\dot{\omega}_\sigma = \tfrac{1}{2}(I\dot{\omega})_\alpha, \end{aligned} \tag{21}$$

[1] There are subtleties involved in identifying the three b_α with an actual vector, but we shall discuss these in Section 2.

where we have written

$$I_{\alpha\sigma} = \delta_{\alpha\sigma} \operatorname{tr} K - K_{\alpha\sigma}, \tag{22}$$

defining the *inertial matrix* I with elements $I_{\alpha\sigma} = I_{\sigma\alpha}$. It requires only some shuffling of indices to show that the second term on the left-hand side of (16) is equal to the first.[2] Similar calculations for the third and fourth terms lead to

$$-\tfrac{1}{2}\epsilon_{\alpha\beta\gamma}[\Omega^2 K - K\Omega^2]_{\beta\gamma} = -\epsilon_{\alpha\mu\rho}\omega_\mu K_{\rho\sigma}\omega_\sigma = -(\omega \times K\omega)_\alpha,$$

where $K\omega$ is the column vector obtained from ω by operating on it with the matrix K. Because $\omega \times \omega = 0$, we may write

$$\omega \times K\omega = \omega \times [(K - \lambda\mathbb{1})\omega],$$

where λ is any number. In particular, by choosing $\lambda = \operatorname{tr} K$, we obtain

$$-\omega \times K\omega = -\omega \times [(K - \mathbb{1} \operatorname{tr} K)\omega] = \omega \times I\omega.$$

Finally, we add this to twice [remember the second term in Eq. (16)] Eq. (21), arriving at

$$I\dot{\omega} + \omega \times I\omega = n. \tag{23}$$

This is the column-vector form of the antisymmetric matrix equation (16), called *Euler's equations for the motion of a rigid body.* Before discussing it in detail we shall show that n does indeed represent the torque vector in $\mathscr{B}$.

From (17) and (19), we obtain

$$n_\gamma = -\tfrac{1}{2}\epsilon_{\alpha\beta\gamma}\left(\frac{\partial V}{\partial D_{\rho\alpha}} D_{\rho\beta} - D_{\rho\alpha}\frac{\partial V}{\partial D_{\rho\beta}}\right) = -\epsilon_{\gamma\alpha\beta}\frac{\partial V}{\partial D_{\rho\alpha}} D_{\rho\beta}.$$

Consider a rotation of the body through an angle θ_γ about the body γ-axis. The component τ_γ of the torque along this axis is defined as the negative rate of change of the potential energy due to such a rotation:

$$\tau_\gamma = -\frac{\partial V}{\partial \theta_\gamma}.$$

Since V is expressed explicitly in terms of D we can write

$$\tau_\gamma = -\frac{\partial V}{\partial D_{\rho\alpha}}\frac{\partial D_{\rho\alpha}}{\partial \theta_\gamma}. \tag{24}$$

In order to calculate τ_γ we must be able to write down the matrix elements of D as functions of θ_γ. Recall what $D(\theta_\gamma)$ is: For each value of θ_γ it tells

[2] This is a reflection of the fact that multiplication by $\epsilon_{\alpha\beta\gamma}$ annihilates the symmetric part of a matrix. The first two terms on the left form just twice the antisymmetric part of $K\dot{\Omega}$.

us how to find the space components x' of a point x in the body if we know its fixed body components x: $x' = D(\theta_\gamma)x$. As θ_γ is varied, $D(\theta_\gamma)$ varies in the following way. For $\theta = 0$ it gives some fixed reference orientation of the body and hence a certain initial orientation of $\mathscr{B}$: call it $\mathscr{B}_{\mathrm{i}}$. Then as θ_γ changes, it is simply a further rotation of the body about the body γ-axis. Thus we can first find the components of the point x in the initial fixed reference frame by transforming back to $\mathscr{B}_{\mathrm{i}}$, and then, knowing the components of x in this frame, we can transform from $\mathscr{B}_{\mathrm{i}}$ to $\mathscr{S}$ and thus find the space components x'. Let us call the matrix for the first transformation $R(\theta_\gamma)$, and the matrix for the second one $D(0) = U$. Then $x' = UR(\theta_\gamma)x$, or

$$D(\theta_\gamma) = UR(\theta_\gamma).$$

In this way we have isolated the θ_γ dependence of $D(\theta_\gamma)$ in the second factor, which is simply the matrix for a rotation about the γ axis. Such matrices are well known. For instance if $\gamma = 3$, then

$$R(\theta_3) = \begin{vmatrix} \cos\theta_3 & -\sin\theta_3 & 0 \\ \sin\theta_3 & \cos\theta_3 & 0 \\ 0 & 0 & 1 \end{vmatrix}. \tag{25}$$

We can now calculate $\partial D/\partial\theta_\gamma$. The result is

$$\frac{\partial D}{\partial\theta_\gamma} = U\frac{\partial R}{\partial\theta_\gamma} = URR^T\frac{\partial R}{\partial\theta_\gamma} = D(\theta_\gamma)R^T\frac{\partial R}{\partial\theta_\gamma}. \tag{26}$$

The combination $R^T\partial R/\partial\theta_\gamma$ is easily calculated and is found to be independent of θ_γ. For example, for rotations about the 3-axis we have

$$R^T\frac{\partial R}{\partial\theta_3} = \begin{vmatrix} 0 & -1 & 0 \\ 1 & 0 & 0 \\ 0 & 0 & 0 \end{vmatrix}.$$

More generally, the result is

$$\left(R^T\frac{\partial R}{\partial\theta_\gamma}\right)_{\alpha\beta} = -\epsilon_{\alpha\beta\gamma}.$$

Thus Eqs. (24) and (26) lead to

$$\begin{aligned} \tau_\gamma &= -\frac{\partial V}{\partial D_{\rho\alpha}} D_{\rho\sigma}\left(R^T\frac{\partial R}{\partial\theta_\gamma}\right)_{\sigma\alpha} = \frac{\partial V}{\partial D_{\rho\alpha}} D_{\rho\sigma}\epsilon_{\sigma\alpha\gamma} \\ &= -\epsilon_{\gamma\alpha\sigma}\frac{\partial V}{\partial D_{\rho\alpha}} D_{\rho\sigma} = n_\gamma, \end{aligned}$$

as asserted.

Let us now return to Eq. (23). In what sense is it an equation of motion? If we solve it, we obtain ω, and hence $\Omega = D^T\dot{D}$, as a function of t. This still leaves us with the first-order equation $\dot{D} = D\Omega$ to solve for D as a function of t. It will be useful, then, to discuss (23), and in particular $\omega(t)$, in more detail.

V-2 ROTATIONAL KINEMATICS

Equation (23) is the dynamical equation for the rotation of a rigid body about either its center of mass or a stationary point in the body, if there is one. In order to solve it we must become better acquainted with the quantities which appear in it.

We begin by showing that the ω_α given by Eq. (20) are the components of a vector $\boldsymbol{\omega}$ in the body system. What do we mean by this? It is, of course, always possible to specify a vector $\boldsymbol{\omega}$ by giving its components ω in one system and calculating them in any other: $\bar{\omega} = U\omega$, where U is the transformation matrix. But we would like more. We would like the components $\bar{\omega}$ of $\boldsymbol{\omega}$ in the other system to be given by Eq. (20) in terms of the matrix elements $\bar{\Omega}$ of the operator $\boldsymbol{\Omega}$ in that other system. Now we know how the matrix of $\boldsymbol{\Omega}$ transforms from system to system: if in one system it is Ω, in another it is

$$\bar{\Omega} = U\Omega U^T,$$

where U is the same orthogonal transformation matrix between the two systems. What we must show then is that

$$\bar{\omega} = U\omega \tag{27}$$

can be obtained by applying $-\frac{1}{2}\epsilon_{\alpha\beta\gamma}$ to $\bar{\Omega}$. Accordingly, Eq. (20) gives

$$\begin{aligned}\bar{\omega}_\gamma &= -\tfrac{1}{2}\epsilon_{\gamma\alpha\beta}\bar{\Omega}_{\alpha\beta} \\ &= -\tfrac{1}{2}\epsilon_{\gamma\alpha\beta}U_{\alpha\rho}\Omega_{\rho\tau}U_{\beta\tau} \\ &= \tfrac{1}{2}\epsilon_{\gamma\alpha\beta}\epsilon_{\sigma\rho\tau}U_{\alpha\rho}U_{\beta\tau}\omega_\sigma.\end{aligned}$$

Now from Eq. (11) of Chapter IV we have

$$\epsilon_{\sigma\rho\tau}U_{\lambda\sigma}U_{\alpha\rho}U_{\beta\tau} = \epsilon_{\lambda\alpha\beta}\det U,$$

or

$$\epsilon_{\sigma\rho\tau}U_{\alpha\rho}U_{\beta\tau} = \epsilon_{\lambda\alpha\beta}U_{\lambda\sigma}\det U,$$

since U is orthogonal. Thus

$$\begin{aligned}\bar{\omega}_\gamma &= \tfrac{1}{2}\epsilon_{\gamma\alpha\beta}\epsilon_{\lambda\alpha\beta}U_{\lambda\sigma}(\det U)\omega_\sigma \\ &= U_{\gamma\sigma}\omega_\sigma \det U,\end{aligned}$$

or

$$\bar{\omega} = (\det U)U\omega, \tag{28}$$

where we have used the relation $\epsilon_{\gamma\alpha\beta}\epsilon_{\lambda\alpha\beta} = 2\delta_{\gamma\lambda}$.

What we have obtained is not quite (27), for the determinant of U may be $+1$ or -1. Equation (28) is the same as (27) only if $\det U = 1$, that is, for proper rotations. If, therefore, we choose to define ω in every coordinate system by Eq. (20), it will form the components of a vector as long as we restrict our coordinate systems to those connected only by proper rotations. Under a reflection the vector $\boldsymbol{\omega}$ corresponding to ω will change to $-\boldsymbol{\omega}$, a new vector. Such a geometrical object is called a *pseudovector*. However, since we shall be dealing almost exclusively with proper rotations, we shall treat $\boldsymbol{\omega}$ as a vector.

It turns out that $\boldsymbol{\omega}$ is a very useful vector. We have seen that its components in $\mathscr{B}$ are given by the elements of $\Omega = D^T\dot{D}$, so that its components in $\mathscr{S}$ are given by the elements of $\Omega' = D\Omega D^T = \dot{D}D^T$. For any point in the body with position vector $\mathbf{x}$, Eq. (6) gives

$$\dot{x}' \equiv \frac{d}{dt} x' = \dot{D}x = \dot{D}D^T x' = \Omega' x',$$

or

$$\dot{x}'_\alpha = -\epsilon_{\alpha\beta\gamma}\omega_\gamma x'_\beta,$$

which may be written

$$\dot{x}' = \omega' \times x'. \tag{29}$$

This equation relates $\dot{x}'$ to x'. Is there some way of relating $\dot{x}'$ to $\dot{x}$? One might think, for instance, that $\dot{x}'$ could be obtained somehow from $D\dot{x}$. But this is certainly not possible, for $\dot{x} = 0$. In fact $\dot{x}'$ and $\dot{x}$ are not, as might be thought, the components of any given vector in the two different frames. This is a reflection of the fact that the time derivative $d\mathbf{s}/dt$ of any vector $\mathbf{s}(t)$ is ambiguous when we are dealing with coordinate systems moving with respect to each other. In fact the derivatives ds'/dt and ds/dt of the components of $\mathbf{s}$ in the space and body systems are not just the components of a single vector in the two coordinate systems, but are components of two different vectors in the two systems. To prove this, we calculate the vectors in two ways and proceed to see whether they are the same or different.

Let us write $\dot{\mathbf{s}}_{\mathscr{S}}$ for the vector whose components in $\mathscr{S}$ are $ds'/dt \equiv \dot{s}'_{\mathscr{S}}$, and $\dot{\mathbf{s}}_{\mathscr{B}}$ for the other vector whose components in $\mathscr{B}$ are $ds/dt \equiv \dot{s}_{\mathscr{B}}$. Although $\dot{\mathbf{s}}_{\mathscr{S}}$ and $\dot{\mathbf{s}}_{\mathscr{B}}$ will be seen to be different vectors, they are defined by the time derivatives of s and s', which are the representations in $\mathscr{B}$ and $\mathscr{S}$ of the same vector $\mathbf{s}$. Since $s' = Ds$, we have

$$\begin{aligned} \dot{s}'_{\mathscr{S}} &= ds'/dt = \dot{D}s + D(ds/dt) \\ &= \Omega' s' + D\dot{s}_{\mathscr{B}}. \end{aligned}$$

We may write, in the usual way, $\dot{s}'_{\mathscr{B}} = D\dot{s}_{\mathscr{B}}$ for the representation of $\dot{\mathbf{s}}_{\mathscr{B}}$ in the coordinate system $\mathscr{S}$. Then, noting that $\Omega' s' = \omega' \times s'$, we have

$$\dot{s}'_{\mathscr{S}} = \dot{s}'_{\mathscr{B}} + \omega' \times s'.$$

This is a relation between the representations in $\mathscr{S}$ of the vectors $\dot{\mathbf{s}}_{\mathscr{B}}$, $\dot{\mathbf{s}}_{\mathscr{S}}$, $\boldsymbol{\omega}$ and $\mathbf{s}$. It is important that it involves only the representations in $\mathscr{S}$, for a vector relation of this type is then just as true in any other coordinate system. Consequently, we may write

$$\dot{\mathbf{s}}_{\mathscr{S}} = \dot{\mathbf{s}}_{\mathscr{B}} + \boldsymbol{\omega} \times \mathbf{s}. \tag{30}$$

This shows not only that $\dot{\mathbf{s}}_{\mathscr{S}}$ and $\dot{\mathbf{s}}_{\mathscr{B}}$ are different, but how to get one from the other. As an example, let $\mathbf{s}$ be the position vector $\mathbf{x}$; then Eq. (30) becomes

$$\dot{\mathbf{x}}_{\mathscr{S}} = \boldsymbol{\omega} \times \mathbf{x} \tag{31}$$

since $\dot{\mathbf{x}}_{\mathscr{B}} = 0$. We see then that Eq. (29) is just a special case of (30) and that in order to state (29) unambiguously we should have written $\dot{x}'_{\mathscr{S}}$ instead of simply $\dot{x}'$.

From (31) we see that all position vectors which (at some instant) are parallel to $\boldsymbol{\omega}$ are (at that instant) stationary in $\mathscr{S}$. Thus at any instant there is a line in the body passing through the origin which is at rest in $\mathscr{S}$ (and, like all lines in the body, also in $\mathscr{B}$). Call this line, parallel to $\boldsymbol{\omega}$, the *axis of rotation.* If $\mathbf{x}$ is any position vector from the origin which is not along this line, we may write

$$\mathbf{x} = \mathbf{x}_{\parallel} + \mathbf{x}_{\perp},$$

where $\mathbf{x}_{\parallel}$ is parallel to $\boldsymbol{\omega}$ and $\mathbf{x}_{\perp}$ is perpendicular. Then the cross product tells us that $\dot{\mathbf{x}}_{\mathscr{S}}$ is perpendicular to $\boldsymbol{\omega}$ and to $\mathbf{x}_{\perp}$, and that its magnitude is

$$|\dot{\mathbf{x}}_{\mathscr{S}}| = |\boldsymbol{\omega}||\mathbf{x}_{\perp}|.$$

Thus, instantaneously the position vector $\mathbf{x}$ of any point in the body is rotating about the axis of rotation with angular velocity $|\boldsymbol{\omega}|$, as seen in Fig. 19. The vector $\boldsymbol{\omega}$, called angular velocity vector, gives the instantaneous axis of rotation and the angular velocity.

This then is the meaning of $\boldsymbol{\omega}(t)$. The orientation of the body develops by constant rotation about $\boldsymbol{\omega}$ as $\boldsymbol{\omega}$ migrates through the body. As we have said, however, when we know $\boldsymbol{\omega}(t)$, we know not D, but $D^T\dot{D}$, so we must go further to obtain the actual motion. Before doing so, however, we would like to show how these results about $\boldsymbol{\omega}$ can be used to derive the Euler equations in a more conventional way.

The angular momentum of the body about the origin of coordinates is

$$\mathbf{L} = \int \rho(\mathbf{x})\mathbf{x} \times \dot{\mathbf{x}}_{\mathscr{S}}\, d^3x.$$

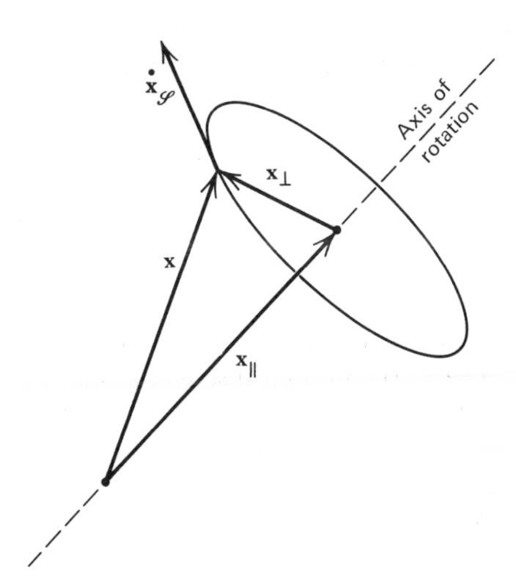

Figure 19. The motion of a point **x** in a rotating body.

By using Eq. (31) we can write this in terms of components in the space system

$$L' = \int \rho(x')x' \times (\omega' \times x')\, d^3x'.$$

Thus we have

$$\begin{aligned} L'_\alpha &= \int \rho(x')\epsilon_{\alpha\beta\gamma}x'_\beta\epsilon_{\gamma\lambda\sigma}\omega'_\lambda x'_\sigma\, d^3x' \\ &= \int \rho(x')[\delta_{\alpha\lambda}\,|x'|^2 - x'_\lambda x'_\alpha]\omega'_\lambda\, d^3x' \\ &= I'_{\alpha\lambda}\omega'_\lambda, \end{aligned}$$

where the $I'_{\alpha\lambda}$ are the elements of the inertial tensor **I** in the space system. The equation is the representation in $\mathscr{S}$ of the vector equation

$$\mathbf{L} = \mathbf{I}\boldsymbol{\omega}, \tag{32}$$

which may be written in any system, including $\mathscr{B}$. Note that according to (32) **L** and $\boldsymbol{\omega}$ are not parallel. See Problem 6 for a simple example.

In Chapter I we saw that $\dot{\mathbf{L}}_{\mathscr{S}}$ is equal to the total torque $\boldsymbol{\tau}$ on the body, provided that **L** is the angular momentum either about a fixed (inertial) point or about the center of mass of the body. In our present discussion the common origin of $\mathscr{B}$ and $\mathscr{S}$ is always one of these points. In the body system, however, we would take a different time derivative, in accord with (30), so to write this statement in $\mathscr{B}$ we must first write

$$\dot{\mathbf{L}}_{\mathscr{S}} = \dot{\mathbf{L}}_{\mathscr{B}} + \boldsymbol{\omega} \times \mathbf{L} = \boldsymbol{\tau}. \tag{33}$$

This is a vector relation, true in any coordinate system, including $\mathscr{B}$. But in $\mathscr{B}$ it is particularly convenient, because in $\mathscr{B}$ the inertial tensor I is a constant, so that

$$\dot{\mathbf{L}}_{\mathscr{B}} = dL/dt = d(I\omega)/dt = I\dot{\omega}.$$

Then by using (32) we may write (33) in $\mathscr{B}$ in the form

$$I\dot{\omega} + \omega \times (I\omega) = \tau. \tag{34}$$

This is just Eq. (23) with τ written in place of n. (We have already shown that $n = \tau$; the present discussion may be looked on as an alternate proof.)

Euler's equations can be further simplified by properly choosing the orientation of $\mathscr{B}$. So far all we have specified about $\mathscr{B}$ is that its origin must be either the center of mass or some fixed point in the body (if there is one). We are still free to orient the body axes in any way that simplifies the problem. Since I is a constant symmetric matrix, the body axes can be oriented in the body so that I is diagonal. These axes are called *principal axes* for the body; an example of how to find them is given in Section 4(b). From now on we shall always assume the body axes to be the principal axes, so that I has the form

$$I = \begin{vmatrix} I_1 & 0 & 0 \\ 0 & I_2 & 0 \\ 0 & 0 & I_3 \end{vmatrix},$$

where I_1, I_2, I_3 are called the *principal moments* of the body. The dynamical properties of any rigid body can be described in terms of its mass M and its three principal moments. (It is easily shown that I_γ is the moment of inertia of the body about the γ-axis.) In terms of the components of ω and τ along the principal axes, Eq. (34) takes the form

$$\begin{aligned} I_1\dot{\omega}_1 + (I_3 - I_2)\omega_2\omega_3 &= \tau_1, \\ I_2\dot{\omega}_2 + (I_1 - I_3)\omega_3\omega_1 &= \tau_2, \\ I_3\dot{\omega}_3 + (I_2 - I_1)\omega_1\omega_2 &= \tau_3. \end{aligned} \tag{35}$$

Thus we return again to the Euler equations. We shall discuss their solution for the zero torque case in Section 4(d). But for $\tau \neq 0$ the solution is relatively complicated. In order, in fact, to discuss any solution, it is convenient to pick a set of generalized coordinates for D other than its matrix elements, and to write (35) in those coordinates. In the next section we describe such a set of coordinates.

V-3 THE EULER ANGLES

One might think that having found the equations of motion for a rigid body in terms of the angular velocity vector $\boldsymbol{\omega}$, one need only find generalized coordinates q_α such that $\omega'_\alpha = \dot{q}_\alpha$, and then proceed to solve for the functions $q_\alpha(t)$. For instance in the usual elementary treatment of rotation about a single fixed axis, so-called plane motion, ω' is just the time derivative of the angle of rotation, which is taken as the generalized coordinate. But in three dimensions this is simply not possible: the components of the angular velocity are not the total time derivatives of any functions that depend on the configuration of the rigid body. The ω'_α are, to put it differently, *nonintegrable.*

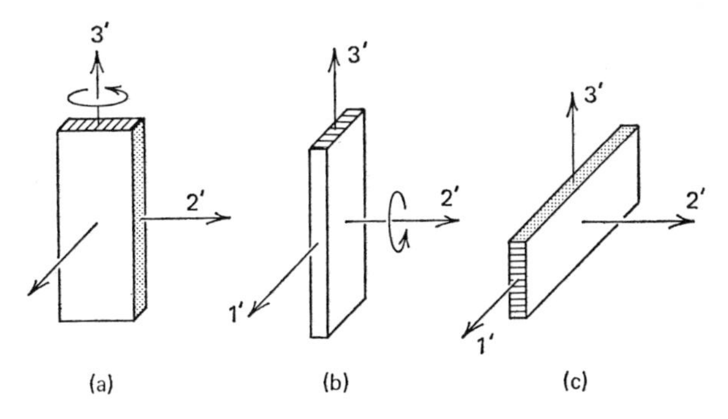

Figure 20. A block rotated through $\pi/2$, first about the 3′-axis and then about the 2′-axis.

This nonintegrability of ω'_α corresponds physically to the fact that the initial and final configurations of a body do not determine the total angle through which the body has been rotated about any given axis and thus do not determine the integrals $q_\alpha = \int \omega'_\alpha \, dt$. A simple illustration of this, not meant to be a proof, is given in Figs. 20 and 21, where a box is taken from a given initial configuration to a given final one in two different ways. We assume that the body and space axes coincide initially. The two ways of getting from the initial to the final orientation are the following. First, Fig. 20: a rotation of $\pi/2$ about the 3′-axis followed by a rotation of $\pi/2$ about the 2′-axis. Thus

$$q_1 = \int \omega'_1 \, dt = 0,$$

$$q_2 = \int \omega'_2 \, dt = \pi/2,$$

$$q_3 = \int \omega'_3 \, dt = \pi/2.$$

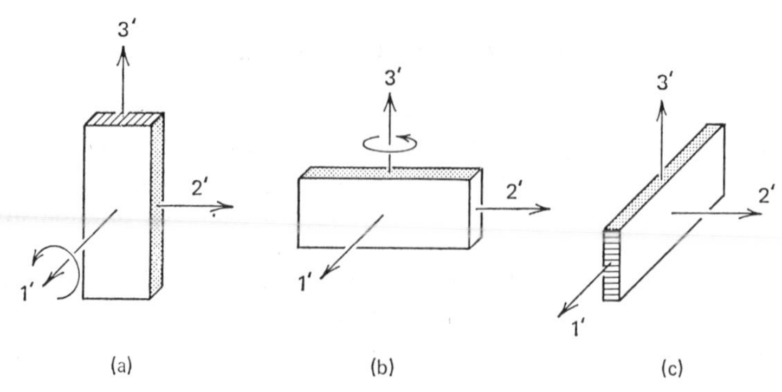

Figure 21. A block rotated through $\pi/2$, first about the 1′-axis and then about the 3′-axis.

Second, Fig. 21: a rotation of $\pi/2$ about the 1′-axis followed by a rotation of $\pi/2$ about the 3′-axis. Thus

$$q_1 = \int \omega_1' \, dt = \pi/2,$$

$$q_2 = \int \omega_2' \, dt = 0,$$

$$q_3 = \int \omega_3' \, dt = \pi/2.$$

It is clear that the final orientation, the same in Figs. 20 and 21, does not determine the q_α uniquely. Hence the q_α cannot be the generalized coordinates of the orientation and the ω_α' are not the time derivatives of suitable coordinates.

Thus we have still to find a suitable set of three unconstrained generalized coordinates. When we find them, the elements of D can be written as functions of these generalized coordinates. Now the orientation can be described by giving in detail a procedure whereby the space axes can be rotated into the body axes, and if this procedure can be specified uniquely by three one-parameter operations, the three parameters will determine D. The procedure of this kind which is most commonly used is the following.

For the purpose of this description, let the body be a sphere with its center fixed. We mark a point on the sphere (say the pole, where the body 3-axis pierces the surface) and draw a short arrow on the surface of the sphere at the point, as in Fig. 22. Then every orientation of the sphere can be given by telling where the pole is (the spherical coordinates of the body 3-axis) and in what direction the arrow is pointing, as in Fig. 22(d). We shall move the sphere from the orientation of Fig. 22(a) to that of Fig. 22(d) in the following

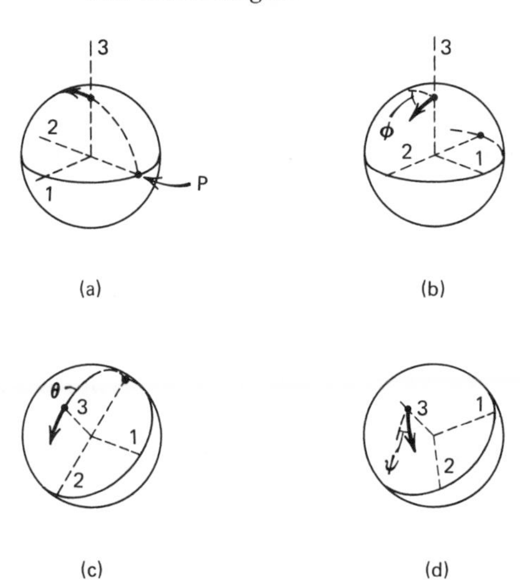

Figure 22. The Euler angles illustrated by a rotated sphere. (a) Starting position. (b) After the first rotation. (c) After the second rotation. (d) Final position after the third rotation.

three operations. First, rotate about the pole (the body 3-axis) until the arrow is pointing at the final position of the pole [Fig. 22(b)]. Second, move the pole in the direction of the arrow into its final position [Fig. 22(c)]. Third, rotate about the new position of the pole (again the body 3-axis) until the arrow is pointing in the proper direction [Fig. 22(d)].

Each one of these operations is a rotation about a certain axis. Let the rotations be through angles ϕ, θ, ψ, respectively. It then follows that by specifying ϕ, θ, and ψ, we specify D. There is only one combination of ϕ, θ, ψ that will take us from the initial to the final orientation, for Fig. 22 shows that the final orientation uniquely determines the direction in which the arrow must point (ϕ), how far it must be moved (θ), and how much it must finally be rotated (ψ). These unique angles are called the *Euler angles.* Their uniqueness results from the specific order in which the rotations are performed. If in the discussion of Figs. 20 and 21 we had specified the order of rotations, we would have obtained uniqueness there, too (see Problem 7).

To analyze D in terms of the Euler angles is our next task. To do this we first specify the positions of the body 1-axis and 2-axis. Since the 3-axis is at the pole, the other two axes lie in the equatorial plane. Let the arrow be pointing away from the positive 2-axis, which intersects the sphere at the point P on Fig. 22(a). Next, let the primed space axes coincide with the initial orientation of the body axes.

The first rotation is a rotation about the 3′-axis through an angle ϕ. The matrix corresponding to this is $D_\phi = R(\phi_3)$, defined by Eq. (25). The second is a rotation about the new position of the 1′-axis (the 1″-axis in Fig. 23) through an angle θ. The matrix corresponding to this is

$$D_\theta = R(\theta_1) = \begin{vmatrix} 1 & 0 & 0 \\ 0 & \cos\theta & -\sin\theta \\ 0 & \sin\theta & \cos\theta \end{vmatrix}. \tag{36}$$

The final rotation is about the new position of the 3′-axis (also its final position, and hence the position of the 3-axis in Fig. 23) through an angle ψ. The matrix corresponding to this is $D_\psi = R(\psi_3)$ defined by Eq. (25). Then the space-system coordinates x' are obtained from the body-system coordinates x by

$$x' = D_\phi D_\theta D_\psi x = Dx, \tag{37}$$

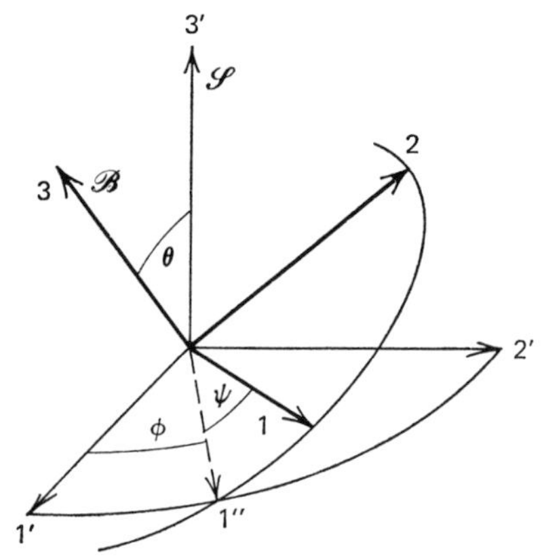

Figure 23. Orientation of the body axes in terms of the Euler angles.

where straightforward calculation yields

$$D = \begin{vmatrix} \cos\phi\cos\psi - \cos\theta\sin\phi\sin\psi & -\cos\phi\sin\psi - \cos\theta\sin\phi\cos\psi & \sin\theta\sin\phi \\ \sin\phi\cos\psi + \cos\theta\cos\phi\sin\psi & -\sin\phi\sin\psi + \cos\theta\cos\phi\cos\psi & -\sin\theta\cos\phi \\ \sin\theta\sin\psi & \sin\theta\cos\psi & \cos\theta \end{vmatrix}. \tag{38}$$

Now that D is known explicitly in terms of the new generalized coordinates ϕ, θ and ψ, the defining formula $\Omega = D^T\dot{D}$ can be used to write down Ω and hence ω ($\boldsymbol{\omega}$ in the body system). By direct but rather tedious calculation one obtains

$$\begin{aligned} \omega_1 &= \dot\theta\cos\psi + \dot\phi\sin\theta\sin\psi, \\ \omega_2 &= -\dot\theta\sin\psi + \dot\phi\sin\theta\cos\psi, \\ \omega_3 &= \dot\psi + \dot\phi\cos\theta. \end{aligned} \tag{39}$$

This result can be obtained by observing that as ϕ, θ, and ψ change, the body is rotating simultaneously about the space 3′-axis, the intermediate 1″-axis, and the body 3-axis with angular velocities $\dot{\phi}$, $\dot{\theta}$, and $\dot{\psi}$. Let the corresponding angular velocity vectors be $\dot{\boldsymbol{\phi}}$, $\dot{\boldsymbol{\theta}}$, and $\dot{\boldsymbol{\psi}}$. It is shown in Problem 8 that angular velocity vectors are additive, so that

$$\boldsymbol{\omega} = \dot{\boldsymbol{\phi}} + \dot{\boldsymbol{\theta}} + \dot{\boldsymbol{\psi}}. \tag{40}$$

We can now read the body components of $\boldsymbol{\omega}$ directly off Fig. 24 by taking the body components of $\dot{\boldsymbol{\phi}}$, $\dot{\boldsymbol{\theta}}$, and $\dot{\boldsymbol{\psi}}$. The space components can be found similarly or directly from the relation

$$\omega' = D\omega. \tag{41}$$

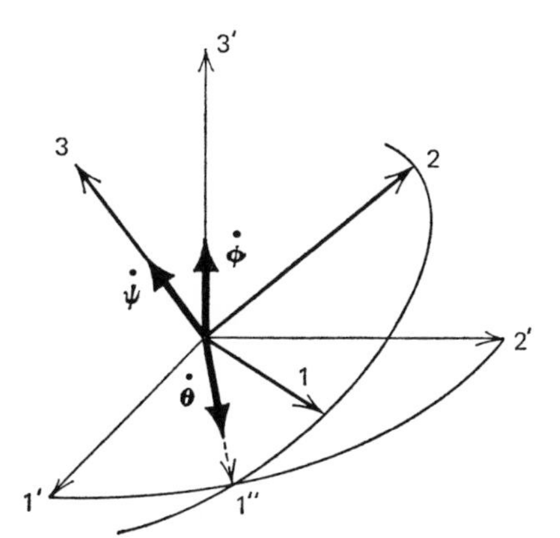

Figure 24. The body components of $\boldsymbol{\omega}$ are the projections of $\dot{\boldsymbol{\phi}}$, $\dot{\boldsymbol{\theta}}$, and $\dot{\boldsymbol{\psi}}$ onto the body axes.

The result is

$$\begin{aligned} \omega_1' &= \dot{\theta} \cos \phi + \dot{\psi} \sin \theta \sin \phi, \\ \omega_2' &= \dot{\theta} \sin \phi - \dot{\psi} \sin \theta \cos \phi, \\ \omega_3' &= \dot{\phi} + \dot{\psi} \cos \theta. \end{aligned} \tag{42}$$

If the components of the external torque on the body can be expressed in terms of the Euler angles, then Eq. (39) can be used in Eq. (35) to obtain three coupled second-order differential equations for ϕ, θ, and ψ. But this may be difficult, and an alternative procedure exists: to express the Lagrangian of the system from the beginning in terms of the Euler angles. In Problem 10 it is shown that the rotational kinetic energy of a rigid body is

$$T = \tfrac{1}{2}\boldsymbol{\omega} \cdot (\mathbf{I}\boldsymbol{\omega}). \tag{43}$$

This is the kinetic energy of the body as measured in the space system, but we are free to express it in terms of the elements of **I** and the components of $\boldsymbol{\omega}$ in any system. The body system is particularly convenient, for in the body system I is a constant. In the system of principal axes, in particular, the kinetic energy is simply

$$T = \tfrac{1}{2}(I_1\omega_1^2 + I_2\omega_2^2 + I_3\omega_3^2). \tag{44}$$

This expression is used with Eq. (39) in Problem 10 to give the kinetic energy in terms of the Euler angles.

If the potential $V(\phi, \theta, \psi)$ is known, the Euler-Lagrange equations of motion can be obtained from the Lagrangian $L = T - V$. These equations will differ from the previous Euler equations, which decompose the

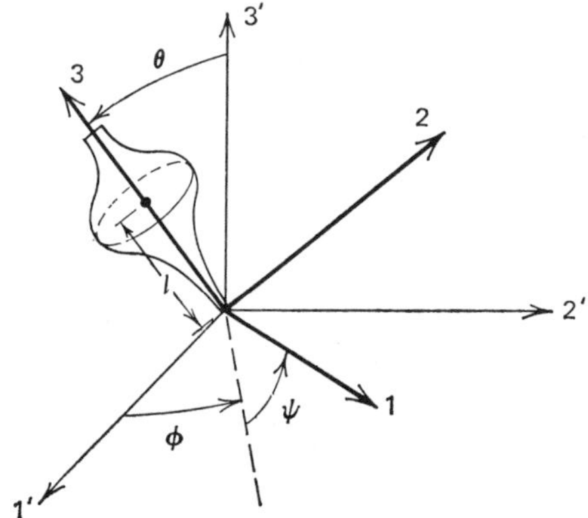

Figure 25. A symmetrical spinning top with one fixed point.

torque along the three body axes, for now we are using the torques $-\partial V/\partial\phi$, $-\partial V/\partial\theta$, $-\partial V/\partial\psi$, which are the components along the nonorthogonal 3'-axis, 1''-axis, and 3-axis of Fig. 24.

As an example, consider the case of a symmetrical spinning top with one point fixed (Fig. 25). We take the 3-axis to be along the axis of symmetry of the top and the 1-axis and 2-axis in any way which will form a right-handed system. Because the top is symmetric, $I_1 = I_2 \equiv I$, and the kinetic energy is then

$$T = \tfrac{1}{2}[I(\dot\theta^2 + \dot\phi^2 \sin^2\theta) + I_3(\dot\psi + \dot\phi\cos\theta)^2].$$

If the center of mass is a distance l above the fixed point, the potential energy due to gravity is

$$V = mgl\cos\theta,$$

where m is the mass of the top. From this the Lagrangian and the Euler-Lagrange equations are easily found. Their integration yields a description of the motion of a top or gyroscope, which has been studied in great detail. An excellent discussion and further bibliography will be found in Goldstein (1950).

V-4 EXAMPLES, APPLICATIONS, AND EXTENSIONS

(a) Acceleration in a Rotating Coordinate System

When the motion of a particle is viewed by an earthbound observer, it is viewed from a noninertial system. In Chapter II, Section 5(c), we used the Lagrangian formalism to derive the equations of motion of a particle in such a rotating noninertial system. These equations can also be derived by using Eq. (30). Thus consider a reference system attached to something, like the Earth, which is rotating at some angular velocity $\boldsymbol{\omega}$. Let the origin of this noninertial *body system* coincide with an inertial space system at the center of mass of the Earth, and let $\mathbf{r}$ be the position vector of a particle moving relative to the common origin. According to Eq. (30) the velocity of such a particle in the space system is related to its velocity in the body system (and hence to its velocity on Earth) by

$$\mathbf{v}_{\mathscr{S}} = \dot{\mathbf{r}} = \dot{\mathbf{r}}_{\mathscr{B}} + \boldsymbol{\omega} \times \mathbf{r} = \mathbf{v}_{\mathscr{B}} + \boldsymbol{\omega} \times \mathbf{r}. \tag{45}$$

We want to find the relation between $\mathbf{a}_{\mathscr{S}}$ and $\mathbf{a}_{\mathscr{B}}$, the acceleration in the space and body systems. (More precisely, $\mathbf{a}_{\mathscr{S}}$ is the vector whose components in the space system are $a'_{\mathscr{S}} = dv'_{\mathscr{S}}/dt$, and $\mathbf{a}_{\mathscr{B}}$ is the vector whose components in the body system are $a_{\mathscr{B}} = dv_{\mathscr{B}}/dt$.) To do this we write (45) in the space system and differentiate, obtaining

$$a'_{\mathscr{S}} = \dot{v}'_{\mathscr{S}} = \dot{v}'_{\mathscr{B}} + \dot{\omega}' \times r' + \omega' \times v'_{\mathscr{S}} \tag{46}$$

where we have used the fact that $v'_{\mathscr{S}} = dr'/dt$. Note two things: first, from Eq. (30) with $\mathbf{s}$ replaced by $\boldsymbol{\omega}$ it follows that $\dot{\boldsymbol{\omega}}_{\mathscr{S}} = \dot{\boldsymbol{\omega}}_{\mathscr{B}}$, so that the subscripts are not needed on $\dot{\boldsymbol{\omega}}$; second, $\dot{v}'_{\mathscr{B}}$ is the derivative of the space-system components of the body-system velocity and so is not itself of physical interest. To eliminate this quantity from (46) we use (30), written in the space system and with $\mathbf{s}$ replaced by $\mathbf{v}_{\mathscr{B}}$. The result is

$$\begin{aligned}\dot{v}'_{\mathscr{B}} &= D\dot{v}_{\mathscr{B}} + \omega' \times v'_{\mathscr{B}} \\ &= a'_{\mathscr{B}} + \omega' \times v'_{\mathscr{B}}.\end{aligned}$$

Consequently (46) becomes

$$a'_{\mathscr{S}} = a'_{\mathscr{B}} + \dot{\omega}' \times r' + \omega' \times v'_{\mathscr{B}} + \omega' \times (v'_{\mathscr{B}} + \omega' \times r'),$$

where use has been made of (45). In this equation all vectors are represented in the same (space) system, and so the equation is easily extended to any system:

$$\mathbf{a}_{\mathscr{S}} = \mathbf{a}_{\mathscr{B}} + \dot{\boldsymbol{\omega}} \times \mathbf{r} + 2\boldsymbol{\omega} \times \mathbf{v}_{\mathscr{B}} + \boldsymbol{\omega} \times (\boldsymbol{\omega} \times \mathbf{r}).$$

For the special case (such as the Earth) in which $\dot{\boldsymbol{\omega}} = 0$ this becomes

$$\mathbf{a}_{\mathscr{S}} = \mathbf{a}_{\mathscr{B}} + 2\boldsymbol{\omega} \times \mathbf{v}_{\mathscr{B}} + \boldsymbol{\omega} \times (\boldsymbol{\omega} \times \mathbf{r}). \tag{47}$$

In Eq. (47) $\mathbf{a}_{\mathscr{B}}$ and $\mathbf{v}_{\mathscr{B}}$ on the right-hand side are the acceleration and velocity of the particle as seeen by a noninertial observer, one attached to the Earth. However, Newton tells us that it is the mass times $\mathbf{a}_{\mathscr{S}}$, the acceleration of the particle relative to the inertial space system, which is equal to the force on the particle. Equation (47) then allows us to write Newton's equation of motion in terms of the acceleration and velocity of the particle as measured by the earthbound observer:

$$\mathbf{F} = m\mathbf{a}_{\mathscr{B}} + 2m\boldsymbol{\omega} \times \mathbf{v}_{\mathscr{B}} + m\boldsymbol{\omega} \times (\boldsymbol{\omega} \times \mathbf{r}).$$

This is equivalent to Eq. (53) of Chapter II, which was derived from the Euler-Lagrange equations.

(b) Principal Moments of a Cube

Let us calculate the inertial matrix of a cube about one of its corners. Let the length of each side of the cube be b, and let its density $\hat{\rho}$ be uniform. According to Eqs. (8) and (22), the inertial matrix elements are in general

$$I_{\alpha\beta} = \int \hat{\rho}(x)(x^2\,\delta_{\alpha\beta} - x_\alpha x_\beta)\,d^3x. \tag{48}$$

For the cube oriented as in Fig. 26, these become

$$I_{11} = I_{22} = I_{33} = \hat{\rho}\int_0^b dx_1 \int_0^b dx_2 \int_0^b dx_3(x_1^2 + x_2^2) = \tfrac{2}{3}\hat{\rho}b^5,$$

$$I_{12} = I_{13} = I_{23} = -\hat{\rho}\int_0^b dx_1 \int_0^b dx_2 \int_0^b dx_3(x_1 x_2) = -\tfrac{1}{4}\hat{\rho}b^5,$$

so that

$$I = \tfrac{2}{3}\hat{\rho}b^5 \begin{vmatrix} 1 & \alpha & \alpha \\ \alpha & 1 & \alpha \\ \alpha & \alpha & 1 \end{vmatrix}$$

where $\alpha = -\frac{3}{8}$.

The principal moments are obtained by diagonalizing I. To do so we find the roots λ of the equation $\det(I - \lambda\mathbb{1}) = 0$. If we write

$$\lambda = \tfrac{2}{3}Mb^2(1 - \alpha\mu),$$

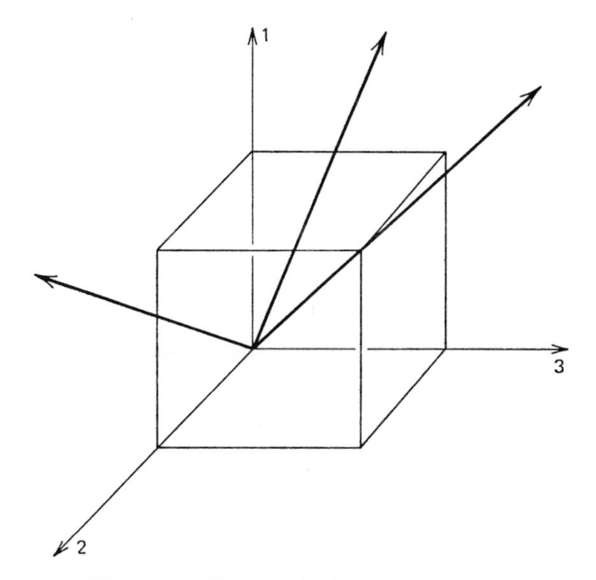

Figure 26. The principal axes of a cube.

where $M = \hat{\rho}b^3$ is the mass of the cube, the equation we must solve is

$$\mu^3 - 3\mu + 2 = 0.$$

The roots of this equation are $\mu = 1, 1, -2$, so that the three eigenvalues are

$$\lambda = \tfrac{11}{12}Mb^2,\ \tfrac{11}{12}Mb^2,\ \tfrac{1}{6}Mb^2.$$

Thus the diagonal form of I is

$$\begin{vmatrix} I_1 & 0 & 0 \\ 0 & I_2 & 0 \\ 0 & 0 & I_3 \end{vmatrix} = \tfrac{1}{12}Mb^2 \begin{vmatrix} 11 & 0 & 0 \\ 0 & 11 & 0 \\ 0 & 0 & 2 \end{vmatrix}.$$

The principal axes corresponding to the principal moments are the eigenvectors of I. The eigenvector belonging to I_3 is easily found to be $(1, 1, 1)$ and is thus directed along the principal diagonal of the cube. Because I is symmetric and hence diagonalized in an orthogonal basis, the eigenvectors belonging to I_2 and I_1 are mutually perpendicular and perpendicular to $(1, 1, 1)$. Furthermore, because $I_1 = I_2$ their directions are otherwise arbitrary. Thus we can take the principal axes to be the direction $(1, 1, 1)$ and any two other perpendicular directions forming an orthogonal system, say $(1, -1, 0)$ and $(\frac{1}{2}, \frac{1}{2}, -1)$ as illustrated in Fig. 26.

(c) The Moment of Inertia about an Arbitrary Axis

In elementary mechanics we learn that if a rigid body is constrained to rotate with angular velocity $\boldsymbol{\omega} = \omega\mathbf{n}$ about a *fixed* axis, then its kinetic energy is

$$T = \tfrac{1}{2}I\omega^2,$$

where

$$I = \int \hat{\rho}(\mathbf{x})\, |\mathbf{x} \times \mathbf{n}|^2\, d^3x \tag{49}$$

is the *moment of inertia* about the axis of rotation. (Note that $|\mathbf{x} \times \mathbf{n}|$ is the distance from $\mathbf{x}$ to the axis.) On the other hand Eq. (43) shows that

$$T = \tfrac{1}{2}\mathbf{n} \cdot (\mathbf{I}\mathbf{n})\omega^2,$$

so that I should be given by

$$I = \mathbf{n} \cdot (\mathbf{I}\mathbf{n}) = n_\alpha I_{\alpha\beta} n_\beta. \tag{50}$$

This is an equation for the moment of inertia about any axis $\mathbf{n}$ in terms of the n_α and the elements of the inertial tensor. That it is consistent with (49) is easily seen. Indeed, (50) says that

$$\begin{aligned} I &= n_\alpha n_\beta \int \hat{\rho}(\mathbf{x})(|\mathbf{x}|^2\, \delta_{\alpha\beta} - x_\alpha x_\beta)\, d^3x \\ &= \int \hat{\rho}(\mathbf{x})(|\mathbf{x}|^2 - (\mathbf{x} \cdot \mathbf{n})^2)\, d^3x \\ &= \int \hat{\rho}(\mathbf{x})\, |\mathbf{x} \times \mathbf{n}|^2\, d^3x, \end{aligned}$$

where we used $n_\alpha \delta_{\alpha\beta} n_\beta = \mathbf{n} \cdot \mathbf{n} = 1$, for $\mathbf{n}$ is a unit vector.

Equation (50) can be given an interesting and useful interpretation by considering all possible directions $\mathbf{n}$ of the axis of rotation. These all lie on a sphere centered about the origin, as shown in Fig. 27. As $\mathbf{n}$ is varied over this sphere the moment of inertia I will vary according to (50). Then if we define for each $\mathbf{n}$ the new vector

$$\mathbf{s} = \frac{\mathbf{n}}{\sqrt{I}} = \frac{\boldsymbol{\omega}}{\sqrt{2T}}, \tag{51}$$

Eq. (50) can be written

$$F(\mathbf{s}) \equiv s_\alpha I_{\alpha\beta} s_\beta = (\mathbf{s}, \mathbf{I}\mathbf{s}) = 1. \tag{52}$$

This is an equation for an ellipsoid; that is, as $\mathbf{n}$ sweeps out the sphere of its possible values, $\mathbf{s}$ sweeps out an ellipsoid, also shown in Fig. 27.

The ellipsoid of possible $\mathbf{s}$ values, called the *inertial ellipsoid*, is a standardized representation of what might be called, for reasons that will become

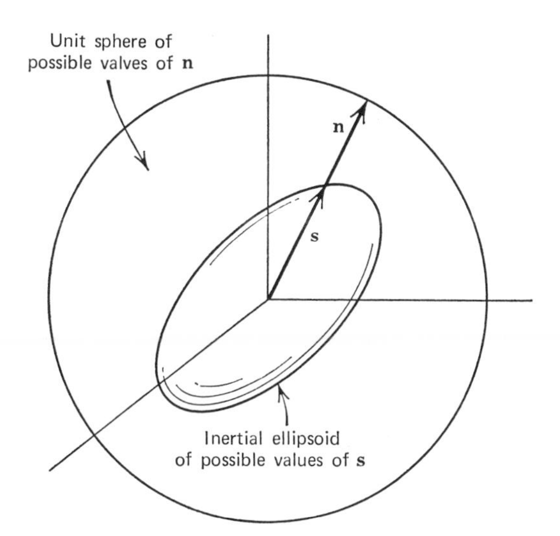

Figure 27. Inertial ellipsoid.

clearer in the next example, the "dynamical shape" of a body. The principal axes of the body correspond to the three principal axes of the ellipsoid, for in terms of these body axes (52) becomes

$$F(\mathbf{s}) = (I_1 s_1^2 + I_2 s_2^2 + I_3 s_3^2) = 1.$$

The lengths of the principal axes are thus seen to be proportional to the $(I_\alpha)^{-1/2}$.

In the next section we shall see how the inertial ellipsoid helps to visualize the motion of an unconstrained rigid body subjected to no external torques.

(d) Torque-Free Motion of a Rigid Body

A special case of particular interest is a rigid body in the absence of external torques. The Earth, artificial satellites, and nonspherical molecules, atoms, and nuclei are all examples of such systems. Although quantum mechanics must be used to describe the behavior properly for the last three systems, the classical description nevertheless provides valuable insight.

In the torque-free case, the Euler equations (35) become

$$\begin{aligned} I_1\dot{\omega}_1 &= (I_2 - I_3)\omega_2\omega_3, \\ I_2\dot{\omega}_2 &= (I_3 - I_1)\omega_3\omega_1, \\ I_3\dot{\omega}_3 &= (I_1 - I_2)\omega_1\omega_2, \end{aligned} \tag{53}$$

which is a set of coupled first-order differential equations for the components of the angular velocity. Under no torques, no work is done on the body as it rotates, and consequently both the angular momentum **L** and the kinetic energy T (relative to the center of mass) are constants of the motion. These results are also derived directly from Eq. (53) in Problem 11. Then Eq. (43), which can be written

$$T = \tfrac{1}{2}\boldsymbol{\omega} \cdot \mathbf{L}, \tag{54}$$

says that the projection of $\boldsymbol{\omega}$ onto **L** is a constant. Since **L** itself is a constant, it follows that the tip of $\boldsymbol{\omega}$ lies always on a fixed plane perpendicular to **L**. This can be described in terms of the inertial ellipsoid. The vector $\mathbf{s} = (2T)^{-1/2}\boldsymbol{\omega}$ of Eq. (51), which goes from the center of the ellipsoid to its surface, lies along $\boldsymbol{\omega}$. On multiplying both sides of (54) by $(2/T)^{1/2}$, we obtain

$$\sqrt{2T} = \mathbf{s} \cdot \mathbf{L},$$

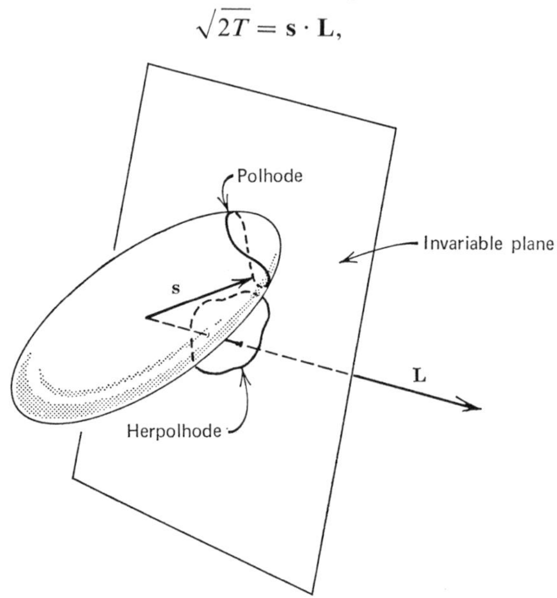

Figure 28. Poinsot's construction. The inertial ellipsoid is rolling on the invariable plane.

so that the tip of **s** also lies always on a fixed plane perpendicular to **L**. Call this plane the *invariable plane*. Its distance from the common origin of **L** and **s**, and hence from the center of the inertial ellipsoid, is the constant projection $\sqrt{2T}/L$ of **s** onto **L**, as illustrated in Fig. 28. The importance of the inertial ellipsoid is that it is tangent to the invariable plane at the point **s** where these two surfaces intersect. This follows at once from Eq. (52) for the inertial ellipsoid: $\mathbf{s} \cdot \mathbf{I}\mathbf{s} = 1$. The normal **p** to the surface of the ellipsoid at **s** is given by

$$\mathbf{p} = \nabla(\mathbf{s} \cdot \mathbf{I}\mathbf{s}) = (\tfrac{1}{2}T)^{-1/2}\mathbf{L},$$

so that **p** is normal also to the invariable plane, implying tangency. Consequently the ellipsoid touches the plane at precisely one point, and this point, since it lies on the axis of rotation, is instantaneously at rest. This, in the usual definition, means that the ellipsoid *rolls* on the plane.

These considerations then lead us to the following description of the torque-free motion of a rigid body (Poinsot's construction). The inertial ellipsoid rolls on the invariable plane, with the center of the ellipsoid fixed at a distance $\sqrt{2T}/L$ from the plane. This determines the instantaneous axis of rotation, which lies along **s**; the magnitude of the angular velocity is determined by $\boldsymbol{\omega} = \mathbf{s}\sqrt{2T}$. In principle this is a complete description of the motion, which one may be able to visualize by studying the curves traced out as in Fig. 28 by the point of tangency on both the inertial ellipsoid and the invariable plane (called the *polhode* and *herpolhode*, respectively). In general, however, the polhode and herpolhode are actually such terribly complicated and nonclosing curves, that even the most exquisite geometrical intuition hardly helps.

In certain special cases, on the other hand, the Euler equations can be solved rather easily and then the Poinsot construction is in fact a useful aid for visualizing the motion. For example, consider a symmetric body in which two of the principal axes are equal. We take $I_1 = I_2 = I$, so that the inertial ellipsoid is a spheroid symmetric about the 3-axis. Equations (53) may be written

$$\begin{aligned} \dot{s}_1 &= -ks_2, \\ \dot{s}_2 &= ks_1, \\ I_3\dot{\omega}_3 &= 0, \end{aligned} \tag{55}$$

where $k = (I_3/I - 1)\omega_3$ is a constant, since according to the last of Eqs. (55) ω_3 is a constant. (We assume specifically that $I_3 \neq I$.) For s_1 we obtain immediately $\ddot{s}_1 + k^2 s_1 = 0$, whose solution is well known:

$$s_1 = a \sin (kt + \varphi),$$

where a and φ are constants. Then

$$s_2 = -a \cos (kt + \varphi),$$

and $s_1^2 + s_2^2 = a^2$ is constant. Thus, as illustrated in Fig. 29, a is the projection of **s** onto the body (1, 2)-plane. This projection (and hence also the projection of $\boldsymbol{\omega}$) is of constant length and rotates with constant angular speed k. Since s_3 is also constant, $s = (s_\alpha s_\alpha)^{1/2}$ is constant, and the **s** vector sweeps out a right circular cone (called the *body cone*) in the body system, whose base is bounded on the ellipsoid by the polhode (see Fig. 29). This cone is swept out with

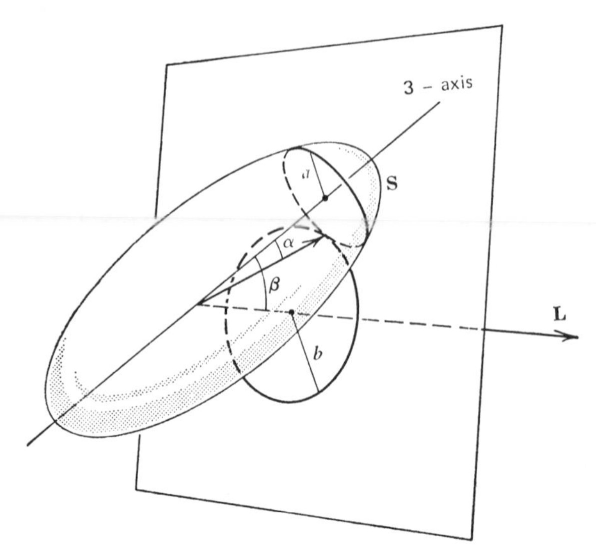

Figure 29. Poinsot's construction for an axially symmetric body.

period

$$P_{\mathscr{B}} = \frac{2\pi}{|k|} = \frac{2\pi}{\omega_3}\left|\frac{I}{I_3 - I}\right|.$$

In the diagram we have assumed that $I_3 < I$, so that the ellipsoid is prolate (cigar shaped).

The herpolhode is also a circle, for the magnitude of **s** is constant. This circle forms the base of a right circular cone which is called the *space cone*, also illustrated in Fig. 29, and at any instant **s** lies on the generator along which the body and space cones are tangent. Since the space cone is fixed on the invariable plane and **s** is instantaneously at rest, the complete motion may be thought of as the body cone rolling on the space cone. If b is the radius of the herpolhode, **s** sweeps out the space cone with period

$$P_{\mathscr{S}} = \frac{b}{a} P_{\mathscr{B}}.$$

Can this approach be applied to the asymmetric case, in which all the principal moments are unequal? Under certain special conditions it can serve as the basis for an approximation. For instance, suppose $\boldsymbol{\omega}$ is initially nearly parallel to one of the principle axes, say the 3-axis. Then ω_1 and ω_2 are initially much less than ω_3, and we may neglect the right-hand side in the last of Eqs. (53), so that ω_3 is approximately a constant, as in the symmetric case. Again we obtain

$$\ddot{\omega}_1 = -k^2\omega_1$$

so long as the approximation is valid, where now

$$k^2 = \frac{(I_3 - I_1)(I_3 - I_2)}{I_1 I_2} \omega_3^2.$$

The approximate solution for ω_1 (and ω_2) will be sinusoidal in t so long as $k^2 < 0$, and thus ω_1 and ω_2 will remain small. But if $k^2 < 0$, ω_1 and ω_2 will not remain small, and the approximation ceases to be valid. This result reflects on the stability of rotation about principal axes; further discussion is left to Problem 14. See also Landau (1960).

(e) A Rolling Sphere

As an illustration both of how difficult a seemingly simple problem is in rigid body motion, and of a relatively easy way to handle it, we consider a sphere of mass M and radius R rolling without slipping on a horizontal plane. The moment of inertial I of a sphere is the same about any diameter, so the kinetic energy (also the Lagrangian) may be written

$$L = T = \tfrac{1}{2}M(\dot{x}_1'^2 + \dot{x}_2'^2) + \tfrac{1}{2}I\omega'^2.$$

Here x_1' and x_2' are Cartesian coordinates in the plane. If we expand $\omega'^2 = \omega_1'^2 + \omega_2'^2 + \omega_3'^2$ in terms of the Euler angles according to Eq. (42) we obtain

$$L = \tfrac{1}{2}M(\dot{x}_1'^2 + \dot{x}_2'^2) + \tfrac{1}{2}I(\dot{\phi}^2 + 2\dot{\phi}\dot{\psi}\cos\theta + \dot{\psi}^2 + \dot{\theta}^2). \tag{56}$$

The nonholonomic constraint equations $R\omega_1' = -\dot{x}_2'$ and $R\omega_2' = \dot{x}_1'$ can be put in the form

$$\begin{aligned} \dot{\theta}\cos\phi + \dot{\psi}\sin\theta\sin\phi + \dot{x}_2'/R &= 0, \\ \dot{\theta}\sin\phi - \dot{\psi}\sin\theta\cos\phi - \dot{x}_1'/R &= 0. \end{aligned} \tag{57}$$

According to Eq. (76) of Chapter IV the Euler-Lagrange equations are then (λ and μ are Lagrange multipliers)

$$M\ddot{x}_1' = -\lambda/R, \tag{58a}$$

$$M\ddot{x}_2' = \mu/R, \tag{58b}$$

$$I\frac{d}{dt}(\dot{\phi} + \dot{\psi}\cos\theta) = I\frac{d\omega_3'}{dt} = 0, \tag{58c}$$

$$I(\ddot{\theta} + \dot{\phi}\dot{\psi}\sin\theta) = \mu\cos\phi + \lambda\sin\phi, \tag{58d}$$

$$I\frac{d}{dt}(\dot{\psi} + \dot{\phi}\cos\theta) = \sin\theta(\mu\sin\phi - \lambda\cos\phi). \tag{58e}$$

Equations (57) and (58) are then seven equations for two Lagrange multipliers, two Cartesian coordinates, and three Euler angles. It is immediately evident from (58c) that ω_3' is constant. Further, we know that the total energy is conserved (see Problem 35, Chapter IV), so that $T = E$ is constant. It can be shown algebraically that these facts and the constraint equations imply that the rotational and translational parts of the kinetic energy are separately constant. But to go further requires some manipulation.

Let us multiply Eq. (58d) by $\cos\phi$ and (58e) by $\sin\phi/\sin\theta$ and add. We then obtain

$$I[\ddot{\theta}\cos\phi + \dot{\phi}\dot{\psi}\sin\theta\cos\phi + \ddot{\psi}\sin\phi/\sin\theta + \ddot{\phi}\sin\phi\cos\theta/\sin\theta - \dot{\theta}\dot{\phi}\sin\phi] = \mu.$$

According to (58c), however, $\ddot{\phi} = -\ddot{\psi}\cos\theta + \dot{\psi}\dot{\theta}\sin\theta$, so that

$$\begin{aligned}\mu &= I[\ddot{\theta}\cos\phi - \dot{\theta}\dot{\phi}\sin\phi + \ddot{\psi}\sin\phi(1-\cos^2\theta)/\sin\theta \\ &\quad + \dot{\phi}\dot{\psi}\sin\theta\cos\phi + \dot{\psi}\dot{\theta}\cos\theta\sin\phi] \\ &= I\frac{d}{dt}(\dot{\theta}\cos\phi + \dot{\psi}\sin\theta\sin\phi) = I\frac{d\omega_1'}{dt}.\end{aligned} \tag{59}$$

Now compare (59) and (58b). We have

$$\mu = I\dot{\omega}_1' = MR\ddot{x}_2'. \tag{60}$$

But the first constraint equation tells us that $\dot{x}_2' = -R\omega_1'$. Thus we have

$$\mu = I\dot{\omega}_1' = -MR^2\dot{\omega}_1'.$$

Since I, M, and R are all positive, we conclude that $\mu = 0$, and hence $\dot{\omega}_1' = 0$ and $\ddot{x}_2' = 0$. Similarly, $\lambda = 0$, $\dot{\omega}_2' = 0$, and $\ddot{x}_1' = 0$. The sphere rolls at constant speed in a straight line and with constant angular velocity (although the angular velocity is not necessarily about a horizontal axis), a trivial result indeed.

Is there some way we could have obtained this result more easily? To answer this question, we look at Eq. (60). Since $M\ddot{x}_2'$ equals the constraint force F_2' in the space 2-direction, μ must be simply $F_2'R$. But $F_2'R$ is the torque τ_1' about the space 1-axis, and this is $\dot{L}_1'$, where $\mathbf{L} = \mathbf{I}\boldsymbol{\omega}$ is the angular momentum. Since the moment of inertia is the same about all axes and hence constant also in the space system $\dot{\mathbf{L}} = \mathbf{I}\dot{\boldsymbol{\omega}}$. Thus Eq. (60) also says $\tau_1' = \dot{L}_1'$. Let us therefore write simply $\mathbf{F} = M\mathbf{a}$ where $\mathbf{F}$ is the constraint force, and $\boldsymbol{\tau} = \mathbf{I}\dot{\boldsymbol{\omega}}$ in the space system and see what we obtain. We have

$$\begin{aligned}F_1' &= M\ddot{x}_1', \\ F_2' &= M\ddot{x}_2', \\ F_1'R &= -I\dot{\omega}_2', \\ F_2'R &= I\dot{\omega}_1'.\end{aligned} \tag{61}$$

The constraint equations are

$$R\omega_1' = -\dot{x}_2',$$
$$R\omega_2' = \dot{x}_1'. \tag{62}$$

Then we have, for instance,

$$F_1' = -I\dot{\omega}_2'/R = M\ddot{x}_1' = MR\dot{\omega}_2',$$

which shows, as before, that $F_1' = 0$.

Thus the problem is very easily solved by using Eqs. (61) and (62) rather than by the Lagrangian formulation. This approach is helpful also in some of the Problems.

Problems

1. Let $\mathscr{F}$ be the internal force per unit volume in a rigid body. That is, the total force $\mathbf{F}_V$ on a volume V of the body is $\mathbf{F}_V = \int_V \mathscr{F}\, d^3x$. Show that in the torque-free case the elements of the symmetric matrix $S = \Lambda + \Lambda^T$ of Eq. (13) are given by

$$S_{\alpha\beta} = \int \mathscr{F}_\beta x_\alpha \, d^3x. \tag{63}$$

   Show that the symmetry of S implies the absence of total torque due to internal forces. Starting from Eq. (63) and the condition that S is symmetric, derive Euler's equations in the absence of external torques. [*Hint:* This last part requires use of Eq. (47).]
2. Derive the result

$$\left(R^T \frac{\partial R}{\partial \theta_\gamma}\right)_{\alpha\beta} = -\epsilon_{\alpha\beta\gamma},$$

   where R is a rotation about the γ axis. [This result is used in Eq. (26)].
3. Prove that if **a** and **b** are vectors, then **a** × **b** is a pseudovector. Prove that if **A** is a vector field, then ∇ × **A** is a pseudovector field.
4. Find the position, as seen by an observer on Earth, of a satellite in a circular orbit passing above the poles. State the position vector in a coordinate system attached to the Earth and with origin at the center of the Earth.
5. A hovering satellite is one placed in a circular orbit above a point on the equator and given a velocity just right to keep it exactly over that point constantly. Suppose a satellite were placed in orbit over a point on the equator with the correct speed (magnitude of velocity) to hover, but at a slightly wrong angle. Describe its motion as seen from Earth.
6. Figure 30 shows two weights, each of mass m, on a light rod of length b. The rod is constrained to rotate at constant angular velocity ω about a vertical axis, maintaining a constant angle θ with the vertical. Calculate (representations of) **L**, **I** and $\boldsymbol{\omega}$ in both the body and space systems, and show that $\mathbf{L} = \mathbf{I}\boldsymbol{\omega}$. Calculate the torque of constraint on the system and

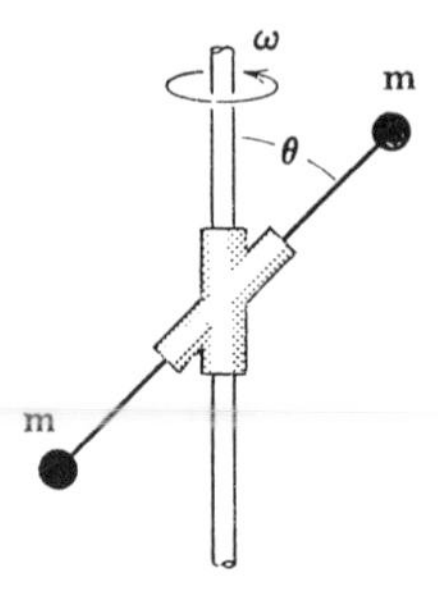

Figure 30. The system of Problem 6.

discuss physically how this torque is provided. Describe the motion which would result if the constraint torque were suddenly turned off.

7. (a) Use Eq. (42) to prove mathematically that there exist no generalized coordinates $q_\alpha(\phi, \theta, \psi)$ such that $\omega'_\alpha = dq_\alpha/dt$. Prove a similar result for the ω_α. (b) Now suppose that instead of the Euler rotations we specified every rotation by the following three operations. *First* rotate through an angle θ_1 about the space 1-axis; *second*, through θ_2 about the space 2-axis; *third*, through θ_3 about the space 3-axis. Show that the θ'_γ are unique for each orientation and that, in this order,

$$D = R(\theta_1)R(\theta_2)R(\theta_3)$$

[notation of Eqs. (25) and (36)]. Find the ω_α in terms of the θ_γ and $\dot{\theta}_\gamma$, obtaining the analogue of Eq. (39). Show explicitly a result we know from part (a) of this problem, namely there exist no generalized coordinates $q_\alpha(\theta_1, \theta_2, \theta_3)$ such that $\omega'_\alpha = dq_\alpha/dt$ or $\omega_\alpha = dq_\alpha/dt$.

8. Let the time-dependent transformation matrix from the body to the space system be written in the form $D = D_b D_a$, where D_a and D_b are also time-dependent rotation matrices. Find Ω (or Ω' in the space system) in terms of $\Omega_a = D_a^T \dot{D}_a$ and $\Omega_b = D_b^T \dot{D}_b$ (or Ω'_a and Ω'_b). Interpret this in terms of superposition of angular velocities, and show that $\boldsymbol{\omega} = \boldsymbol{\omega}_a + \boldsymbol{\omega}_b$.

9. Derive Eq. (39) in the following way. Show from (37) that $\Omega = D^T \dot{D} = (D_\theta D_\psi)^T \Omega_\phi D_\theta D_\psi + D_\psi^T \Omega_\theta D_\psi + \Omega_\psi$. Then from what you know about the relation between ω and Ω you can almost read off Eq. (39).

10. Show that the kinetic energy T of a rigid body rotating about a fixed point may be written

$$T = \tfrac{1}{2}\boldsymbol{\omega} \cdot (\mathbf{I}\boldsymbol{\omega}).$$

From this, prove that the principal moments of inertia are always positive. Use Eq. (39) to write T in terms of the Euler angles, and calculate the generalized momenta p_ϕ, p_θ, p_ψ conjugate to the Euler angles. Show that each of these generalized momenta is just the projection of the angular momentum onto the corresponding angular velocity vector $\dot{\boldsymbol{\phi}}$, $\dot{\boldsymbol{\theta}}$, or $\dot{\boldsymbol{\psi}}$.

11. Show directly from Eq. (53) that the angular momentum **L** and kinetic energy T of a rigid body are both constants of the motion in the torque-free case.
12. For the motion of a symmetric body with no torques, show that the angle β between **L** and the axis of symmetry of the body is given by

$$\tan \beta = \frac{Ia}{I_3 s_3},$$

and that the angle α between **s** and the axis of symmetry is given by

$$\tan \alpha = \frac{I_3}{I} \tan \beta.$$

Here a is the radius of the polhode.
13. For the motion of a symmetric body with no torques, show that the radius b of the herpolhode is given by

$$b = \frac{k}{\sqrt{2T}} \sin \beta,$$

where k is defined at Eq. (55) and β in Problem 12.
14. Try tossing an object with three unequal moments of inertia (a match book, a chalk eraser, an airplane) into the air, flicking it so that it spins about one of its principal axes. It will continue to rotate stably if the axis chosen belongs to the smallest or largest moment of inertia, but almost never otherwise. Explain this behavior [refer to the very end of Section 4(d)].
15. Write down the Euler-Lagrange equations for the motion of the symmetric top. Show that ω_3 is a constant of the motion. Show that p_ϕ and p_ψ are constants of the motion.
16. Find the angular velocity $\boldsymbol{\omega}$, the kinetic energy T, and the angular momentum **L** of a uniform solid right circular cone of mass M, height h and radius R at the base, rolling without slipping on its side on a horizontal plane. Find the total torque $\boldsymbol{\tau}$ on the cone and analyze how gravitational and constraint forces provide this torque. (Assume the cone returns to its initial position with period P.) Find the minimum value of the coefficient of friction for which the cone will not slip. Find the minimum value of P for which the cone will not tilt.
17. Consider the rolling disc problem of Section 5(e), Chapter IV. Find the angular momentum vector **L** (about the center of mass) for the general solution of Eq. (83). From the **L** you obtain, find the torque $\boldsymbol{\tau}$, and analyze how the applied forces produce this torque. [*Hint:* you will have to invent a mechanism to keep the disc vertical. This holonomic constraint was eliminated by proper choice of two generalized coordinates. Which two?]
18. A coin (a flat disc) of radius R and mass M is constrained to roll without slipping on a horizontal plane in a uniform gravitational field. (The coin is not constrained to remain vertical as in Problem 17.) Describe the motion.

19. A sphere of radius R and mass M is constrained to roll without slipping on a horizontal turntable rotating with constant angular speed β. Describe its motion. Consider the limit $\beta = 0$. Compare Section 4(e).
20. A sphere of radius R and mass M is constrained (say, by magnet-like forces) to roll on a vertical plane in a uniform gravitational field. The plane itself rotates about a vertical axis with constant angular speed β. Describe the motion. Consider the limit $\beta = 0$.

CHAPTER VI

The Hamiltonian Formulation

Up to this point we have been describing mechanical systems with n degrees of freedom by using n generalized coordinates q_α and the Lagrangian function L. The Euler-Lagrange equations then give us a set of n second-order differential equations which determine the time dependence of the q_α. Recall that the Lagrangian formalism was introduced partly to allow a mechanical system to be described by any convenient set of coordinates; it allows transformations from one set of generalized coordinates to another.

In this chapter we formulate the Hamiltonian approach to mechanics, in which the n variables q_α are replaced by a set of $2n$ variables, and L is replaced by a new function, the Hamiltonian H. The n second-order Euler-Lagrange equations are then replaced by $2n$ first-order equations. This Hamiltonian formulation is equivalent to the Lagrangian one, for it predicts the same motions for any mechanical system, but it has a number of advantages.

Perhaps its principal advantage (and the only one we mention now) is that by doubling the number of variables, the Hamiltonian formulation allows a much broader set of general transformations, among which the coordinate transformations appear as a special class. An important aspect of this chapter will be the study and classification of this broader set. Later, in the next chapter, we shall see how these transformations can be used in solving actual problems.

VI-1 PHASE SPACE, THE CANONICAL EQUATIONS, AND POISSON BRACKETS

We return now to the Euler-Lagrange equations for a mechanical system in generalized coordinates:

$$\frac{d}{dt}\frac{\partial L}{\partial \dot{q}_\alpha} - \frac{\partial L}{\partial q_\alpha} = 0. \tag{1}$$

We assume that all constraints have been removed by appropriate choice of the generalized coordinates (assuming also that there are no nonholonomic constraints). These n second-order equations can be transformed trivially into $2n$ first-order equations by treating the $\dot{q}_\alpha$ and q_α as $2n$ independent variables s_α, $\alpha = 1, \ldots, 2n$. Thus we may write

$$s_\alpha = q_\alpha \quad \text{for} \quad \alpha = 1, \ldots, n,$$
$$s_\alpha = \dot{q}_{\alpha-n} \quad \text{for} \quad \alpha = n+1, \ldots, 2n. \tag{2}$$

Then Eq. (1) becomes

$$\frac{d}{dt}\frac{\partial L}{\partial s_{\alpha+n}} - \frac{\partial L}{\partial s_\alpha} = 0, \qquad \alpha = 1, \ldots, n, \tag{3}$$

which are n equations for the $2n$ functions $s_\alpha(t)$. Clearly we need n more equations. But these are obvious:

$$s_{\alpha+n} = \dot{s}_\alpha, \qquad \alpha = 1, \ldots, n. \tag{4}$$

What we have done is simply to treat (4) not as a definition, but as an equation of motion.[1]

The particular definition of the new variables we have given in Eq. (2) is not, however, canonical.[2] This is because it is not convenient for one of our principal purposes, namely to make general transformations among the $2n$ resulting variables. It is much more fruitful to construct the set of $2n$ variables out of the q_α and the momenta

$$p_\alpha = \frac{\partial L}{\partial \dot{q}_\alpha}. \tag{5}$$

Half of the resulting $2n$ equations of motion, those which like (4) are in fact reinterpretations of a definition, are easily obtained. One need only solve (5) for the $\dot{q}_\alpha$ as functions of q, p, t. This requires, of course, that (5) can be inverted. Now, in general any set of equations of the form $y_\alpha = f_\alpha(x)$ (where $x = \{x_1, \ldots, x_n\}$ and $\alpha = 1, \ldots, n$) can be inverted to yield $x_\alpha = g_\alpha(y)$ only if $\det(\partial f_\alpha/\partial x_\beta) \neq 0$. In our present case this means that $\det(\partial^2 L/\partial \dot{q}_\alpha\, \partial \dot{q}_\beta) \neq 0$, a condition we have already accepted in what seemed a different context (see the Remark at the end of Chapter II, Section 2). The other n equations can be obtained from the Euler-Lagrange equations, for they state that $\dot{p}_\alpha = \partial L/\partial q_\alpha$. However, this can be done in such a way that the inverse of (5) is not explicitly inserted into this differential equation. To do this we turn our attention from L to a new function H of q, p, t in terms of which we shall write the equations of motion.

[1] To show that (3) and (4) together are equivalent to (1) is trivial. In fact, the straightforward way to solve the system of $2n$ equations is to use (4) to eliminate n of the variables from (3), arriving immediately at (1).

[2] By "canonical" we mean generally accepted, standard, conventional. It derives from the meaning "according to the canons." The word is used, as we shall see, quite technically in mechanics.

Let the *Hamiltonian* function $H(q, p, t)$ be defined by

$$H(q, p, t) = \dot{q}_\alpha(q, p, t)p_\alpha - L(q, \dot{q}(q, p, t), t). \tag{6}$$

The independent variables here are q, p, t, not q, $\dot{q}$, t. This means that in order to write down H we must solve (5) for $\dot{q}$ as a function of q, p, t, and insert this solution into the expression for $L(q, \dot{q}, t)$. Then we assert that

$$\begin{aligned} \dot{q}_\alpha &= \frac{\partial H}{\partial p_\alpha}, \\ \dot{p}_\alpha &= -\frac{\partial H}{\partial q_\alpha}. \end{aligned} \tag{7}$$

These equations mean, for instance, that the expression for $\dot{q}_\alpha$ as a function of q, p, t is obtained by calculating $\partial H/\partial p_\alpha$, and that the function so calculated becomes the right-hand side of one of the $2n$ equations of motion.

PROOF. From the definition of H we have (remember, q and p are the independent variables)

$$\frac{\partial H}{\partial p_\beta} = \dot{q}_\beta + p_\alpha \frac{\partial \dot{q}_\alpha}{\partial p_\beta} - \frac{\partial L}{\partial \dot{q}_\gamma}\frac{\partial \dot{q}_\gamma}{\partial p_\beta},$$

for L depends on p only through $\dot{q}$. Then (5) leads immediately to the first of Eqs. (7). To obtain the second, we calculate

$$\frac{\partial H}{\partial q_\beta} = p_\alpha \frac{\partial \dot{q}_\alpha}{\partial q_\beta} - \frac{\partial L}{\partial q_\beta} - \frac{\partial L}{\partial \dot{q}_\gamma}\frac{\partial \dot{q}_\gamma}{\partial q_\beta}.$$

This time the first and third terms cancel according to (5), and Eq. (1) tells us that

$$\frac{\partial L}{\partial q_\beta} = \frac{d}{dt}\frac{\partial L}{\partial \dot{q}_\beta} = \frac{d}{dt}p_\beta = \dot{p}_\beta.$$

Thus we obtain the second of Eqs. (7).

Equations (7) are $2n$ equations of first order, which, with the definitions (5) and (6), are equivalent to the Euler-Lagrange equations. They are called *Hamilton's canonical equations of motion*, and the $2n$ variables q, p are called *canonical variables*. Their symmetry is evident and, as we shall see, useful. If we were as clever in writing down Hamiltonians as we have become in writing down Lagrangians, we could start any problem with (7) rather than with (1). Actually no great cleverness is required, for the Hamiltonian is in many ways a more natural function to write down than the Lagrangian. This is because the Hamiltonian possesses several remarkable properties that make it easy to write down.

First, we show that

$$\frac{\partial H}{\partial t} = -\frac{\partial L}{\partial t}. \tag{8}$$

In fact from (6) we have [remember: $\dot{q}_\alpha = \dot{q}_\alpha(q, p, t)$]

$$\frac{\partial H}{\partial t} = p_\alpha \frac{\partial \dot{q}_\alpha}{\partial t} - \frac{\partial L}{\partial t} - \frac{\partial L}{\partial \dot{q}_\gamma}\frac{\partial \dot{q}_\gamma}{\partial t}.$$

As usual the first and third terms cancel according to (5), and we arrive at (8).

Second, we show that

$$\frac{dH}{dt} = \frac{\partial H}{\partial t}. \tag{9}$$

Indeed, we have

$$\frac{dH}{dt} = \frac{\partial H}{\partial q_\alpha}\dot{q}_\alpha + \frac{\partial H}{\partial p_\alpha}\dot{p}_\alpha + \frac{\partial H}{\partial t}.$$

In this equation the first and second terms cancel according to (7), and we arrive at (9).

From those two properties, we see that if L does not depend explicitly on t, neither does H, and that H is then a constant of the motion.

Third, let us consider a standard Lagrangian of the form $L = T - V$, where the potential energy V is independent of the $\dot{q}_\alpha$, and the kinetic energy T is homogeneous quadratic in the $\dot{q}_\alpha$. Then, still taking q, p, t as the independent variables, we see that

$$\begin{aligned} H &= \frac{\partial L}{\partial \dot{q}_\alpha}\dot{q}_\alpha - L \\ &= \frac{\partial T}{\partial \dot{q}_\alpha}\dot{q}_\alpha - (T - V) \\ &= 2T - (T - V) \\ &= T + V. \end{aligned}$$

Here we have used the fact that if T is homogeneous quadratic in the $\dot{q}_\alpha$, then $\dot{q}_\alpha \partial T/\partial \dot{q}_\alpha = 2T$. (See Problem 2 of Chapter III.) Thus H is the total energy.

It is clear from this last property of H that to write it down is not very difficult. One need only write down the total energy in terms of the momentum and position variables. [See Section 6(a).]

Remark. The Hamiltonian is a constant of the motion and equal to the energy only if $L(q, \dot{q}, t)$ has no explicit time dependence and if $L = T - V$ with T homogeneous quadratic in the $\dot{q}_\alpha$. Other possibilities are that H may be a constant of the motion but not the energy (if T is not homogeneous quadratic), that H may be the energy but not constant (if $\partial L/\partial t \neq 0$), or that H may be neither. See Problems 1 and 2.

Let us now proceed, as we did in going over from the q_α, $\dot{q}_\alpha$ to the s_α, to introduce a set of $2n$ variables ξ_α on *phase space*, which is the name given the $2n$-dimensional space of q's and p's. We do this by making the following definition

$$\begin{aligned} \xi_\alpha &= q_\alpha \qquad \text{for} \qquad \alpha = 1, \ldots, n, \\ \xi_\alpha &= p_{\alpha-n} \qquad \text{for} \qquad \alpha = n+1, \ldots, 2n. \end{aligned} \tag{10}$$

In order also to write the canonical equations in simple form we introduce the $2n \times 2n$ matrix

$$\Gamma = \begin{vmatrix} 0_n & \mathbb{1}_n \\ -\mathbb{1}_n & 0_n \end{vmatrix} \tag{11}$$

with matrix elements $\gamma_{\alpha\beta}$ (by $\mathbb{1}_n$ we mean the $n \times n$ unit matrix, and by 0_n the $n \times n$ null matrix). Then if we write $H(\xi, t)$ for the Hamiltonian, Eqs. (7) become

$$\dot{\xi}_\alpha = \gamma_{\alpha\beta} \frac{\partial H}{\partial \xi_\beta} . \tag{12}$$

Remarks. 1. We shall often switch back and forth from the ξ-notation to the (q, p)-notation. It will be useful for the reader, as he proceeds, to practice going from one system to the other. In using the summation convention, as we switch back and forth, it is understood that the indices on the q_α and p_α run from 1 to n, while those on the ξ_α run from 1 to $2n$. *2.* Hamilton's canonical equations can be obtained from a variational principle, as are the Euler-Lagrange equations. This is shown in Section 6(d).

The matrix Γ has several important properties which are easily proved. We state them without the proof, which we leave to the student (Problem 3).

1. Γ is orthogonal, that is,

$$\Gamma^T\Gamma = \mathbb{1} \qquad \text{or} \qquad \gamma_{\alpha\beta}\gamma_{\alpha\rho} = \delta_{\beta\rho}.$$

2. Γ is antisymmetric, that is,

$$\Gamma^T + \Gamma = 0 \qquad \text{or} \qquad \gamma_{\alpha\beta} + \gamma_{\beta\alpha} = 0.$$

3. Γ is unimodular, that is,

$$\det \Gamma = 1.$$

Let us now summarize the new, *Hamiltonian* approach, in which we view mechanics in the following way. A mechanical system is completely specified at any time t by giving all of the q_α and $\dot{q}_\alpha$ or, equivalently, all of the q_α and

p_α. Thus the state of a system is specified by a point in phase space, namely a set of values for the $2n$ variables ξ_α; the problem of mechanics is to find how this point moves in time, or *the motion in phase space*. The initial conditions tell us where in phase space the system starts, but it is the Hamiltonian which tells us, through the canonical equations (12), how it proceeds from there. Or we may invert the order of these two ideas: the Hamiltonian determines all of the possible motions the system can perform in phase space, the initial conditions picking out the particular motion which is the solution to a particular problem. Thus with a Hamiltonian we associate a set of motions. The problem of mechanics is then to exhibit this association: to find the motions associated with each given Hamiltonian.

It is possible to use the Hamiltonian directly to determine something about how other functions vary along motions before actually calculating the motions themselves. Let $F(\xi, t)$ be any *dynamical variable*, that is, any function of $q, \dot{q}, t$ or, equivalently, of q, p, t. Then

$$\begin{aligned}\frac{dF}{dt} &= \frac{\partial F}{\partial \xi_\alpha}\dot{\xi}_\alpha + \frac{\partial F}{\partial t}\\ &= \frac{\partial F}{\partial \xi_\alpha}\gamma_{\alpha\beta}\frac{\partial H}{\partial \xi_\beta} + \frac{\partial F}{\partial t}.\end{aligned} \tag{13}$$

[The reader should obtain this expression in the (q, p)-notation.] Thus if we know the Hamiltonian, we know immediately how dF/dt depends on the coordinates in phase space, or how F varies along a motion. Of course, to use this to calculate dF/dt as a function of the time we must know the coordinates as functions of the time. But for one special case (13) is useful without this additional information, namely, if F is a constant of the motion. In this case $dF/dt = 0$, and it thus follows that F is a constant of the motion if and only if the right-hand side of (13) vanishes.

To see how this works, let us take the familiar example of the two-dimensional, symmetric simple harmonic oscillator. (More complicated examples will be found in Section 6 and among the Problems.) The kinetic energy is $\frac{1}{2}m(\dot{q}_1^2 + \dot{q}_2^2)$ and the potential energy is $\frac{1}{2}k(q_1^2 + q_2^2)$. In order to simplify the algebra, we set $m = k = 1$. Then $L = \frac{1}{2}(\dot{q}_1^2 + \dot{q}_2^2) - \frac{1}{2}(q_1^2 + q_2^2)$, and $p_\alpha = \dot{q}_\alpha$. Straightforward calculation yields

$$H = \tfrac{1}{2}(p_1^2 + p_2^2) + \tfrac{1}{2}(q_1^2 + q_2^2)$$

or, in the ξ-notation,

$$H = \tfrac{1}{2}\xi_\alpha\xi_\alpha.$$

This expression for H could also have been obtained by noting that $T = \frac{1}{2}(\dot{q}_1^2 + \dot{q}_2^2)$ is homogeneous quadratic in the $\dot{q}_\alpha$, so that H is the total energy. The canonical equations of motion are

$$\dot{q}_\alpha = p_\alpha \qquad \text{and} \qquad \dot{p}_\alpha = -q_\alpha$$

or, in the ξ-notation,

$$\dot{\xi}_\rho = \gamma_{\rho\sigma}\xi_\sigma.$$

Note that half of the canonical equations are just the defining equations for the p_α, or rather their solutions for the $\dot{q}_\alpha$.

Let us solve these equations in their second form. We take the derivative with respect to time [how would you solve them in the (q, p)-notation?] to obtain

$$\ddot{\xi}_\rho = \gamma_{\rho\sigma}\dot{\xi}_\sigma = \gamma_{\rho\sigma}\gamma_{\sigma\tau}\xi_\tau = -\delta_{\rho\tau}\xi_\tau = -\xi_\rho.$$

The solution may be written

$$\xi_\rho = \Delta_\rho \cos(t + \varphi_\rho),$$

where the constants Δ_ρ and φ_ρ are connected by the four first-order equations. Thus although ρ runs from 1 to 4, there are not eight but four independent constants of integration.

Equation (13) can be used to show that the angular momentum

$$l = q_1 p_2 - q_2 p_1$$

is a constant of the motion. We have, in the (q, p)-notation,

$$\begin{aligned} \frac{dl}{dt} &= \frac{\partial l}{\partial q_\alpha}\frac{\partial H}{\partial p_\alpha} - \frac{\partial l}{\partial p_\alpha}\frac{\partial H}{\partial q_\alpha} \\ &= p_2 p_1 - p_1 p_2 + q_2 q_1 - q_1 q_2 = 0, \end{aligned}$$

so that l is indeed a constant of the motion.

The first term on the right-hand side of (13) is seen to be an important expression; it is called the *Poisson bracket* of F with H. In general, the Poisson bracket $[S, R]$ of the dynamical variable S with the dynamical variable R is defined as

$$\begin{aligned} [S, R] &= \frac{\partial S}{\partial \xi_\alpha}\gamma_{\alpha\beta}\frac{\partial R}{\partial \xi_\beta} \\ &= \frac{\partial S}{\partial q_\alpha}\frac{\partial R}{\partial p_\alpha} - \frac{\partial S}{\partial p_\alpha}\frac{\partial R}{\partial q_\alpha}. \end{aligned} \tag{14}$$

(As usual, the sum in the first line is from $\alpha, \beta = 1$ to $\alpha, \beta = 2n$, and in the second line it is from $\alpha, \beta = 1$ to $\alpha, \beta = n$.) In terms of such brackets we may write (12) and (13) in the form

$$\begin{aligned} \dot{\xi}_\alpha &= [\xi_\alpha, H], \\ F &= [F, H] + \frac{\partial F}{\partial t}. \end{aligned} \tag{15}$$

Note that the first of these equations is merely the special case of the second in which F is chosen to be ξ_α.

The Poisson bracket has the following important properties, which the reader can easily prove (Problem 4).

Property 1. *Linearity.* If α and β are numbers, then

$$[\alpha S + \beta R, T] = \alpha[S, T] + \beta[R, T].$$

Property 2. *Antisymmetry.*

$$[S, R] = -[R, S].$$

Property 3. *Product rule.*

$$[S, RT] = [S, R]T + R[S, T].$$

(To remember this, think of $[S, \ ]$ as being a derivative operating on the product RT.)

Property 4. *The Jacobi indentity.*

$$[R, [S, T]] + [S, [T, R]] + [T, [R, S]] = 0.$$

Property 5. *Chain rule.* Suppose the dynamical variables S and R are given not in terms of the ξ_α, but in terms of some other set η_α of general coordinates on phase space. We assume that there exist equations of the form

$$\eta_\alpha = \eta_\alpha(\xi, t) \tag{16}$$

connecting these two sets of coordinates. Then

$$[R, S] = \frac{\partial R}{\partial \eta_\alpha} [\eta_\alpha, \eta_\beta] \frac{\partial S}{\partial \eta_\beta} . \tag{17}$$

Let us prove this fifth property. We have

$$[R, S] = \frac{\partial R}{\partial \xi_\rho} \gamma_{\rho\sigma} \frac{\partial S}{\partial \xi_\sigma} = \frac{\partial R}{\partial \eta_\alpha} \frac{\partial \eta_\alpha}{\partial \xi_\rho} \gamma_{\rho\sigma} \frac{\partial \eta_\beta}{\partial \xi_\sigma} \frac{\partial S}{\partial \eta_\beta}$$
$$= \frac{\partial R}{\partial \eta_\alpha} [\eta_\alpha, \eta_\beta] \frac{\partial S}{\partial \eta_\beta} .$$

Property 6. *Fundamental brackets.* For the particular case $R = \xi_\alpha$, $S = \xi_\beta$, we have directly from (14)

$$[\xi_\alpha, \xi_\beta] = \gamma_{\alpha\beta}, \tag{18}$$

or, in the (q, p)-notation

$$[q_\alpha, q_\beta] = [p_\alpha, p_\beta] = 0, \qquad [q_\alpha, p_\beta] = \delta_{\alpha\beta}.$$

We have now stated, through Eqs. (15), the main results of this section in terms of Poisson brackets. The properties of these brackets are very important to the structure of mechanics. The next section proves a theorem which establishes this importance.

VI-2 THE POISSON BRACKET THEOREM

Given some Hamiltonian $H(\xi, t)$, we say that *the time development* or set of motions of the resulting mechanical system *is generated by* H according to the canonical equations (12) or (15). It is possible, on the other hand, to write down a time development or a set of motions in phase space generated by no Hamiltonian $H(\xi, t)$. This could be done in several ways. For instance, by giving $2n$ sufficiently well-behaved functions $\Psi_\alpha(\xi, t)$ one could write down a set of coupled ordinary differential equations $\dot{\xi}_\alpha = \Psi_\alpha(\xi, t)$. Their solutions $\xi_\alpha = f_\alpha(t, \xi^0)$ may then be called the motions $\xi_\alpha(t)$. Note that they depend on the initial point $\xi_\alpha^0 = f_\alpha(0, \xi^0)$, a dependence we will usually suppress. Alternatively, one could give the $\xi_\alpha(t) = f_\alpha(t, \xi^0)$ directly, defining paths through phase space, being careful that the paths do not intersect. Then by calculating $\dot{\xi}_\alpha = \partial f_\alpha/\partial t$ at each point on every path, one obtains a set of functions $\Psi_\alpha(\xi, t) = \dot{\xi}_\alpha$. In either case, to say that no Hamiltonian generates these motions is to say that there exists no function $H(\xi, t)$ such that $\Psi_\alpha(\xi, t) = \gamma_{\alpha\beta}\partial H/\partial \xi_\beta$. The fundamental dynamical property of the Poisson brackets can then be stated in the form of the following theorem.

POISSON BRACKET THEOREM. Let $\xi_\alpha(t)$ be the time development of a system on phase space. This development is generated by some Hamiltonian $H(\xi, t)$ if and only if every pair of dynamical variables $R(\xi, t)$, $S(\xi, t)$ satisfies the relation

$$\frac{d}{dt}[R, S] = [\dot{R}, S] + [R, \dot{S}]. \tag{19}$$

PROOF. (*a*) *Necessity*. Assume that the development is generated by $H(\xi, t)$. Then according to (15),

$$\frac{d}{dt}[R, S] = [[R, S], H] + \frac{\partial}{\partial t}[R, S].$$

The first term can be handled by the Jacobi identity and Property 2:

$$[[R, S], H] = [[R, H], S] + [R, [S, H]].$$

For the second term we have

$$\begin{aligned}\frac{\partial}{\partial t}[R, S] &= \frac{\partial}{\partial t}\left(\frac{\partial R}{\partial \xi_\alpha}\gamma_{\alpha\beta}\frac{\partial S}{\partial \xi_\beta}\right) \\ &= \frac{\partial^2 R}{\partial \xi_\alpha\, \partial t}\gamma_{\alpha\beta}\frac{\partial S}{\partial \xi_\beta} + \frac{\partial R}{\partial \xi_\alpha}\gamma_{\alpha\beta}\frac{\partial^2 S}{\partial \xi_\beta\, \partial t} \\ &= \left[\frac{\partial R}{\partial t}, S\right] + \left[R, \frac{\partial S}{\partial t}\right].\end{aligned}$$

Combining these results, we obtain

$$\frac{d}{dt}[R, S] = \left[[R, H] + \frac{\partial R}{\partial t}, S\right] + \left[R, [S, H] + \frac{\partial S}{\partial t}\right]$$

$$= [\dot{R}, S] + [R, \dot{S}],$$

as asserted.

(*b*) *Sufficiency*. What we wish to prove is that a function $H(\xi, t)$ exists such that

$$\frac{\partial H}{\partial \xi_\beta} = \dot{\xi}_\alpha \gamma_{\alpha\beta},$$

or that these equations, treated as a set of partial differential equations for H, have a solution. This will be done by showing that the right-hand sides $\chi_\beta = \dot{\xi}_\alpha \gamma_{\alpha\beta}$, when written as functions of ξ, t, satisfy the integrability condition $\partial \chi_\beta / \partial \xi_\alpha = \partial \chi_\alpha / \partial \xi_\beta$.

Remark. We have used this technique before (Chapter III, Section 4) and we will use it again. It strictly applies only in *simply connected* regions of the variable space, that is, in regions in which every closed path can be continuously distorted to a point. (This means that there are no essential holes in the region.) We assume then that we deal only with simply connected regions of phase space.

Assume, then, that Eq. (19) is satisfied for every pair R, S. Let $\dot{\xi}_\alpha = \Psi_\alpha(\xi, t)$ be given at every point in phase space, or let it be calculated from the given time development $\xi_\alpha(t)$. Now if (19) holds for every pair R, S, it certainly holds for the pair ξ_α, ξ . We have

$$\frac{d}{dt}[\xi_\alpha, \xi_\beta] = \frac{d}{dt}\gamma_{\alpha\beta} = 0 = [\dot{\xi}_\alpha, \xi_\beta] + [\xi_\alpha, \dot{\xi}_\beta]$$

$$= \frac{\partial \Psi_\alpha}{\partial \xi_\mu}\gamma_{\mu\lambda}\frac{\partial \xi_\beta}{\partial \xi_\lambda} + \frac{\partial \xi_\alpha}{\partial \xi_\mu}\gamma_{\mu\lambda}\frac{\partial \Psi_\beta}{\partial \xi_\lambda}$$

$$= \frac{\partial \Psi_\alpha}{\partial \xi_\mu}\gamma_{\mu\beta} + \gamma_{\alpha\lambda}\frac{\partial \Psi_\beta}{\partial \xi_\lambda}.$$

Multiply both sides by $\gamma_{\alpha\rho}\gamma_{\beta\sigma}$. This yields

$$\frac{\partial \chi_\sigma}{\partial \xi_\rho} - \frac{\partial \chi_\rho}{\partial \xi_\sigma} = 0,$$

where, as before, $\chi_\sigma = \Psi_\beta \gamma_{\beta\sigma} = \dot{\xi}_\beta \gamma_{\beta\sigma}$. Thus the integrability conditions are satisfied, which completes the proof.

Examples of motions which are generated by a Hamiltonian are found abundantly in nature. We have already given the example of the simple harmonic oscillator, for instance; others will be found in Section 6 and among the Problems. Consider, on the other hand, the motion whose equations are

$$\dot{q} = pq \qquad \text{and} \qquad \dot{p} = -pq.$$

If this were generated by a Hamiltonian H, then $\dot{q}$ would be $\partial H/\partial p = pq$, and $-\dot{p}$ would be $\partial H/\partial q = pq$. The second partials will then not agree, so there exists no Hamiltonian. These equations, incidentally, can be solved easily. Their solutions are

$$q(t) = q^0 \frac{Ce^{Ct}}{p^0 + q^0 e^{Ct}}, \qquad p(t) = p^0 \frac{C}{p^0 + q^0 e^{Ct}},$$

where $C = p^0 + q^0 = p + q$ is a constant of the motion. Let us show that the Poisson Bracket Theorem is correct in this example. We choose $R = q$, $S = p$. First, we have

$$\frac{d}{dt}[q, p] = \frac{d}{dt} 1 = 0.$$

Second, we have

$$[\dot{q}, p] + [q, \dot{p}] = [qp, p] - [q, qp] = [q, p]p - [q, p]q = p - q.$$

Thus $(d/dt)[q, p] \neq [\dot{q}, p] + [q, \dot{p}]$, in accordance with the theorem. We shall return to this example again in Section 6(e).

From the Poisson Bracket Theorem we obtain the following important corollary.

COROLLARY (Poisson's theorem). If R and S are constants of a Hamiltonian-generated motion, so is $[R, S]$. Indeed, (19) implies that if $\dot{R} = \dot{S} = 0$, then $d[R, S]/dt = 0$.

Remark. Poisson's theorem provides us with a way of generating constants of the motion by taking Poisson brackets of pairs of others. There is, however, no guarantee that a constant so generated will be independent of the original two. We have discussed independence of the constants of the motion in Chapter 3. There we extablished the condition that $2n$ constants $C_1, \ldots, C_{2n}$ are independent if and only if the Jacobian has the property

$$0 < \left| \frac{\partial(q, \dot{q})}{\partial C_\alpha} \right| < \infty,$$

or, equivalently, if and only if

$$0 < \left| \frac{\partial C_\alpha}{\partial(q, \dot{q})} \right| < \infty.$$

This is equivalent to the condition that $0 < |\det (\partial C_\alpha/\partial \xi_\beta)| < \infty$.

Let us go back to our example of the symmetric harmonic oscillator. It is easily shown that

$$R = q_1 \cos t - p_1 \sin t$$

is a constant of the motion. In fact, consider

$$[R, H] + \frac{\partial R}{\partial t} = \tfrac{1}{2}[q_1 \cos t - p_1 \sin t, q_1^2 + q_2^2 + p_1^2 + p_2]$$
$$+ \frac{\partial}{\partial t}(q_1 \cos t - p_1 \sin t).$$

Let us take the Poisson bracket term by term. Using Properties 1–3 and Eqs. (18), we have

$$\tfrac{1}{2}[q_1 \cos t, q_1^2 + q_2^2 + p_1^2 + p_2^2]$$
$$= \tfrac{1}{2} \cos t(2[q_1, q_1]q_1 + 2[q_1, q_2]q_2 + 2[q_1, p_1]p_1 + 2[q_1, p_2]p_2).$$

All terms but the third vanish, and we obtain p_1 cost for this term of the Poisson bracket. The rest is left to the reader: R is a constant of the motion. We have already seen that $l = q_1p_2 - q_2p_1$ is also a constant of the motion. By a calculation similar to the above we obtain the new constant of the motion

$$[l, R] = q_2 \cos t - p_2 \sin t.$$

VI-3 CANONICAL TRANSFORMATIONS

Now that we are treating mechanics as the study of motions on phase space rather than on configuration space, it is interesting to discuss something like generalized coordinates on phase space. Thus, let

$$\eta_\alpha = \eta_\alpha(\xi, t) \tag{20}$$

be a transformation (always invertible) from one set of coordinates ξ_α to another set η_α on phase space.

There are many advantages that can be gained by changing in this way to new variables. For instance, the differential equations for the motions may be simplified, especially if some of the η_α are constants of the motion. We would like, however, to preserve in general the very simple form of the canonical equations, and so we shall restrict the class of transformations to those that do just that.

To be explicit, we suppose that the motion, that is the $\dot{\xi}_\alpha$, is generated by a Hamiltonian H in accordance with Eq. (12). Then given a transformation, we can calculate the functions

$$\dot{\eta}_\alpha(\xi, t) = [\eta_\alpha, H] + \frac{\partial \eta_\alpha}{\partial t}$$

and can write these in terms of the η_α by using the inverse of (20). Thus we have

$$\dot{\eta}_\alpha = \varphi_\alpha(\eta, t).$$

We shall require that the transformation (20) have the property that there exists a function $K(\eta, t)$ such that

$$\varphi_\alpha(\eta, t) = \dot{\eta}_\alpha = \gamma_{\alpha\beta}\frac{\partial K}{\partial \eta_\beta}. \tag{21}$$

It is evident that this will not be true for all generalized coordinates on phase space, that is, for all transformations (20). When an invertible transformation has this property, we shall call it *canonoid* (i.e., not quite canonical) *with respect to H.*

It is inherent in this definition that a transformation which is canonoid with respect to one Hamiltonian need not be so with respect to another. Let us give an example in one degree of freedom (then phase space has two dimensions). We write q, p as usual for ξ_1 and ξ_2, and Q, P for η_1 and η_2. Consider the transformation

$$\begin{aligned} Q &= q, \\ P &= p^{1/2} - q^2, \end{aligned} \tag{22}$$

which inverts to give

$$\begin{aligned} p &= (P + Q^2)^2, \\ q &= Q. \end{aligned}$$

This transformation is canonoid with respect to the (unit mass) free-particle Hamiltonian $H = \frac{1}{2}p^2$. The equations of motion for this case are $\dot{p} = 0$ and $\dot{q} = p$, so that

$$\begin{aligned} \dot{Q} &= \dot{q} = p = (P + Q^2)^2, \\ \dot{P} &= \tfrac{1}{2}p^{-1/2}\dot{p} - 2q\dot{q} = -2Q(P + Q^2)^2. \end{aligned}$$

To prove that a "new Hamiltonian" $K(Q, P)$ exists, we merely exhibit it. It is left to the reader to verify that $K = \frac{1}{3}(P + Q^2)^3$ will do the trick.

On the other hand, the transformation of (22) is not canonoid with respect to the simple harmonic oscillator Hamiltonian $H = \frac{1}{2}p^2 + \frac{1}{2}q^2$. In this case the equations of motion are $\dot{q} = p$ and $\dot{p} = -q$, so that

$$\begin{aligned} \dot{Q} &= (P + Q^2)^2, \\ \dot{P} &= -\tfrac{1}{2}Q(P + Q^2)^{-1} - 2Q(P + Q^2)^2. \end{aligned}$$

That no K exists in this case can be seen by first writing $\dot{P} = -\partial K/\partial Q$ and using this to calculate $\partial^2 K/\partial P\,\partial Q$, and then writing $\dot{Q} = \partial K/\partial P$ and using this to calculate $\partial^2 K/\partial Q\,\partial P$. If these two expressions are unequal, no K exists. The reader will find that they are indeed unequal, differing by $\frac{1}{2}Q(P + Q^2)^{-2}$. [This example illustrates the coordinate dependence of the Poisson Bracket Theorem; see Section 6(e).]

There are, however, transformations which are canonoid with respect to all Hamiltonians H. One such example is

$$Q = p,$$
$$P = -q. \tag{23}$$

It is not difficult to show that (23) satisfies these rather grandiose claims, but we put that off until we have discussed the problem in more generality. The reader may check, however, that (23) is canonoid at least with respect to the free particle and the simple harmonic oscillator.

A transformation canonoid with respect to all Hamiltonians is called *canonical*.[3] We want to discuss such canonical transformations in some detail, and we start with the following theorem.

CANONICAL TRANSFORMATION THEOREM (Currie 1971). Let R and S be dynamical variables and let ζ_α be a set of general coordinates on phase space. We shall write

$$[R, S]^\zeta = \frac{\partial R}{\partial \zeta_\alpha} \gamma_{\alpha\beta} \frac{\partial S}{\partial \zeta_\beta}$$

(note the superscript ζ) for the Poisson bracket of R with S taken with respect to the ζ_α.

Let $\eta_\alpha = \eta_\alpha(\xi, t)$ be an invertible transformation on phase space (actually on a simply connected region of phase space; see the Remark in the proof of the Poisson Bracket Theorem). Then the following three statements are equivalent.

(A) η_α is canonical.

(B) There exists a nonzero[4] constant z such that for all dynamical variables R, S, we have

$$[R, S]^\eta = z[R, S]^\xi. \tag{24}$$

(C) η_α is canonoid with respect to all quadratic Hamiltonians of the form

$$H = C + c_\alpha \xi_\alpha + \tfrac{1}{2}\omega_{\alpha\beta}\xi_\alpha\xi_\beta, \tag{25}$$

where C, the c_α, and the $\omega_{\alpha\beta} = \omega_{\beta\alpha}$ are constants.

PROOF. That (A) implies (C) follows immediately from the definition of a canonical transformation.

Now we show that (B) implies (A). According to (B),

$$\frac{d}{dt}[R, S]^\eta = z \frac{d}{dt}[R, S]^\xi.$$

Assume that $\xi_\alpha(t)$ is a time development generated by some Hamiltonian $H(\xi, t)$

[3] There is some lack of uniformity about the terminology. Usually, however, what we call *restricted canonical* is what is called canonical in other texts. (See Section 4.)

[4] We specify $z \neq 0$ because otherwise, as is easily shown, the Jacobian of the transformation is zero and the transformation is not invertible.

Then we can apply the Poisson Bracket Theorem to the right-hand side, obtaining

$$\frac{d}{dt}[R, S]^\eta = z([\dot{R}, S]^\xi + [R, \dot{S}]^\xi).$$

Now we can apply (B) to the right-hand side of this equation, arriving at

$$\frac{d}{dt}[R, S]^\eta = [\dot{R}, S]^\eta + [R, \dot{S}]^\eta,$$

which according to the Poisson bracket theorem implies that $\eta_\alpha(t)$ is a time development generated by some Hamiltonain $K(\eta, t)$. Thus K exists for arbitrary H, and $\eta_\alpha(\xi, t)$ is a canonical transformation.

Finally we show that (C) implies (B), completing the proof that all three statements are equivalent. To begin, first consider the case $c_\alpha = 0$, $\omega_{\alpha\beta} = 0$ in Eq. (25), so that $H = C$. Then the equations of motion read $\dot{\xi}_\alpha = 0$. Since, according to (C), η_α is canonoid with respect to this H, the time development of $\eta_\alpha(t)$ is generated by a Hamiltonian, and the Poisson bracket theorem tells us that

$$\frac{d}{dt}[\xi_\alpha, \xi_\beta]^\eta = [\dot{\xi}_\alpha, \xi_\beta]^\eta + [\xi_\alpha, \dot{\xi}_\beta]^\eta = 0 \tag{26}$$

(equal to zero because $\dot{\xi}_\alpha = 0$).

Let us write the $[\xi_\alpha, \xi_\beta]^\eta$ as functions of ξ, t:

$$[\xi_\alpha, \xi_\beta]^\eta = \mu_{\alpha\beta}(\xi, t) = -\mu_{\beta\alpha}(\xi, t). \tag{27}$$

Then (26) can be written (recall that $H = C$)

$$\frac{d\mu_{\alpha\beta}}{dt} = [\mu_{\alpha\beta}, C]^\xi + \frac{\partial \mu_{\alpha\beta}}{\partial t} = 0,$$

or

$$\frac{\partial \mu_{\alpha\beta}}{\partial t} = 0. \tag{28}$$

Note that Eq. (28) does not depend on the form of H, for the $\mu_{\alpha\beta}(\xi, t)$ depend only on the functions $\eta_\alpha(\xi, t)$ and these are independent of the Hamiltonian.

Second, consider the case $C = 0$, $\omega_{\alpha\beta} = 0$ in Eq. (25), so that $H = c_\alpha \xi_\alpha$. Then the equations of motion are $\dot{\xi}_\alpha = \gamma_{\alpha\beta} c_\beta$. Since η_α is canonoid with respect to this H, we arrive again at (26) (equal to zero now because $[\gamma_{\alpha\nu} c_\nu, \xi_\beta]^\eta = 0$). Therefore

$$\frac{d\mu_{\alpha\beta}}{dt} = [\mu_{\alpha\beta}, c_\lambda \xi_\lambda]^\xi + \frac{\partial \mu_{\alpha\beta}}{\partial t} = 0,$$

and by calculating the Poisson bracket and using (28) we arrive at

$$\frac{\partial \mu_{\alpha\beta}}{\partial \xi_\lambda} \gamma_{\lambda\nu} c_\nu = 0.$$

This is to be true for arbitrary c_ν. Since Γ is a nonsingular matrix, this means that

$$\frac{\partial\mu_{\alpha\beta}}{\partial\xi_\lambda} = 0. \tag{29}$$

Consequently the $\mu_{\alpha\beta}$ are independent both of t and the ξ_λ, and thus are constants, forming an antisymmetric constant matrix M.

Third, consider the case $C = 0$, $c_{\alpha\beta} = 0$ in Eq. (25), so that $H = \frac{1}{2}\omega_{\alpha\beta}\xi_\alpha\xi_\beta$. The equations of motion are $\dot\xi_\alpha = \gamma_{\alpha\beta}\omega_{\beta\nu}\xi_\nu$. Since η_α is canonoid with respect to this H, the Poisson Bracket Theorem yields

$$\frac{d}{dt}[\xi_\alpha, \xi_\beta]^\eta = \gamma_{\alpha\rho}\omega_{\rho\nu}[\xi_\nu, \xi_\beta]^\eta + [\xi_\alpha, \xi_\nu]^\eta\gamma_{\beta\rho}\omega_{\rho\nu}.$$

This may be written in terms of the $\mu_{\alpha\beta}$. Since the $\mu_{\alpha\beta}$ are constants, the left-hand side is zero and we have

$$0 = \gamma_{\alpha\rho}\omega_{\rho\nu}\mu_{\nu\beta} + \mu_{\alpha\nu}\gamma_{\beta\rho}\omega_{\rho\nu},$$

or in matrix notation

$$\Gamma\Omega M = M\Omega\Gamma,$$

where Ω is the arbitrary symmetric matrix of the $\omega_{\alpha\beta}$. When we multiply on the right by Γ and on the left by $-\Gamma$, this becomes

$$\Omega M\Gamma = \Gamma M\Omega = (M\Gamma)^T\Omega. \tag{30}$$

This is to be true for arbitrary symmetric Ω. If we choose $\Omega = \mathbb{1}$, we obtain

$$M\Gamma = (M\Gamma)^T,$$

so that (30) becomes

$$\Omega(M\Gamma) = (M\Gamma)\Omega.$$

Thus $M\Gamma$ is a symmetric matrix that commutes with all symmetric matrices. It can then be shown (see Problem 11) that $M\Gamma$ is a constant multiple of the unit matrix. (The proof is similar to the proof of Problem 32, Chapter IV, that if a matrix K commutes with all rotation matrices, K is a multiple of $\mathbb{1}$.) We write $M\Gamma = -z\mathbb{1}$. Now multiplying on the right by $-\Gamma$, we arrive at $M = z\Gamma$, or

$$\mu_{\alpha\beta} = [\xi_\alpha, \xi_\beta]^\eta = z\gamma_{\alpha\beta}. \tag{31}$$

Finally let R, S be any dynamical variables. Then by Eq. (17)

$$[R, S]^\eta = \frac{\partial R}{\partial\xi_\mu}[\xi_\mu, \xi_\nu]^\eta\frac{\partial S}{\partial\xi_\nu} = z[R, S]^\xi.$$

This completes the proof.

Let us return to our previous example. At Eq. (23) we had claimed that $Q = p, P = -q$ is canonoid with respect to all Hamiltonians, or is canonical. In the ξ-notation, (23) may be written

$$\eta_\alpha = \gamma_{\alpha\beta}\xi_\beta \qquad \text{or} \qquad \xi_\alpha = -\gamma_{\alpha\beta}\eta_\beta, \tag{32}$$

where the η_α are Q and P, and the ξ_α are q and p. To prove that the transformation is canonical we need only check the Poisson bracket. From (32) we have

$$\begin{aligned}[\xi_\alpha, \xi_\beta]^\eta &= \frac{\partial \xi_\alpha}{\partial \eta_\mu} \gamma_{\mu\nu} \frac{\partial \xi_\beta}{\partial \eta_\nu} \\ &= \gamma_{\alpha\mu}\gamma_{\mu\nu}\gamma_{\beta\nu} \\ &= \gamma_{\alpha\mu}\delta_{\mu\beta} = \gamma_{\alpha\beta}.\end{aligned}$$

This satisfies the criterion with $z = 1$.

In short, we have established now that *a transformation is canonical* in accordance with our definition *if and only if it preserves Poisson brackets to within a constant factor* z.

It will be important later that the canonical transformations form a group. For our purposes this means essentially that the product of two canonical transformations is canonical and that the inverse of a canonical transformation is canonical. Let us prove these assertions.

Product. Let the transformations

$$\eta_\alpha = \eta_\alpha(\xi, t) \qquad \text{and} \qquad \zeta_\alpha = \zeta_\alpha(\eta, t)$$

each be canonical. Then the transformation from ξ to ζ, namely

$$\zeta_\alpha = \zeta_\alpha(\eta(\xi, t), t)$$

is also canonical. To prove this, we use (24) and (31):

$$[\xi_\alpha, \xi_\beta]^\zeta = z_2[\xi_\alpha, \xi_\beta]^\eta = z_1 z_2 \gamma_{\alpha\beta},$$

where z_1 is the constant associated with the transformation from ξ to η, and z_2 with the transformation from η to ζ. This proves the assertion.

Inverse. We wish to show that if the transformation $\eta_\alpha = \eta_\alpha(\xi, t)$ is canonical, so is its inverse $\xi_\alpha = \xi_\alpha(\eta, t)$. To do this we again use (24), now with $R = \eta_\alpha$, $S = \eta_\beta$:

$$[\eta_\alpha, \eta_\beta]^\eta = \gamma_{\alpha\beta} = z[\eta_\alpha, \eta_\beta]^\xi,$$

or

$$[\eta_\alpha, \eta_\beta]^\xi = z^{-1}\gamma_{\alpha\beta}.$$

This proves the assertion and shows that if a canonical transformation is associated with a constant z, its inverse is associated with $1/z$.

The group of canonical transformations will play an important part in the remainder of our discussion of mechanics. In the next sections we shall try to gain a clearer view of these transformations, how they relate to each other and to the structure of mechanics itself.

VI-4 RESTRICTED CANONICAL TRANSFORMATIONS AND GENERATING FUNCTIONS

The principal result of Section 3 is the Canonical Transformation Theorem: a transformation is canonical if and only if it preserves Poisson brackets up to a constant factor z. We would like now to attack the problem of classifying the canonical transformations, that is of specifying them in some minimal way. Recall, for instance, the group of rotations. Each rotation could be specified by giving three numbers, say the axis of rotation $\hat{\mathbf{n}}$ (the normalized eigenvector) and the angle of rotation θ (essentially the trace of the operator). Is there some similarly condensed way to specify a canonical transformation?

As a start, we can give z to specify the transformation at least partially. In fact, every canonical transformation can be compounded of two (see Problem 13), the first of which is of the form

$$\eta_\alpha = |z|^{-1/2}\omega_{\alpha\beta}\xi_\beta$$

(where $\omega_{\alpha\beta}$ is a certain constant matrix), and the second of which is a canonical transformation with $z = 1$. Once this is established, we are left only with the problem of classifying the ones with $z = 1$.

Let us agree to call a canonical transformation $\eta_\alpha = \eta_\alpha(\xi, t)$ *restricted* if $z = 1$. Thus a transformation is *restricted canonical* if and only if for every pair of dynamical variables R, S

$$[R, S]^\xi = [R, S]^\eta. \tag{33}$$

For example, the transformation

$$\eta_\alpha = z^{1/2}\xi_\alpha, \qquad 0 < z \neq 1,$$

is canonical, but not restricted canonical, as is easily verified. Henceforth we shall confine our discussion of canonical transformations to the restricted ones, but we shall ordinarily omit the word "restricted." It should be noted that in the literature the term "canonical" is often used to denote transformations which we call restricted canonical.

Restricted canonical transformations are easily shown to have Jacobian equal to ± 1. In fact let $\eta_\alpha(\xi, t)$ be restricted canonical. Then from (33) we have

$$[\eta_\alpha, \eta_\beta]^\xi = \frac{\partial \eta_\alpha}{\partial \xi_\mu}\gamma_{\mu\nu}\frac{\partial \eta_\beta}{\partial \xi_\nu} = \gamma_{\alpha\beta},$$

and by taking determinants we have

$$[\det(\partial\eta_\alpha/\partial\xi_\mu)]^2 \det\Gamma = \det\Gamma,$$

or

$$\det(\partial\eta_\alpha/\partial\xi_\mu) = \pm 1.$$

Consider then the transformation $\eta_\alpha = \eta_\alpha(\xi, t)$. To classify the (restricted) canonical transformations we will study the expression

$$\lambda_{\mu\nu}\dot{\xi}_\mu\xi_\nu - \lambda_{\mu\nu}\dot{\eta}_\mu\eta_\nu, \tag{34}$$

where the $\lambda_{\mu\nu}$ are the elements of the $2n \times 2n$ matrix

$$\Lambda = \begin{Vmatrix} 0_n & \mathbb{1}_n \\ 0_n & 0_n \end{Vmatrix}. \tag{35}$$

(Note that $\Lambda - \Lambda^T = \Gamma$ or $\lambda_{\mu\nu} - \lambda_{\nu\mu} = \gamma_{\mu\nu}$.) We will see that when (34) is written out as a function of the ξ_α and $\dot{\xi}_\alpha$, it tells us much about the transformation. Let us write $\dot{\eta}_\alpha = (\partial\eta_\alpha/\partial\xi_\mu)\dot{\xi}_\mu + \partial\eta_\alpha/\partial t$, and insert this into (34). We then have

$$\lambda_{\mu\nu}\dot{\xi}_\mu\xi_\nu - \lambda_{\mu\nu}\dot{\eta}_\mu\eta_\nu = \varphi_\mu\dot{\xi}_\mu + \psi, \tag{36}$$

where we have put

$$\varphi_\mu = \lambda_{\mu\nu}\xi_\nu - \lambda_{\alpha\beta}\frac{\partial\eta_\alpha}{\partial\xi_\mu}\eta_\beta, \tag{37}$$

$$\psi = -\lambda_{\alpha\beta}\frac{\partial\eta_\alpha}{\partial t}\eta_\beta. \tag{38}$$

By differentiating (37) we get

$$\frac{\partial\varphi_\mu}{\partial\xi_\nu} - \frac{\partial\varphi_\nu}{\partial\xi_\mu} = \gamma_{\mu\nu} - \{\xi_\mu, \xi_\nu\}^\eta, \tag{39}$$

where we have written

$$\{\xi_\mu, \xi_\nu\}^\eta = \frac{\partial\eta_\alpha}{\partial\xi_\mu}\gamma_{\alpha\beta}\frac{\partial\eta_\beta}{\partial\xi_\nu}$$

for the so-called *Lagrange bracket* of the ξ_μ with respect to the η_α. It is shown in Problem 14 that a transformation is canonical if and only if

$$\{\xi_\mu, \xi_\nu\}^\eta = \gamma_{\mu\nu}. \tag{40}$$

Thus (39) becomes

$$\frac{\partial\varphi_\mu}{\partial\xi_\nu} - \frac{\partial\varphi_\nu}{\partial\xi_\mu} = 0$$

so, in a way that we have often seen, there exists a function $F(\xi, t)$ such that

$$\varphi_\mu = \frac{\partial F}{\partial\xi_\mu}. \tag{41}$$

By suitable differentiation of (37) and (38) we similarly find that

$$\frac{\partial \varphi_\mu}{\partial t} - \frac{\partial \psi}{\partial \xi_\mu} = \frac{\partial \eta_\alpha}{\partial t} \gamma_{\alpha\beta} \frac{\partial \eta_\beta}{\partial \xi_\mu}. \tag{42}$$

The right-hand side can be simplified. Let us suppose that the mechanical system we are dealing with has ξ-Hamiltonian $H(\xi, t)$ and new η-Hamiltonian $K(\eta, t)$, that is, that

$$\dot{\xi}_\alpha = \gamma_{\alpha\lambda} \frac{\partial H}{\partial \xi_\lambda} \quad \text{and} \quad \dot{\eta}_\alpha = \gamma_{\alpha\lambda} \frac{\partial K}{\partial \eta_\lambda}. \tag{43}$$

On the other hand we also have

$$\begin{aligned} \dot{\eta}_\alpha = \frac{d}{dt} \eta_\alpha &= \frac{\partial \eta_\alpha}{\partial \xi_\rho} \dot{\xi}_\rho + \frac{\partial \eta_\alpha}{\partial t} \\ &= \frac{\partial \eta_\alpha}{\partial \xi_\rho} \gamma_{\rho\lambda} \frac{\partial H}{\partial \xi_\lambda} + \frac{\partial \eta_\alpha}{\partial t}, \end{aligned}$$

so that

$$\frac{\partial \eta_\alpha}{\partial t} = \gamma_{\alpha\lambda} \frac{\partial K}{\partial \eta_\lambda} - \frac{\partial \eta_\alpha}{\partial \xi_\rho} \gamma_{\rho\lambda} \frac{\partial H}{\partial \xi_\lambda}.$$

Thus the right-hand side of (42) is

$$\frac{\partial \eta_\alpha}{\partial t} \gamma_{\alpha\beta} \frac{\partial \eta_\beta}{\partial \xi_\mu} = - \frac{\partial \eta_\alpha}{\partial \xi_\rho} \gamma_{\alpha\beta} \frac{\partial \eta_\beta}{\partial \xi_\mu} \gamma_{\rho\lambda} \frac{\partial H}{\partial \xi_\lambda} + \gamma_{\alpha\lambda} \gamma_{\alpha\beta} \frac{\partial \eta_\beta}{\partial \xi_\mu} \frac{\partial K}{\partial \eta_\lambda} = \frac{\partial}{\partial \xi_\mu} (-H + K),$$

where we have used (40) and the fact that $\gamma_{\rho\mu}\gamma_{\rho\lambda} = \delta_{\mu\lambda}$. On inserting this into (42) and using (41) we obtain

$$\begin{aligned} 0 &= \frac{\partial}{\partial \xi_\mu} [\psi - H + K] - \frac{\partial \varphi_\mu}{\partial t} \\ &= \frac{\partial}{\partial \xi_\mu} \left[\psi - H + K - \frac{\partial F}{\partial t} \right]. \end{aligned}$$

This implies that the expression in brackets is a function of t alone. However, (41) determines F only up to an additive function $f(t)$, and we can therefore choose $f(t)$ so that

$$\psi - H + K = \frac{\partial F}{\partial t}. \tag{44}$$

On inserting (41) and (44) into (36), (37), and (38) we arrive at

$$\lambda_\mu \dot{\xi}_\mu \xi_\nu - H - (\lambda_{\mu\nu} \dot{\eta}_\mu \eta_\nu - K) = \frac{\partial F}{\partial \xi_\mu} \dot{\xi}_\mu + \frac{\partial F}{\partial t} \equiv \frac{dF}{dt}, \tag{45}$$

and

$$\lambda_{\mu\nu}\xi_\nu - \lambda_{\alpha\beta}\frac{\partial\eta_\alpha}{\partial\xi_\mu}\eta_\beta = \frac{\partial F}{\partial\xi_\mu}, \tag{46a}$$

$$-H + K - \lambda_{\alpha\beta}\frac{\partial\eta_\alpha}{\partial t}\eta_\beta = \frac{\partial F}{\partial t}. \tag{46b}$$

Thus if $\eta_\alpha(\xi, t)$ is canonical, there exists a function F satisfying Eq. (46*a*), and then H and K are related through F by (46*b*). Equivalently, there exists an F satisfying (45).

These two statements are equivalent because the $\dot{\xi}_\mu$ appear linearly on both sides of (45), and we can therefore equate their coefficients. This leads exactly to (46*a*) and (46*b*). The converse is trivial. But (45) must be treated cautiously, for it makes two separate assertions. The first, contained in (46*a*), is a set of differential equations for F. The second, contained in (46*b*), is a relationship between K and H. We shall return to this second assertion later.

Let us first look at (46*a*). What it tells us is that every canonical transformation determines an $F(\xi, t)$ uniquely up to an additive $f(t)$. The converse uniqueness does not hold, however, for two different canonical transformations may yield the same F. Or to put it differently, given F, Eqs. (46*a*) have several sets of solutions η_α. This means that to a certain extent, but not entirely, the $F(\xi, t)$ functions classify the canonical transformations. This function $F(\xi, t)$ will be called the *generating function* of the canonical transformation $\eta_\alpha(\xi, t)$. The partial classification it yields is then that to every (restricted) canonical transformation corresponds a generating function. But unfortunately, to every generating function corresponds, as we have pointed out, an entire class of canonical transformations.

Now let us turn to (46*b*). Suppose we know H and that the transformation $\eta_\alpha(\xi, t)$ is canonical. All we had proven previously is that then an η-Hamiltonian K exists. But now (46*b*) tells us actually how to calculate K in terms of H and η, provided that we know $F(\xi, t)$. Note that since F is determined only up to an additive $f(t)$, K is determined only up to its derivative $f'(t)$. But this is as it should be, for $f'(t)$ will not enter into the canonical equations and so will not affect the motion: in fact K can in principle be determined only up to an additive function of the time alone.

Section 6(f) is an example of the relation between a transformation and its generating function.

Remarks. *1.* In the (p, q)-notation Eqs. (45) and (46) become

$$\dot{q}_\mu p_\mu - H - (\dot{Q}_\mu P_\mu - K) = \frac{dF}{dt}, \tag{45'}$$

and

$$\left.\begin{aligned} p_\mu - \frac{\partial Q_\alpha}{\partial q_\mu} P_\alpha &= \frac{\partial F}{\partial q_\mu}, \\ -\frac{\partial Q_\alpha}{\partial p_\mu} P_\alpha &= \frac{\partial F}{\partial p_\mu}, \end{aligned}\right\} \tag{46'a}$$

$$-H + K - \frac{\partial Q_\alpha}{\partial t} P_\alpha = \frac{\partial F}{\partial t}. \tag{46'b}$$

In these equations, as usual, the indices run from 1 to n, not from 1 to $2n$.
2. If the $\eta_\alpha(\xi, t)$ are independent of t, that is if $\partial\eta_\alpha/\partial t = 0$, the generating function can also be chosen so that $\partial F/\partial t = 0$. With this choice Eq. (46*b*) becomes $K = H$. This means that the η-Hamiltonian is obtained from the ξ-Hamiltonian simply by using the inverse transformation $\xi_\alpha(\eta)$ to write the Hamiltonian in terms of η instead of ξ.

We now have a third way to characterize canonicity (restricted) of a transformation.[5] A transformation $\eta_\alpha(\xi, t)$ is canonical if and only if there exists an $F(\xi, t)$ such that (46*a*) is an identity.

In Section 6(d) it is shown that Hamilton's canonical equations for the ξ_α can be obtained from the variational principle in phase space by using the Lagrangian $L(\xi, \dot{\xi}, t) = \lambda_{\mu\nu}\dot{\xi}_\mu\xi_\nu - H$. The canonical equations for the η_α can of course similarly be obtained by using the Lagrangian $L'(\eta, \dot{\eta}, t) = \lambda_{\mu\nu}\dot{\eta}_\mu\eta_\nu - K$. The ξ-equations are

$$\left(\frac{d}{dt}\frac{\partial}{\partial\dot{\xi}_\alpha} - \frac{\partial}{\partial\xi_\alpha}\right)L(\xi, \dot{\xi}, t) = 0, \tag{47}$$

and the η-equations are

$$\left(\frac{d}{dt}\frac{\partial}{\partial\dot{\eta}_\alpha} - \frac{\partial}{\partial\eta_\alpha}\right)L'(\eta, \dot{\eta}, t) = 0. \tag{48}$$

What Eq. (45) tells us is that L and L' differ by the total time derivative of some function $F(\xi, t)$, which means (see the theorem toward the end of Section 2 of Chapter II, as well as Problem 8 in that chapter) that (47) and the equations

$$\left(\frac{d}{dt}\frac{\partial}{\partial\dot{\xi}_\alpha} - \frac{\partial}{\partial\xi_\alpha}\right)L'(\eta, \dot{\eta}, t) = 0 \tag{49}$$

are exactly the same, term for term. Note that it is not (47) and (48) that are the same, a fact which is sometimes misunderstood in the literature.

[5] The other two are (1) preservation of the Poisson brackets and (2) existence of an η-Hamiltonian $K(\eta, t)$ for every ξ-Hamiltonian $H(\xi, t)$. This last characterization, actually our definition, includes also the unrestricted canonicals.

That (47) and (49) are exactly the same is remarkable. It is certainly not self-evident from the fact that (47) and (48) are equations for the same set of motions on phase space; we know of this equality only because we have established the existence of $F(\xi, t)$. It is possible, however, to reverse the procedure, that is, to calculate (49) and use the fact that $\eta_\alpha(\xi, t)$ is a canonical transformation, to show that (47) and (49) are exactly the same (see Problem 15). In this way the existence of a function $F(\xi, t)$ satisfying (45) can be established. In the literature this alternate procedure is sometimes attempted.

Remark. We have seen that the inverse of a canonical transformation is also canonical. Let us write everything now as a function of η and $\dot{\eta}$ instead of ξ and $\dot{\xi}$. Then Eq. (45), which we write in the form

$$\lambda_{\mu\nu}\dot{\eta}_\mu\eta_\nu - K - (\lambda_{\mu\nu}\dot{\xi}_\mu\xi_\nu - H) = -\frac{dF}{dt},$$

shows us immediately that $-F$ is the generating function of the inverse transformation $\xi_\alpha(\eta, t)$. (Of course we must insert this inverse transformation into F to write it as a function of η rather than ξ.)

We have now taken the first step in the classification of the (restricted) canonicals: to every $F(\xi, t)$ corresponds a class of such transformations. To go further at this point we would have to classify within each such class. Transformations within each class, for which the generating function is zero (*Mathieu transformations*), are discussed in Section 6(g). Another simpler approach to classification is discussed in the next section and in Section 6(h).

VI-5 CANONICAL TRANSFORMATIONS OF TYPE

To complete the classification of the canonical transformations by assigning to each generating function $F(\xi, t)$ a standard canonical transformation in its class and then classifying the Mathieu transformations is a difficult procedure. Therefore we shall proceed in a different way.

A point in phase space can be specified by any set of $2n$ independent functions of the ξ_μ, for instance by the η_μ obtained from the ξ_μ by a canonical (or any other invertible) transformation. We will call such a set of $2n$ independent functions a *coordinate system on phase space*. Now it is possible that a given canonical transformation has the property that the original q_α and the new Q_α form a coordinate system on phase space. Because this is not a property common to all canonicals, it classifies them to some extent. We shall call those that have this property *canonical transformations of type 1*. For example, the transformation

$$Q_\alpha = -p_\alpha, \qquad P_\alpha = q_\alpha$$

is of type 1, whereas the identity transformation is not.

If the transformation is of type 1, every function $F(\xi, t)$ on phase space can be written as a function of the q_α and Q_α. Of course it is a different function of the q_α and Q_α than of the q_α and p_α (i.e., of the ξ_α), and we shall therefore write

$$F(\xi, t) = F^1(q, Q, t),$$

the superscript 1 indicating that this is the case of type 1. The way to obtain F^1 is to replace the p_α by their expressions in terms of q_α, Q_α,

$$p_\alpha = p_\alpha^1(q, Q, t),$$

and to substitute these in $F(\xi, t) = F(q, p, t)$:

$$F^1(q, Q, t) = F(q, p^1(q, Q, t), t).$$

Mathematically, this superscript 1 indicates the independent variables for differentiation. For instance,

$$\frac{\partial F^1}{\partial q_\beta} = \frac{\partial F}{\partial q_\beta} + \frac{\partial F}{\partial p_\mu}\frac{\partial p_\mu^1}{\partial q_\beta}. \tag{50}$$

If we apply this to $Q_\alpha^1 \equiv Q_\alpha$, we obtain a useful identity:

$$\frac{\partial Q_\alpha^1}{\partial q_\beta} = 0 = \frac{\partial Q_\alpha}{\partial q_\beta} + \frac{\partial Q_\alpha}{\partial p_\mu}\frac{\partial p_\mu^1}{\partial q_\beta},$$

$$\frac{\partial Q_\alpha}{\partial q_\beta} = -\frac{\partial Q_\alpha}{\partial p_\mu}\frac{\partial p_\mu^1}{\partial q_\beta}. \tag{51}$$

Now let F be the generating function of a transformation of type 1. Then we may use (50) and (51) to obtain an expression for the transformation in terms of F^1: first multiply the second of Eqs. (46′*a*) by $\partial p_\mu^1/\partial q_\beta$ and use (51) to get

$$\frac{\partial Q_\alpha}{\partial q_\beta}P_\alpha = \frac{\partial F}{\partial p_\mu}\frac{\partial p_\mu^1}{\partial q_\beta},$$

and then insert this into (50):

$$\frac{\partial F^1}{\partial q_\beta} = \frac{\partial F}{\partial q_\beta} + \frac{\partial Q_\alpha}{\partial q_\beta}P_\alpha.$$

Finally, from the first of (46′*a*) it follows that[6]

$$p_\beta = \frac{\partial F^1}{\partial q_\beta}. \tag{52}$$

[6] Actually we should write p_β^1 on the left-hand side of this equation, for it gives p_β as a function of q_α, Q_α, t. But we are using the superscripts mostly as mnemonics for differentiation, and they may be dropped whenever the meaning is clear. This applies also to P_β in Eq. (53).

Similarly, the second of (46′a) gives

$$\frac{\partial F^1}{\partial Q_\beta} = \frac{\partial F}{\partial p_\mu}\frac{\partial p^1_\mu}{\partial Q_\beta} = -\frac{\partial Q_\alpha}{\partial p_\mu} P_\alpha \frac{\partial p^1_\mu}{\partial Q_\beta},$$

and with

$$\frac{\partial Q^1_\alpha}{\partial Q_\beta} = \delta_{\alpha\beta} = \frac{\partial Q_\alpha}{\partial p_\mu}\frac{\partial p^1_\mu}{\partial Q_\beta},$$

this becomes

$$P_\beta = -\frac{\partial F^1}{\partial Q_\beta}. \tag{53}$$

Remark. This calculation can be performed in a simpler way: Write out the right-hand side of (45′) in terms of q, Q, t:

$$\dot{q}_\mu p_\mu - H - \dot{Q}_\mu P_\mu + K = \frac{\partial F^1}{\partial q_\mu}\dot{q}_\mu + \frac{\partial F^1}{\partial Q_\mu}\dot{Q}_\mu + \frac{\partial F^1}{\partial t}. \tag{54}$$

Now equate coefficients of $\dot{q}_\mu$ and $\dot{Q}_\mu$ on both sides. This gives (52) and (53).

Summarizing, if we know that the variables q_α and Q_α form a coordinate system on phase space, and if we know the generating function F, but now as a function of these two particular variables, then (52) and (53) give us the canonical transformation. All we need do is solve (52) algebraically for the Q_β as functions of q, p, t and insert these solutions into (53). This classifies the canonicals of type 1. To each such transformation corresponds its own particular $F^1(q, Q, t)$. No other canonical corresponds to the same F^1. Indeed, given any differentiable $F^1(q, Q, t)$, Eqs. (52) and (53) determine the transformation uniquely. And given the transformation, Eqs. (52) and (53) are differential equations for F^1 that determine it uniquely (actually, as usual, only up to an additive function of the time alone).

This, in a sense, is the first part of the reward you get at the end of this chapter: a simple way to produce canonical transformations. For every function F^1 of $2n + 1$ variables there is one canonical of type 1. Note that to specify $F^1(q, Q, t)$ is not the same as to specify the generating function $F(\xi, t)$. It would not be enough to specify the latter, for as we have seen $F(\xi, t)$ does not uniquely determine the transformation. Further, not all functions $F^1(q, Q, t)$ are acceptable. For instance, it must be possible to solve (52) for the Q_α, so that $\det(\partial^2 F^1/\partial q_\alpha \partial Q_\beta)$ must be finite. If the transformation is to be differentiable, F^1 must be twice differentiable. We shall say that such an F^1 is a *generating function of type 1*.

The relation between K and H can also be written in terms of F^1, either by using (54) (the simplest way) or by going through (46′b). In either way one arrives at

$$K = H + \frac{\partial F^1}{\partial t}. \tag{55}$$

Of course not all canonicals are of type 1. For instance $Q = q, P = p + q$ is canonical, but q and Q are trivially and obviously not independent. However, in this case q and P are independent, as are p and Q or p and P. If the q_α and P_β are independent, we shall say the transformation is of type 2. If the p_α and Q_β are independent, it is of type 3 and if the p_α and P_β are independent, it is of type 4. The example we have just given is simultaneously of types 2, 3, and 4.

In the case of each of these types a *generating function of* the required type can be defined. Consider, for instance, type 2, and let us use (45′) as we did in the Remark after Eq. (53). We write both sides of (45′) as functions of q, P, t (now using the superscript 2 as we did the superscript 1):

$$\dot{q}_\mu p_\mu - H - \frac{\partial Q_\alpha^2}{\partial q_\mu} P_\alpha \dot{q}_\mu - \frac{\partial Q_\alpha^2}{\partial P_\mu} P_\alpha \dot{P}_\mu - \frac{\partial Q_\alpha^2}{\partial t} P_\alpha + K$$
$$= \frac{\partial F'^2}{\partial q_\mu} \dot{q}_\mu + \frac{\partial F'^2}{\partial P_\mu} \dot{P}_\mu + \frac{\partial F'^2}{\partial t}.$$

[The prime on $F'^2(q, P, t) \equiv F(q, p^2(q, P, t), t)$ will soon be explained.] The coefficients of $\dot{q}_\mu$ give

$$p_\mu = \frac{\partial Q_\alpha^2}{\partial q_\mu} P_\alpha + \frac{\partial F'^2}{\partial q_\mu} = \frac{\partial}{\partial q_\mu}(F'^2 + Q_\alpha^2 P_\alpha)$$

(remember that q, P are the independent variables, or that $P_\alpha^2 \equiv P_\alpha$). The coefficients of $\dot{P}_\mu$ give

$$0 = \frac{\partial Q_\alpha^2}{\partial P_\mu} P_\alpha + \frac{\partial F'^2}{\partial P_\mu} = \frac{\partial}{\partial P_\mu}(F'^2 + Q_\alpha^2 P_\alpha) - Q_\mu^2.$$

The remainder is

$$K = H + \frac{\partial Q_\alpha^2}{\partial t} P_\alpha + \frac{\partial F'^2}{\partial t}$$
$$= H + \frac{\partial}{\partial t}(F'^2 + Q_\alpha^2 P_\alpha).$$

Let us write (here is where that prime comes in)

$$F^2 = F'^2 + Q_\alpha^2 P_\alpha, \tag{56}$$

and let us call F^2 the *generating function of type 2*. Then, dropping the superscript on Q_μ^2 (see footnote 6), we have

$$\begin{aligned} p_\mu &= \frac{\partial F^2}{\partial q_\mu}, \\ Q_\mu &= \frac{\partial F^2}{\partial P_\mu}, \\ K &= H + \frac{\partial F^2}{\partial t}. \end{aligned} \tag{57}$$

Note that $F^2(q, P, t)$ is not simply $F(\xi, t)$ with the p_α written as functions of the q_α, P_α. That gives F'^2, and then F^2 is obtained from (56).

Similar equations for the other two types of transformations will be obtained in the Problems. When all of these four cases are codified, we have both a classification and a procedure by which to produce canonical transformations of all four types. It can be shown that this takes care of all canonical transformations [see Section 6(h)], so that it is, in fact, the reward we spoke of earlier. To each function of $2n + 1$ variables there corresponds a set of four transformations, one of each type. Since the types overlap, however, this is a redundant classification, but this will be discussed in Section 6(h).

The Hamiltonian formulation and canonical-transformation theory will be used in the next chapter to discuss such things as conservation laws, symmetry, and the motion of a system itself.

VI-6 EXAMPLES, APPLICATIONS, AND EXTENSIONS

(a) Two Special Hamiltonians

Two Hamiltonians are of special interest: that for a particle in a central field and that for a particle in an electromagnetic field. We shall calculate both.

First the central force problem. Let the force be given by the potential $V(r)$, so that in spherical polar coordinates the Lagrangian is

$$L = \tfrac{1}{2}m(\dot{r}^2 + r^2\dot{\theta}^2 + r^2\dot{\varphi}^2 \sin^2 \theta) - V(r).$$

Then we have

$$\begin{aligned} p_r &= \frac{\partial L}{\partial \dot{r}} = m\dot{r}, \\ p_\theta &= \frac{\partial L}{\partial \dot{\theta}} = mr^2\dot{\theta}, \\ p_\varphi &= \frac{\partial L}{\partial \dot{\varphi}} = mr^2\dot{\varphi} \sin^2 \theta. \end{aligned} \tag{58}$$

The Hamiltonian may be calculated by finding $H = p_\alpha \dot{q}_\alpha - L$, but it is easier to recall that H is the total energy. Since $L = T - V$, we have

$$\begin{aligned} H &= T + V = L + 2V \\ &= \frac{p_r^2}{2m} + \frac{p_\theta^2}{2mr^2} + \frac{p_\varphi^2}{2mr^2 \sin^2 \theta} + V(r). \end{aligned} \tag{59}$$

Since φ does not appear in the Hamiltonian, we obtain

$$\dot{p}_\varphi = -\frac{\partial H}{\partial \varphi} = 0,$$

or p_φ is a constant of the motion. The other canonical equations are easy to obtain. Three of them are just (58) solved for $\dot{r}$, $\dot{\theta}$, $\dot{\varphi}$, two of them are

$$\dot{p}_r = -\frac{\partial H}{\partial r} = \frac{p_\theta^2}{mr^3} + \frac{p_\varphi^2}{mr^3 \sin^2 \theta} - V'(r),$$

$$\dot{p}_\theta = -\frac{\partial H}{\partial \theta} = \frac{p_\varphi^2 \cos \theta}{mr^2 \sin^3 \theta},$$

and the last, as mentioned, is $\dot{p}_\varphi = 0$.

Now the electromagnetic field. According to Eq. (58) of Chapter II, the Lagrangian is

$$L = \tfrac{1}{2} m\dot{x}^2 - e\Phi + \frac{e}{c}\dot{\mathbf{x}} \cdot \mathbf{A}.$$

Then we have

$$p_\alpha = m\dot{x}_\alpha + \frac{e}{c} A_\alpha, \tag{60}$$

which is easily solved for $\dot{x}_\alpha$. When this is inserted into $H = p_\alpha \dot{q}_\alpha - L$, we obtain

$$H = \frac{1}{2m}\left(p_\alpha - \frac{e}{c} A_\alpha\right)\left(p_\alpha - \frac{e}{c} A_\alpha\right) + e\Phi. \tag{61}$$

This Hamiltonian looks just like the Hamiltonian of a particle in the potential field $e\Phi$, except that the momentum $\mathbf{p}$ is replaced by $\mathbf{p} - (e/c)\mathbf{A}$. Note that the *canonical* momentum defined by Eq. (60) is not the *dynamical* momentum $m\dot{\mathbf{x}}$.

It is left to the reader to derive the canonical equations.

(b) Small Vibrations in the Hamiltonian Formulation

In Problem 7 you are asked to use the Hamiltonian formalism to find the normal modes and frequencies for a system of small vibrations. Here we want to solve the equations in a different way, ignoring normal modes.

Suppose we are given the Hamiltonian in the form

$$H = \tfrac{1}{2}\omega_{\alpha\beta}\xi_\alpha\xi_\beta,$$

where the $\omega_{\alpha\beta}$ form the $2n$-dimensional matrix

$$\Omega = \begin{vmatrix} K & 0_n \\ 0_n & M^{-1} \end{vmatrix}$$

where K and M^{-1} are the n-dimensional symmetric matrices of the $\kappa_{\alpha\beta}$ and $\sigma_{\alpha\beta}$, respectively, of Problem 7. We shall assume in this treatment that not only M^{-1} but also K is a positive definite matrix.

The canonical equations are

$$\dot{\xi}_\alpha = \gamma_{\alpha\mu}\omega_{\mu\nu}\xi_\nu,$$

or

$$\dot{x} = \Gamma\Omega x, \tag{62}$$

where x is the vector with components ξ_μ, and Ω is the matrix whose elements are $\omega_{\alpha\beta}$.

Now a formal solution of (62) is possible immediately:

$$x = e^{\Gamma\Omega t}x_0, \tag{63}$$

where we define the exponential by

$$e^{\Gamma\Omega t} = \sum_{j=0}^{\infty}\frac{(\Gamma\Omega t)^j}{j!}. \tag{64}$$

The solution of Eq. (63) is more than formal only if we can show that the series converges to an actual matrix. This means that each element of the matrix must converge to some limit, or specifically that

$$\lim_{N\to\infty}\left[\sum_{j=0}^{N}\frac{(\Gamma\Omega t)^j}{j!}\right]_{\alpha\beta}$$

exists. To prove this existence, we exhibit the form of the matrix and prove that in one basis the limit exists. It then follows that it exists in all bases.

Let us define a new $2n$-dimensional matrix R by writing

$$RR \equiv R^2 = \begin{vmatrix} M^{-1}K & 0_n \\ 0_n & KM^{-1} \end{vmatrix}.$$

It can be shown by techniques used in Chapter IV, Section 5(b) that $M^{-1}K$ and KM^{-1} are diagonalizable and positive definite. But for any diagonalizable positive definite matrix A we can define the square-root matrix $A^{1/2}$ in terms of the (positive) eigenvalues of A: each eigenvector of $A^{1/2}$ is also an eigenvector of A, but the corresponding eigenvalue of $A^{1/2}$ is the positive square root of the eigenvalue of A. Clearly A and $A^{1/2}$ are simultaneously diagonalizable, by construction. Then square-root matrices can be found for $M^{-1}K$ and KM^{-1}, and thus R exists.

Then direct calculation according to (64) yields

$$\begin{aligned} e^{\Gamma\Omega t} &= \mathbb{1} - \frac{R^2t^2}{2!} + \frac{R^4t^4}{4!} - \frac{R^6t^6}{6!} + \cdots \\ &\quad + \Gamma \begin{vmatrix} M & 0_n \\ 0_n & K^{-1} \end{vmatrix} R\left(Rt - \frac{R^3t^3}{3!} + \cdots\right) \\ &\equiv \cos Rt + \Gamma \begin{vmatrix} M & 0_n \\ 0_n & K^{-1} \end{vmatrix} R \sin Rt, \end{aligned} \tag{65}$$

where we have defined the sine and cosine of a matrix by power series. (We used the extra assumption that K is positive definite when we assumed that K^{-1} exists.) Now we need to show that the power series expressions for $\cos Rt$ and $\sin Rt$ converge. But this becomes obvious when we consider the basis in which R is diagonal, for then the power series define diagonal matrices whose elements are just the sines and cosines of the (diagonal) elements of Rt. Thus the series do in fact converge.

On inserting (65) into (63) we arrive at

$$x(t) = \left(\cos Rt + \Gamma \begin{vmatrix} M & 0_n \\ 0_n & K^{-1} \end{vmatrix} R \sin Rt\right) x^0. \tag{66}$$

The vector x^0 contains the initial conditions: its elements are the initial q_α and p_α. As indicated by the sine and cosine functions, the solution $x(t)$ has oscillatory properties. This form of the solution does not, however, explicitly exhibit the normal modes and frequencies; to exhibit them you would have to diagonalize R by a canonical transformation (i.e., leaving Γ invariant). We will not go through this procedure, but see Problem 25.

An interesting canonical transformation, useful in quantum mechanics, can be performed for the small vibrations problem. We illustrate it on the one-dimensional case, or the one-dimensional simple harmonic oscillator. As we have done once before, let $m = k = 1$, so that

$$H = \tfrac{1}{2}(q^2 + p^2).$$

Now perform the *complex* canonical transformation

$$\begin{aligned} Q &= \frac{1}{\sqrt{2}}(q + ip), \\ P &= \frac{i}{\sqrt{2}}(q - ip) \end{aligned} \tag{67}$$

(the reader may check that this is indeed canonical). The Hamiltonian now becomes (since the transformation is time independent, $K = H$)

$$K = H = -iQP,$$

and the canonical equations become

$$\dot{Q} = -iQ, \qquad \dot{P} = iP. \tag{68}$$

Notice that Q and P are now uncoupled. On the other hand, $P = iQ^*$ by definition, so that the second canonical equation follows from the first by

$$\dot{P} = i\dot{Q}^* = i(iQ^*) = iP.$$

The solution of these equations is left to the reader.

Equation (67) provides an interesting example of the richness introduced by the Hamiltonian formalism, for this canonical transformation altered the Hamiltonian to the point that it cannot be obtained from a Lagrangian. In fact, if one tries in a simple-minded way to form the Lagrangian $L(Q, \dot{Q})$ corresponding to the Hamiltonian of Eq. (68), one obtains $L = 0$ to the extent that one obtains anything at all. (See Problem 27. Compare Problem 6.)

On the other hand, this Hamiltonian can be obtained from the variational principle in phase space, as in Section 6(d). Then the "Lagrangian" function to be varied is

$$L(Q, P, \dot{Q}, \dot{P}) = P\dot{Q} - H = P\dot{Q} + iPQ.$$

Note that L is indeed zero when the equations of motion are inserted.

In quantum mechanics one sometimes treats this problem by starting with this Lagrangian. Moreover, one ordinarily ignores the fact that the variation is in phase space, treating P and Q as configuration-space variables, called $a^\dagger$ and a, respectively. Then the Lagrangian is

$$L = a^\dagger \dot{a} + ia^\dagger a, \tag{69}$$

and the Euler-Lagrange equations are

$$\dot{a}^\dagger = ia^\dagger \qquad \text{and} \qquad \dot{a} = -ia.$$

Now in fact it is impossible to construct the Hamiltonian from this configuration-space Lagrangian, for

$$p = \frac{\partial L}{\partial \dot{a}} = a^\dagger \qquad \text{and} \qquad p^\dagger = \frac{\partial L}{\partial \dot{a}^\dagger} = 0$$

are not equations that can be inverted to give $\dot{a}$ and $\dot{a}^\dagger$ in terms of p and $p^\dagger$. In spite of this if one proceeds blindly, ignoring this impossibility, one obtains

$$H = \frac{\partial L}{\partial \dot{a}}\,\dot{a} + \frac{\partial L}{\partial \dot{a}^\dagger}\,\dot{a}^\dagger - L = -ia^\dagger a. \tag{70}$$

Then two of the canonical equations are

$$\dot{p} = ia^\dagger \qquad \text{and} \qquad \dot{p}^\dagger = ia.$$

Comparison of this and the Euler-Lagrange equations shows that up to initial conditions

$$p = a^\dagger, \qquad p^\dagger = -a.$$

Thus the Hamiltonian may be written

$$H = -ipa, \tag{71}$$

and we are essentially back to Eq. (68). Incidentally, since for Eq. (68) we had $P = iQ^*$ or $p = ia^*$, we find that $a^\dagger = ia^*$.

(c) Poisson Brackets of the Angular Momentum

For a single particle the angular momentum $\mathbf{l}$ is defined by

$$l_\alpha = \epsilon_{\alpha\beta\gamma} x_\beta p_\gamma.$$

The Poisson brackets of the angular momentum with various types of dynamical variables have interesting properties.

First consider scalar functions of the p_α and x_α. These can depend only on the magnitudes of $\mathbf{x}$ and $\mathbf{p}$ and on the angle between them, or on x^2, p^2, and $\mathbf{p} \cdot \mathbf{x}$. Thus if we want to know the Poisson bracket of some scalar function f with l_α, we need know only the Poisson brackets of x^2, p^2, and $\mathbf{p} \cdot \mathbf{x}$ with l_α. These are easily calculated:

$$\begin{aligned}
[x^2, l_\alpha] &= \epsilon_{\alpha\beta\gamma}[x_\mu x_\mu, x_\beta p_\gamma] = 2\epsilon_{\alpha\beta\gamma} x_\mu x_\beta [x_\mu, p_\gamma] \\
&= 2\epsilon_{\alpha\beta\gamma} x_\mu x_\beta \delta_{\mu\gamma} = 2\epsilon_{\alpha\beta\gamma} x_\beta x_\gamma = 0; \\
[p^2, l_\alpha] &= 2\epsilon_{\alpha\beta\gamma} p_\mu p_\gamma [p_\mu, x_\beta] = 0; \\
[\mathbf{x} \cdot \mathbf{p}, l_\alpha] &= \epsilon_{\alpha\beta\gamma}[x_\mu p_\mu, x_\beta p_\gamma] = \epsilon_{\alpha\beta\gamma}\{x_\mu p_\gamma [p_\mu, x_\beta] + p_\mu x_\beta [x_\mu, p_\gamma]\} \\
&= \epsilon_{\alpha\beta\gamma}(-x_\beta p_\gamma + p_\gamma x_\beta) = 0.
\end{aligned}$$

Thus every scalar dynamical variable *commutes* (that is, has zero Poisson bracket) with the angular momentum, or if $f = f(x^2, p^2, \mathbf{x} \cdot \mathbf{p})$, then

$$[f, \mathbf{l}] = 0. \tag{72}$$

Second, consider vector dynamical variables. The only vector other than $\mathbf{x}$ and $\mathbf{p}$ themselves that can be formed from $\mathbf{x}$ and $\mathbf{p}$ is $\mathbf{x} \times \mathbf{p}$. Therefore every vector dynamical variable must be of the form $\mathbf{f} = f_1\mathbf{x} + f_2\mathbf{p} + f_3\mathbf{x} \times \mathbf{p}$, where f_1, f_2, and f_3 are scalar dynamical variables. Since we already know that $\mathbf{l}$ commutes with scalars, we need know in addition only its Poisson brackets with $\mathbf{x}$, $\mathbf{p}$, and $\mathbf{x} \times \mathbf{p} = \mathbf{l}$. These are easily calculated:

$$[x_\mu, l_\alpha] = \epsilon_{\alpha\beta\gamma} x_\beta [x_\mu, p_\gamma] = \epsilon_{\mu\alpha\beta} x_\beta;$$

$$[p_\mu, l_\alpha] = \epsilon_{\alpha\beta\gamma} p_\gamma [p_\mu, x_\beta] = \epsilon_{\mu\alpha\gamma} p_\gamma;$$

$$\begin{aligned}
[l_\mu, l_\alpha] &= \epsilon_{\alpha\beta\gamma}\epsilon_{\mu\nu\lambda}(x_\nu p_\gamma [p_\lambda, x_\beta] + p_\lambda x_\beta [x_\nu, p_\gamma]) \\
&= -\epsilon_{\alpha\beta\gamma}\epsilon_{\mu\nu\beta} x_\nu p_\gamma + \epsilon_{\alpha\beta\gamma}\epsilon_{\mu\gamma\lambda} p_\lambda x_\beta \\
&= -(\delta_{\alpha\nu}\delta_{\mu\gamma} - \delta_{\alpha\mu}\delta_{\nu\gamma}) x_\nu p_\gamma + (\delta_{\alpha\lambda}\delta_{\beta\mu} - \delta_{\alpha\mu}\delta_{\beta\lambda}) p_\lambda x_\beta \\
&= x_\mu p_\alpha - x_\alpha p_\mu = \epsilon_{\mu\alpha\nu} l_\nu.
\end{aligned}$$

Combining these results, we have

$$[f_\mu, l_\alpha] = \epsilon_{\mu\alpha\beta} f_\beta \tag{73}$$

for any vector dynamical variable $\mathbf{f} = \mathbf{e}_\mu f_\mu$, where $\mathbf{e}_\mu$ is the unit vector in the μ direction. When multiplied by $\mathbf{e}_\mu$ and summed over μ, Eq. (73) becomes

$$[\mathbf{f}, l_\alpha] = \mathbf{e}_\alpha \times \mathbf{f}. \tag{74}$$

If, in particular, we take $\mathbf{f}$ to be $\mathbf{l}$ itself, we obtain

$$[l_1, l_2] = l_3, \qquad [l_2, l_3] = l_1, \qquad [l_3, l_1] = l_2. \tag{75}$$

This is often taken to mean that the components of $\mathbf{l}$ cannot be chosen as the momenta of a set of canonical variables. Actually what it means is that the transformation from the Cartesian canonical variables x_α, p_α to the θ_α, l_α (where the θ_α are any functions of x_α, p_α) is not a canonical transformation, for the brackets in (75) are with respect to the Cartesian variables. There may, however, exist noncanonical transformations, which do not preserve the Poisson brackets, such that the components of $\mathbf{l}$ become the canonical momenta.

(d) Variational Principle

Hamilton's canonical equations can be obtained from a variational principle, as are the Euler-Lagrange equations. Now, however, the variation is over paths in phase space, which has $2n$ dimensions, twice the n dimensions of configuration space. The function involved is, interestingly enough, again the Lagrangian L, but now considered a function of the ξ_α and $\dot{\xi}_\alpha$:

$$L(\xi, \dot{\xi}, t) = p_\alpha \dot{q}_\alpha - H = \lambda_{\alpha\beta}\dot{\xi}_\alpha \xi_\beta - H, \tag{76}$$

where we are using the matrix Λ defined by Eq. (35). We wish to obtain the extremum of the integral

$$\int_{t_0}^{t_1} L(\xi, \dot{\xi}, t)\, dt$$

taken along paths $\xi_\alpha(t, \varepsilon)$ starting and ending at the same points:

$$\xi_\alpha(t_0, \varepsilon) = \xi_\alpha(t_0, 0),$$
$$\xi_\alpha(t_1, \varepsilon) = \xi_\alpha(t_1, 0).$$

Note that now we require the ξ_α, that is the q's and the p's both to be fixed at the end points. In the configuration-space variational principle we required only the q's to be fixed at the end points. This emphasizes again that the present variational principle is different from the earlier one in configuration space. As in the configuration-space case, we choose $\varepsilon = 0$ to be the physical path, and then precisely as in that case, we arrive at

$$\frac{d}{dt}\frac{\partial L}{\partial \dot{\xi}_\rho} - \frac{\partial L}{\partial \xi_\rho} = 0.$$

With L given by (76), this becomes

$$\frac{d}{dt}(\lambda_{\rho\beta}\xi_\beta) - \lambda_{\alpha\rho}\dot{\xi}_\alpha + \frac{\partial H}{\partial \xi_\rho} = 0,$$

or

$$(\lambda_{\rho\alpha} - \lambda_{\alpha\rho})\dot{\xi}_\alpha = \gamma_{\rho\alpha}\dot{\xi}_\alpha = -\frac{\partial H}{\partial \xi_\rho}.$$

On multiplying both sides by $\gamma_{\rho\sigma} = -\gamma_{\sigma\rho}$ and using the orthogonality of Γ, we arrive at Hamilton's canonical equations

$$\dot{\xi}_\sigma = \gamma_{\sigma\rho}\frac{\partial H}{\partial \xi_\rho}.$$

(e) Noncanonical Transformations

Consider a time development $\xi_\alpha(t)$ which is generated by no Hamiltonian $H(\xi, t)$, that is, such that Eq. (19), the Poisson Bracket Theorem, does not hold for every R and S. When we proved the Poisson Bracket Theorem we were not considering the possibility of transforming to new coordinates on phase space, and we now ask the following question. Is it possible to transform from the ξ_α to new coordinates $\eta_\alpha(\xi, t)$ for which a Hamiltonian $K(\eta, t)$ exists, that is, such that

$$\dot{\eta}_\alpha = \gamma_{\alpha\beta}\frac{\partial K}{\partial \eta_\beta}?$$

If such a transformation exists, its inverse will, of course, not be canonoid with respect to K (for no H exists).

Sometimes it is indeed possible to save the situation, so to speak, by such a transformation, and we shall give an instance. Recall the example of Section 1 in which we showed that $\dot{q} = pq$ and $\dot{p} = -qp$ is generated by no

Hamiltonian. But consider the new coordinates

$$Q = \ln (q/p), \qquad P = q + p.$$

Then since $q + p$ is, as we have seen, a constant of the motion, we obtain $\dot{P} = 0$; a simple calculation yields $\dot{Q} = P$. A new Hamiltonian is obvious:

$$K(Q, P) = \tfrac{1}{2}P^2.$$

By definition, however, the transformation from Q, P to q, p is not canonoid with respect to K, and hence not canonical. The reader may verify that $[Q, P]^{q,p} \neq 1$.

(f) A Generating Function and Its Transformation

We now give a detailed example of the relation between a generating function and its transformation. Consider the canonical transformation in one degree of freedom:

$$\begin{aligned} Q &= p, \\ P &= \frac{1}{m}pt - q. \end{aligned} \tag{77}$$

(Verify the canonicity by checking the only Poisson bracket that does not vanish by antisymmetry.)

First, calculating F. Equations (46′a) are

$$p = \frac{\partial F}{\partial q}, \qquad -\frac{1}{m}pt + q = \frac{\partial F}{\partial p},$$

and their solution is

$$F = pq - \frac{p^2}{2m}t + f(t),$$

where f is an arbitrary function.

Second, the connection between H and K. Equation (46′b) becomes

$$K = H - \frac{p^2}{2m} + f'(t), \tag{78}$$

where $f' = df/dt$. Given H we can then find K from (78), and can put it in terms of Q and P by using (77). Note that $f'(t)$, and hence also $f(t)$, are irrelevant in K, for they do not enter the canonical equations. Therefore we shall set $f = 0$.

Third, special forms of H. Let H be of the form $H = T + V(q)$, where $T = p^2/2m$ is the kinetic energy. Then with $f = 0$, Eq. (78) becomes $K = V$, which shows that the canonical transformation of (77) "eliminates" the kinetic energy from the Hamiltonian. This is not in general useful, for

$V(q) = V(Qt/m - P)$ is not simply a function of Q, but for one particular case this becomes very neat. Indeed, let $V = 0$. This is then the free-particle problem. Now (78) becomes $K = 0$, and the canonical equations for Q and P are

$$\dot{Q} = 0 \qquad \text{and} \qquad \dot{P} = 0.$$

Q and P are hence constants of the motion. Of course! Just take a look at (77) and remember that this is a free particle. But perhaps this suggests an interesting method for solving the general dynamical problem: find a canonical transformation such that $K = 0$. We shall return to this method again in the next chapter.

Fourth, reversing the procedure. Suppose we are given the generating function $F(\xi, t) = pq - p^2t/2m$; let us find all corresponding canonical transformations (as mentioned already, the result will not be unique). The differential equations for the transformation, namely for $Q(q, p, t)$ and $P(q, p, t)$, are

$$\frac{\partial F}{\partial q} = p = p - \frac{\partial Q}{\partial q} P,$$

$$\frac{\partial F}{\partial p} = q - pt/m = -\frac{\partial Q}{\partial p} P.$$

The general solution of these equations is

$$Q = \Phi(p, t),$$

$$P = \frac{pt/m - q}{\partial \Phi/\partial p}, \tag{79}$$

where Φ is an arbitrary function such that $\partial\Phi/\partial p \neq 0$. Equations (79) give the entire class of canonical transformations corresponding to our generating function. If, as is true in general, $\partial\Phi/\partial t \neq 0$, then (78) is no longer the equation for K. On the other hand, for any Φ such that $\partial\Phi/\partial t = 0$, (78) still holds, and then for the free-particle Hamiltonian H, both Q and P will be constants of the motion. The reader may verify this explicitly. He will discover that it means that for the free particle, not only p but, of course, every function Φ of p alone is a constant of the motion.

(g) Successive Canonical Transformations and Mathieu Transformations

Let $\eta_\alpha(\xi, t)$ be a (restricted) canonical transformation whose generating function is $F(\xi, t)$, and let $\zeta_\alpha(\eta, t)$ be one whose generating function is $G(\eta, t)$. Then what is the generating function of the transformation from the ξ_α to the ζ_α?

To answer this question we use (46a) to write

$$\begin{aligned} \lambda_{\mu\nu}\xi_\nu - \lambda_{\alpha\beta}\frac{\partial \eta_\alpha}{\partial \xi_\mu}\eta_\beta &= \frac{\partial F}{\partial \xi_\mu}, \\ \lambda_{\mu\nu}\eta_\nu - \lambda_{\alpha\beta}\frac{\partial \zeta_\alpha}{\partial \eta_\mu}\zeta_\beta &= \frac{\partial G}{\partial \eta_\mu}. \end{aligned} \tag{80}$$

Multiply the second of these by $\partial\eta_\mu/\partial\xi_\rho$. This yields

$$\lambda_{\mu\nu}\eta_\nu \frac{\partial \eta_\mu}{\partial \xi_\rho} = \lambda_{\alpha\beta}\frac{\partial \zeta_\alpha}{\partial \xi_\rho}\zeta_\beta + \frac{\partial G}{\partial \xi_\rho},$$

which, after some changes in indices, may be inserted into the second term of the first of Eqs. (80). The result is

$$\lambda_{\mu\nu}\xi_\nu - \lambda_{\alpha\beta}\frac{\partial \zeta_\alpha}{\partial \xi_\mu}\zeta_\beta = \frac{\partial}{\partial \xi_\mu}(F + G.) \tag{81}$$

This shows that $F + G$ written as a function of the ξ_α, that is, $F(\xi, t) + G(\eta(\xi, t), t)$, generates the transformation from the ξ_α to the ζ_α. An arbitrary function of the time can be added to $F + G$.

Let us extend this to a canonical of type 1. Let $F^1(q, Q, t)$ generate the transformation from q, p to Q, P according to Eqs. (52) and (53), and let $G^1(Q, Q', t)$ generate the transformation from Q, P to Q', P'. Then

$$p_\alpha = \frac{\partial F^1}{\partial q_\alpha}; \qquad P_\alpha = \frac{\partial G^1}{\partial Q_\alpha}$$

$$P_\alpha = -\frac{\partial F^1}{\partial Q_\alpha}; \qquad P'_\alpha = -\frac{\partial G^1}{\partial Q'_\alpha}.$$

Let

$$J^1(q, Q', t) = F^1(q, Q(q, Q'), t) + G^1(Q(q, Q'), Q', t). \tag{82}$$

Then we have

$$\begin{aligned} \frac{\partial J^1}{\partial q_\alpha} &= \frac{\partial F^1}{\partial q_\alpha} + \frac{\partial F^1}{\partial Q_\beta}\frac{\partial Q_\beta}{\partial q_\alpha} + \frac{\partial G^1}{\partial Q_\beta}\frac{\partial Q_\beta}{\partial q_\alpha} \\ &= p_\alpha - P_\beta\frac{\partial Q_\beta}{\partial q_\alpha} + P_\beta\frac{\partial Q_\beta}{\partial q_\alpha} = p_\alpha, \end{aligned}$$

and

$$\begin{aligned} \frac{\partial J^1}{\partial Q'_\alpha} &= \frac{\partial F^1}{\partial Q_\beta}\frac{\partial Q_\beta}{\partial Q'_\alpha} + \frac{\partial G^1}{\partial Q_\beta}\frac{\partial Q_\beta}{\partial Q'_\alpha} + \frac{\partial G^1}{\partial Q'_\alpha} \\ &= -P_\beta\frac{\partial Q_\beta}{\partial Q'_\alpha} + P_\beta\frac{\partial Q_\beta}{\partial Q'_\alpha} - P'_\alpha = -P'_\alpha. \end{aligned}$$

Thus for transformations of type 1 also, the result of two successive canonicals is generated by the sum of the generating functions of type 1. Only in this case we must replace the intermediate Q_α by functions of other variables not in one, but in both of the initial generating functions.

From Eq. (81) it follows that if $-F = G$, that is if $\xi_\alpha(\eta, t)$ and $\zeta_\alpha(\eta, t)$ are in the class of transformations all of which are generated by the same function, then $\zeta_\alpha(\xi, t)$, the transformation between two variables in this class, is generated by zero. Such canonical transformations, with generating function zero, are called *Mathieu transformations.* For them we have the following theorem.

THEOREM. If $\eta_\alpha(\xi, t)$, or equivalently $Q_\alpha(q, p, t)$, $P_\alpha(q, p, t)$ is a Mathieu transformation, then Q_α is homogeneous of degree zero[7] in the p_α, and P_α is homogeneous of first degree in the p_α.

PROOF. If $F = 0$, Eqs. (46′*a*) become

$$P_\alpha \frac{\partial Q_\alpha}{\partial q_\beta} = p_\beta, \qquad P_\alpha \frac{\partial Q_\alpha}{\partial p_\beta} = 0. \tag{83}$$

Multiplying the first of these by $\partial P_\gamma / \partial p_\beta$, and using the Poisson bracket relation

$$[Q_\alpha, P_\gamma] = \delta_{\alpha\gamma} = \frac{\partial Q_\alpha}{\partial q_\beta}\frac{\partial P_\gamma}{\partial p_\beta} - \frac{\partial Q_\alpha}{\partial p_\beta}\frac{\partial P_\gamma}{\partial q_\beta},$$

we arrive at

$$p_\beta \frac{\partial P_\gamma}{\partial p_\beta} = P_\alpha \delta_{\alpha\gamma} + P_\alpha \frac{\partial Q_\alpha}{\partial p_\beta}\frac{\partial P_\gamma}{\partial q_\beta}.$$

But the second term on the right-hand side of this equation vanishes according to the second of Eqs. (83). Consequently we have

$$p_\beta \frac{\partial P_\gamma}{\partial p_\beta} = P_\gamma. \tag{84}$$

This, according to Euler's theorem for homogeneous functions, is a necessary and sufficient condition that P_γ be homogeneous of first degree in the p_β.

[7] Recall that a function $f(x_1, \ldots, x_n)$ is called homogeneous of degree k if

$$f(\lambda x_1, \lambda x_2, \ldots, \lambda x_n) = \lambda^k f(x_1, \ldots, x_n).$$

It does not follow, incidentally, that every homogeneous function of first degree is linear or that every homogeneous function of degree zero is independent of the variables. Common examples in three dimensions are the magnitude of the position vector (homogeneous of first degree) and any component of the unit vector pointing toward the origin (homogeneous of degree zero).

To prove the rest of the theorem, multiply the first of Eqs. (83) now by $\partial Q_\gamma / \partial p_\beta$, and use the Poisson bracket relation

$$[Q_\alpha, Q_\gamma] = 0 = \frac{\partial Q_\alpha}{\partial q_\beta} \frac{\partial Q_\gamma}{\partial p_\beta} - \frac{\partial Q_\alpha}{\partial p_\beta} \frac{\partial Q_\gamma}{\partial q_\beta},$$

arriving at

$$p_\beta \frac{\partial Q_\gamma}{\partial p_\beta} = P_\alpha \frac{\partial Q_\alpha}{\partial p_\beta} \frac{\partial Q_\gamma}{\partial q_\beta}.$$

Again the right-hand side vanishes by the second of Eqs. (83), so that

$$p_\beta \frac{\partial Q_\gamma}{\partial p_\beta} = 0. \tag{85}$$

This is a necessary and sufficient condition that Q_γ be homogeneous of degree zero in the p_β.

The proof of the converse, namely, that for a canonical transformation (84) and (85) imply (83), is left to the reader.

(h) Some Remarks About Classifying the Canonical Transformations

We have not yet solved the problem of classifying the canonical transformations. Our first approach involved classifying by finding a standard transformation for each generating function $F(\xi, t)$ and then classifying the Mathieu transformations. This is a difficult method and has not, to our knowledge, ever been carried through.

Our second approach involved transformations of the four types. To each such transformation corresponds a unique generating function of one of the four types. The trouble with this approach is twofold. First, a given transformation may be simultaneously of more than one type. Second, it is not yet clear that every canonical transformation is of one of the four types. In fact, clearly a transformation may be of one type in one degree of freedom and of another type in a second degree of freedom. Let us call such transformations generally *of type* or of *mixed type*. Then are all canonical transformations of type?

We shall not prove the result here, but shall merely state it (Caratheodory, 1965). We define an *elementary canonical transformation* (it is in general of mixed type) as one in which k (where $0 \leq k \leq n$) of the q_α, p_α are merely relabeled according to

$$Q_\alpha = p_\alpha, \qquad P_\alpha = -q_\alpha,$$

and all the others q_β, p_β are left alone. Then every canonical transformation can be composed of an elementary one followed by a transformation of type 1. This completely classifies the canonicals. In the proof, incidentally,

Caratheodory shows that the Jacobian of every canonical is equal to $+1$. (Previously we have shown only that it is ± 1.)

Problems

1. A particle in a uniform gravitational field is constrained to the surface of a sphere centered at the origin. The radius of the sphere varies in time: $r = r(t)$ with $r(t)$ a given function. Obtain the Hamiltonian and the canonical equations. Discuss energy conservation. Is the Hamiltonian the total energy?
2. Consider the Hamiltonian

$$H = \frac{p^2}{2m} e^{-q/a}$$

in one degree of freedom. Solve for $q(t)$ and $p(t)$. The equation for $\ddot{q}$ may look at first like the equation for a particle subjected to a retarding force proportional to $\dot{q}^2$, but it is not. For positive $\dot{q}$ it is a retarding force, while for negative $\dot{q}$ it increases the speed. This is an illustration of the kind of trouble one gets into if one tries to write a Hamiltonian for a dissipative problem. But consider positive $\dot{q}$. Then we may think of this as a one-dimensional dissipative system in which q is a Cartesian coordinate. Take $q(0) = 0$ and $\dot{q}(0) > 0$. Is there a terminal velocity? At what rate is energy dissipated? What is conserved? What term must be added to the Hamiltonian to provide an additional constant (positive) force? For the new problem with the additional positive force answer the above three questions, now taking $\dot{q}(0) \geq 0$.
3. Show that the matrix Γ of Eq. (11) is orthogonal, antisymmetric, and unimodular. Show that if $A\Gamma = \Gamma A$, then A is of the form

$$A = \begin{vmatrix} U_n & V_n \\ -V_n & U_n \end{vmatrix}$$

where U_n and V_n are $n \times n$ matrices.
4. Prove Properties 1, 2, 3, and 4 of the Poisson brackets.
5. Let $\mathbf{A}$ be the Runge-Lenz vector for the Kepler problem [see Chapter III, Section 5(c)]. Calculate $[A_\alpha, l_\rho]$ and $[A_\alpha, H]$ explicitly, where $\mathbf{l}$ is the angular momentum. According to (19) these should be constants of the motion. Are they independent of $\mathbf{A}$, $\mathbf{l}$, and H? (Note: very few derivatives have to be taken to perform these calculations. Use the properties of the Poisson brackets.)
6. Given the Hamiltonian

$$H = q_1 p_1 - q_2 p_2 - a q_1^2 + b q_2^2,$$

where a and b are constants. Show that

$$F_1 = (p_2 - b q_2)/q_1,$$
$$F_2 = q_1 q_2,$$
$$F_3 = q_1 e^{-t}$$

are constants of the motion. Discuss their independence and whether or not there exist other independent constants of the motion. If such exist, exhibit some until you have the maximum number of independent ones. Show explicitly that the $[F_i, F_j]$ are constants of the motion. [This Hamiltonian is pathological in that it comes from no Lagrangian. In fact, one of the equations of motion is $\dot{q}_1 = q_1$, which looks like a velocity-dependent constraint. See Problem 27. Compare Section 6(b).]

7. Consider the Hamiltonian $H = T + V$, where

$$T = \tfrac{1}{2}\sigma_{\alpha\beta}p_\alpha p_\beta$$

and

$$V = \tfrac{1}{2}\kappa_{\alpha\beta}q_\alpha q_\beta.$$

Assume that $T > 0$ for $p \neq 0$, and $V \geq 0$ for $q \neq 0$. This is the small vibrations problem in Hamiltonian form. Solve it within the Hamiltonian formalism; that is, find the normal modes and frequencies.

8. Consider a particle moving in a plane under the influence of the generalized potential

$$V = \frac{1}{r}(1 + \dot{r}^2),$$

where r is the distance from the origin. Write down the Lagrangian in polar coordinate r, θ, and from it find p_r, p_θ, and H. Obtain the canonical equations and show that angular momentum is conserved. Reduce the problem for r to a single first-order differential equation for r.

9. A particle moves in a coordinate system fixed to the Earth, which is rotating with fixed angular velocity $\boldsymbol{\omega} = \mathbf{e}_3\omega$ relative to an inertial frame. Assume that the potential (e.g., gravitational) is independent of t in the rotating Earth's system. Find the Lagrangian in the Earth's system. (Remember that you know L originally in the inertial system.) Find momenta conjugate to the Cartesian coordinates x_1, x_2, x_3 of the Earth's system, and calculate the Hamiltonian in this system. Show that H is not the total energy, but that it is conserved.

10. Find a canonical transformation such that the Hamiltonian of a freely falling body in one degree of freedom becomes $K(P, Q) = P$. Solve the problem in terms of Q and P and transform back to obtain the usual solution.

11. Let A be a symmetric linear operator on an n-dimensional vector space S such that $A\Omega = \Omega A$ for all symmetric operators Ω on S. Prove that A is a multiple of $\mathbb{1}$.

12. Let

$$Q_1 = q_1^2, \qquad P_1 = P_1(q_1, q_2, p_1, p_2),$$

$$Q_2 = q_1 + q_2, \qquad P_2 = P_2(q_1, q_2, p_1, p_2)$$

be a canonical transformation. Complete the transformation by finding the most general expressions for P_1 and P_2. Show that a particular choice for P_1 and P_2 will reduce

$$H = \left(\frac{p_1 - p_2}{2q_1}\right)^2 + p_2 + (q_1 + q_2)^2$$

to

$$K = P_1^2 + P_2.$$

Use this to solve for q_1 and q_2 as functions of the time.

13. Let S be the matrix

$$S = \begin{vmatrix} 1_n & 0_n \\ 0_n & -1_n \end{vmatrix}.$$

Show that any nonrestricted canonical transformation $\eta_\alpha(\xi, t)$ can be written in the form

$$\eta_\alpha = (S^\epsilon)_{\alpha\beta}\, |z|^{-1/2} \zeta_\beta(\xi, t),$$

where $\epsilon = 0$ if z is positive and $\epsilon = 1$ if z is negative, and $\zeta_\beta(\xi, t)$ is a restricted canonical transformation determined uniquely by $\eta_\alpha(\xi, t)$.

14. Let

$$\{\xi_\alpha, \xi_\beta\}^\eta \equiv \frac{\partial \eta_\mu}{\partial \xi_\alpha} \gamma_{\mu\nu} \frac{\partial \eta_\nu}{\partial \xi_\beta}$$

be the *Lagrange bracket* of the ξ_β with respect to the η_β. Show that $\{\xi_\alpha, \xi_\rho\}^\eta$ is the inverse transpose of $[\xi_\alpha, \xi_\beta]^\eta$, that is, that

$$\{\xi_\alpha, \xi_\beta\}^\eta [\xi_\alpha, \xi_\gamma]^\eta = \delta_{\beta\gamma},$$

and therefore that the transformation from ξ_α to η_β is (restricted) canonical if and only if $\{\xi_\alpha, \xi_\beta\}^\eta = \gamma_{\alpha\beta}$. Write these equations in the (q, p)-notation.

15. Let $\eta_\alpha(\xi, t)$ be a (restricted) canonical transformation, and let $L'(\xi, \dot\xi, t) = \lambda_{\mu\nu}\dot\eta_\mu(\xi, t)\eta_\nu(\xi, t) - K$, where K is the η-Hamiltonian expressed as a function of the ξ_α. Show by direct calculation that

$$\frac{d}{dt}\frac{\partial L}{\partial \dot\xi_\alpha} - \frac{\partial L}{\partial \xi_\alpha} = \frac{d}{dt}\frac{\partial L'}{\partial \dot\xi_\alpha} - \frac{\partial L'}{\partial \xi_\alpha},$$

where $L = \lambda_{\mu\nu}\dot\xi_\mu\xi_\nu - H$, thereby establishing the existence of the generating function.

16. Obtain the following equations for canonical transformations of types 3 and 4.

Type 3 (p_α and Q_β independent):

$$q_\alpha = -\frac{\partial F^3}{\partial p_\alpha},$$

$$P_\alpha = -\frac{\partial F^3}{\partial Q_\alpha},$$

$$K = H + \frac{\partial F^3}{\partial t}.$$

Express $F^3(p, Q, t)$ in terms of $F(\xi, t)$.

Type 4 (p_α and P_β independent):

$$q_\alpha = -\frac{\partial F^4}{\partial p_\alpha},$$

$$Q_\alpha = \frac{\partial F^4}{\partial P_\alpha},$$

$$K = H + \frac{\partial F^4}{\partial t}.$$

Express $F^4(p, P, t)$ in terms of $F(\xi, t)$.

17. Assign a type to one of the possible canonical transformations of Problem 12. Obtain the generating function of the transformation and the generating function of the type you assigned. Can the transformation be of some type other than the one you assigned? Can it be of a more general type [see Section 6(h)]?
18. Consider the generating function $F(\xi, t) = pq - p^2t/2m$. The canonicals generated by this $F(\xi, t)$ are given by Eq. (79) with arbitrary $\Phi(p, t)$. Suppose that we take two different Φ functions, $\Phi_A(p, t)$ and $\Phi_B(p, t)$, and that each yields a canonical transformation Q_A, P_A and Q_B, P_B according to (79). Show explicitly that the transformation from Q_A, P_A to Q_B, P_B is Mathieu (i.e., that it is generated by $F = 0$ and that it possesses the correct homogeneity properties).
19. The point transformation from Cartesian x_1, x_2, x_3 coordinates to spherical polar r, θ, φ can be completed to yield a canonical transformation. Show that if one takes the usual definition of momenta conjugate to x_1, x_2, x_3, this completion may be the one given by Eq. (58).
20. Consider the point transformation

$$Q = \text{arc tan}\,(\lambda q/p).$$

Complete it to form a canonical transformation, showing that

$$P = \tfrac{1}{2}\left(\frac{p^2}{\lambda + \lambda q^2}\right) + R(q, p, t),$$

where R is an arbitrary function homogeneous of degree zero in the pair of variables q, p. Apply this transformation to the simple harmonic oscillator of mass m and circular frequency $\omega = \lambda/m$. (Choose R to simplify the problem, solve it in terms of P, Q, and transform back.)

21. Consider a particle subject to the force

$$F = -kq - \frac{\alpha}{q^3}.$$

Show that this system is described by the Hamiltonian

$$H = \frac{p^2}{2m} + \frac{k}{2}q^2 + A\frac{p}{q}$$

where A must be a properly chosen constant. Use the canonical transformation of Problem 20 to solve this problem.

22. Show for a particle in an electromagnetic field that a gauge transformation changes the Hamiltonian and the orbits on phase space, but not the orbits on configuration space.
23. In one degree of freedom p may be replaced wherever it appears in the Hamiltonian by $p - f(q)$, where $f(q)$ is arbitrary, without changing the orbits on configuration space. Prove this statement and discuss its relation to Problem 22. By completing the square appropriately in the Hamiltonian of Problem 21, eliminate the term Ap/q, and show that the Hamiltonian may therefore be written $H = T + V$, where $-V'$ is the given force F.
24. Consider a Mathieu transformation of type 1. Then $F^1 = F = 0$, and Eqs. (52) and (53) yield $p_\alpha = 0$, $P_\alpha = 0$. What is wrong? Prove it!
25. Apply the treatment of Section 6(b), Eq. (66) to solve for the motion of the two-dimensional harmonic oscillator with

$$K = \begin{vmatrix} 2k & -k \\ -k & k \end{vmatrix}, \qquad M = m\mathbb{1}.$$

Calculate R, check convergence. See how diagonalizing R exhibits the normal modes. [This problem corresponds to a mass m suspended from a spring of constant k, from which is suspended another equal mass from a similar spring.]

26. Just before stating the Poisson Bracket Theorem we discuss placing a time development on phase space by giving $\xi_\alpha(t) = f_\alpha(t, \xi^0)$, "being careful that the paths do not intersect." Prove that it is then possible to write $\dot{\xi}_\alpha = \psi_\alpha(\xi, t)$ uniquely with no ξ^0 dependence.
27. Given a Hamiltonian, how do you find the Lagrangian? Show that the Euler-Lagrange equations follow from Hamilton's. Obtain necessary and sufficient conditions for the transition from the Hamiltonian to the Lagrangian.
28. We have proven that if $\eta_\alpha(\xi, t)$ is canonical, a function F exists satisfying (46a). Prove the converse.
29. Let $\eta_\alpha(\xi)$ be a (nonrestricted) canonical transformation with $z \neq 1$. Let $H(\xi, t)$ be the ξ-Hamiltonian. Find the η-Hamiltonian $K(\eta, t)$.

CHAPTER VII

Continuous Families of Canonical Transformations

In Chapter III continuous families of transformations in configuration space were used to relate constants of the motion to symmetries of the Lagrangian. In the last chapter we saw that configuration-space transformations are a subclass of the much larger class of canonical transformations. In this chapter we shall use continuous families of canonical transformations to relate constants of the motion to symmetries of the Hamiltonian. The theory of continuous families of canonical transformations will lead, in addition, to a new view of the equations of motion and to new methods for solving them (Hamilton-Jacobi theory and canonical perturbation theory).

VII-1 FAMILIES OF CANONICAL TRANSFORMATIONS. INFINITESIMAL GENERATORS

In Chapter VI we considered transformations between the variables ξ_α and η_α. Here we change notation slightly and consider a time-dependent canonical transformation between the variables ξ_α^0 and ξ_α:

$$\xi_\alpha = Z_\alpha(\xi^0, t; \theta). \tag{1}$$

This transformation depends on a parameter θ so chosen that the transformation is continuous and differentiable in θ for all values of θ in some finite interval which includes $\theta = 0$. Equation (1) thus defines a *continuous family of canonical transformations*. We let $\theta = 0$ correspond to the identity transformation, so that

$$\xi_\alpha^0 = Z_\alpha(\xi^0, t; 0).$$

Being canonical, these transformations are in vertible: there exist functions Z_α^0 such that

$$\xi_\alpha^0 = Z_\alpha^0(\xi, t; \theta). \tag{2}$$

The canonical transformation for each value of θ has a generating function $F^0(\xi^0, t; \theta)$, also continuous and differentiable in θ, unique up to an arbitrary function of θ and t. (We write F^0 to emphasize that it is a function of ξ^0 rather than of ξ.)

We wish to study how the ξ_α depend on θ for fixed ξ^0_α and t. To do this we shall derive a differential equation for the Z_α functions. We start with Eq. (46a) of Chapter VI, which now reads

$$\lambda_{\alpha\rho}\xi^0_\rho = \lambda_{\mu\nu}\frac{\partial Z_\mu}{\partial \xi^0_\alpha} Z_\nu + \frac{\partial F^0}{\partial \xi^0_\alpha},$$

and differentiate with respect to θ, holding the ξ^0_α and t fixed. Direct calculation yields

$$0 = \lambda_{\mu\nu}\frac{\partial Z_\mu}{\partial \xi^0_\alpha}\frac{\partial Z_\nu}{\partial \theta} + \lambda_{\mu\nu}\frac{\partial^2 Z_\mu}{\partial \xi^0_\alpha\,\partial \theta} Z_\nu + \frac{\partial^2 F^0}{\partial \xi^0_\alpha\,\partial \theta}$$

$$= \lambda_{\mu\nu}\frac{\partial Z_\mu}{\partial \xi^0_\alpha}\frac{\partial Z_\nu}{\partial \theta} - \lambda_{\mu\nu}\frac{\partial Z_\mu}{\partial \theta}\frac{\partial Z_\nu}{\partial \xi^0_\alpha} + \frac{\partial}{\partial \xi^0_\alpha}\left[\lambda_{\mu\nu}\frac{\partial Z_\mu}{\partial \theta} Z_\nu + \frac{\partial F^0}{\partial \theta}\right].$$

We define the term in brackets as

$$G^0(\xi^0, t; \theta) = \frac{\partial F^0}{\partial \theta} + \lambda_{\mu\nu}\frac{\partial Z_\mu}{\partial \theta} Z_\nu, \tag{3}$$

and then by interchanging μ and ν in the second term of the previous equation we get

$$(\lambda_{\mu\nu} - \lambda_{\nu\mu})\frac{\partial Z_\mu}{\partial \xi^0_\alpha}\frac{\partial Z_\nu}{\partial \theta} + \frac{\partial G^0}{\partial \xi^0_\alpha} = 0. \tag{4}$$

Equations (1) and (2) show that

$$\frac{\partial Z_\mu}{\partial \xi^0_\alpha}\frac{\partial Z^0_\alpha}{\partial \xi_\beta} = \delta_{\mu\beta},$$

so by multiplying (4) by $\gamma_{\beta\rho}\partial Z^0_\alpha/\partial \xi_\beta$ and using the fact that $\lambda_{\mu\nu} - \lambda_{\nu\mu} = \gamma_{\mu\nu}$ we arrive at

$$\frac{\partial Z_\rho}{\partial \theta} = \gamma_{\rho\beta}\frac{\partial G^0}{\partial \xi^0_\alpha}\frac{\partial Z^0_\alpha}{\partial \xi_\beta}. \tag{5}$$

If we now introduce $G(\xi, t; \theta) = G^0(\xi^0, t; \theta) = G^0(Z^0(\xi, t; \theta), t; \theta)$, Eq. (5) becomes, changing the indices,

$$\frac{\partial Z_\nu}{\partial \theta} = \gamma_{\nu\mu}\frac{\partial G}{\partial \theta}. \tag{6}$$

This is the promised differential equation for the Z_α functions.

So far we have been viewing transformations such as (1) and (2) as coordinate transformations on phase space (the passive view), according to which ξ and ξ^0 are the coordinates of a fixed phase point in two different coordinate systems. Let us look at (6), however, from the active view, according to which ξ and ξ^0 are the coordinates of two different phase points in a fixed coordinate system. In this view Eq. (1) tells us that ξ moves as θ varies, and we thus put our attention on what happens to ξ when ξ^0 and t are held fixed.

In the active view[1] Eq. (1) becomes a set of parametric equations

$$\xi_\nu = \xi_\nu(\theta)$$

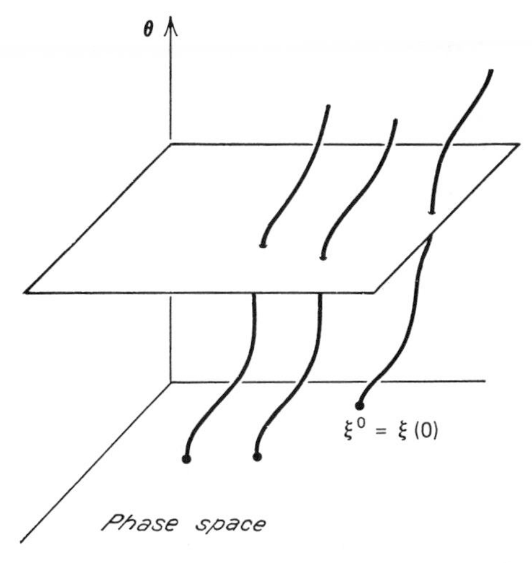

Figure 31. The active view of a continuous family of transformations on phase space.

for a one-dimensional curve which passes through the original point ξ^0 at $\theta = 0$ in the $2n$-dimensional phase space (see Fig. 31). There is a different curve for each different original point and the collection of all these nonintersecting curves is called the set of θ-orbits of the transformation. On a θ-orbit ξ^0 and t are fixed and ξ_ν is a function only of θ, so that we can write

$$\frac{\partial Z_\nu}{\partial \theta} = \frac{d\xi_\nu}{d\theta}, \tag{7}$$

and thus it is seen that Eq. (6) gives the differential equation for each θ-orbit:

$$\frac{d\xi_\nu}{d\theta} = \gamma_{\nu\mu}\frac{\partial G}{\partial \xi_\mu}. \tag{8}$$

[1] In Chapter VI we always viewed canonical transformations passively. In the active view the same mathematical relations are interpreted differently. For instance, for time-independent transformations we have $K(Q, P) = H(q, p)$, which in the active view equates the value of K at the new point Q, P to the value of H at the old point q, p.

These equations are marvelously similar to Hamilton's canonical equations: just replace θ by t and G by H. We shall see in the next section that this similarity has actually more than formal significance.

The interesting thing about these θ-orbits is that every point $\xi(\theta)$ on a given orbit is related to the original point $\xi^0 = \xi(0)$ (and hence to every other point on the orbit) by a canonical transformation. What we have shown is that for each one-parameter family of such canonical transformations there exists a function $G(\xi, t; \theta)$, called the *infinitesimal generator* of the transformation, such that the θ-orbits are given by Eq. (8). However, the usefulness of these generators and θ-orbits comes from the fact that the converse is also true. We state the entire result as a theorem:

THEOREM. Every continuous and differentiable function (dynamical variable) generates a family of canonical transformations through Eq. (8), and for each family of canonical θ-orbits there exists an infinitesimal generator satisfying (8).

PROOF. We have already proved the second part of this theorem. The first part says that for any function G every solution $\xi = \xi(\theta)$ of Eq. (8) is related to its original point $\xi^0 = \xi(0)$ by a canonical transformation for each θ. To prove this we need only show (see Problem 14 of Chapter VI) that the Lagrange bracket of ξ_α^0 and ξ_β^0 is $\gamma_{\alpha\beta}$:[2]

$$\{\xi_\alpha^0, \xi_\beta^0\}^\xi \equiv \frac{\partial \xi_\mu}{\partial \xi_\alpha^0} \gamma_{\mu\nu} \frac{\partial \xi_\nu}{\partial \xi_\beta^0} = \gamma_{\alpha\beta}. \tag{9}$$

The proof is as follows:

Let us write [we omit the time dependence of $Z(\xi^0, t; \theta)$]

$$\{\xi_\alpha^0, \xi_\beta^0\}^\xi = l_{\alpha\beta}(\xi^0, \theta).$$

Then it is clear that

$$l_{\alpha\beta}(\xi^0, 0) = \gamma_{\alpha\beta},$$

or that Eq. (9) holds at $\theta = 0$. How does $l_{\alpha\beta}$ vary with θ? We have

$$\begin{aligned}
\frac{\partial}{\partial \theta} l_{\alpha\beta} &= \frac{\partial}{\partial \theta}\left(\frac{\partial Z_\mu}{\partial \xi_\alpha^0} \gamma_{\mu\nu} \frac{\partial Z_\nu}{\partial \xi_\beta^0}\right) = \frac{\partial}{\partial \xi_\alpha^0}\left(\frac{\partial Z_\mu}{\partial \theta} \gamma_{\mu\nu}\right) \frac{\partial Z_\nu}{\partial \xi_\beta^0} + \frac{\partial Z_\mu}{\partial \xi_\alpha^0} \frac{\partial}{\partial \xi_\beta^0}\left(\gamma_{\mu\nu} \frac{\partial Z_\nu}{\partial \theta}\right) \\
&= \frac{\partial}{\partial \xi_\alpha^0}\left(\frac{\partial G}{\partial \xi_\nu}\right) \frac{\partial Z_\nu}{\partial \xi_\beta^0} - \frac{\partial Z_\mu}{\partial \xi_\alpha^0} \frac{\partial}{\partial \xi_\beta^0}\left(\frac{\partial G}{\partial \xi_\mu}\right) \\
&= \frac{\partial}{\partial \xi_\alpha^0}\left(\frac{\partial G}{\partial \xi_\nu} \frac{\partial \xi_\nu}{\partial \xi_\beta^0}\right) - \frac{\partial G}{\partial \xi_\nu} \frac{\partial^2 Z_\nu}{\partial \xi_\alpha^0 \, \partial \xi_\beta^0} \\
&\quad - \left[\frac{\partial}{\partial \xi_\beta^0}\left(\frac{\partial G}{\partial \xi_\mu} \frac{\partial \xi_\mu}{\partial \xi_\alpha^0}\right) - \frac{\partial G}{\partial \xi_\mu} \frac{\partial^2 Z_\mu}{\partial \xi_\alpha^0 \, \partial \xi_\beta^0}\right] \\
&= \frac{\partial}{\partial \xi_\alpha^0} \frac{\partial G^0}{\partial \xi_\alpha^0} - \frac{\partial}{\partial \xi_\beta^0} \frac{\partial G^0}{\partial \xi_\alpha^0} = 0.
\end{aligned}$$

[2] The theorem can also be proved by showing that the Poisson bracket $[\xi_\alpha, \xi_\beta]^{\xi^0}$ is equal to $\gamma_{\alpha\beta}$. However the proof is then a little more complicated, and it is left to the student as a problem.

Thus $l_{\alpha\beta}$ does not depend on θ, or

$$l_{\alpha\beta}(\xi^0, \theta) = l_{\alpha\beta}(\xi^0, 0) \equiv l_{\alpha\beta}(\xi^0) = \gamma_{\alpha\beta}.$$

This proves the theorem.

The importance of this theorem is that G is an arbitrary function, so that a family of canonical transformations can be generated by any dynamical variable. As an example consider the infinitesimal generator

$$G = -\xi_1 + \tfrac{1}{2}\xi_2^2 = -q + \tfrac{1}{2}p^2$$

for a system with one degree of freedom (two dimensions). Equation (8) now reads

$$\frac{dq}{d\theta} = \frac{\partial G}{\partial p} = p,$$

$$\frac{dp}{d\theta} = -\frac{\partial G}{\partial q} = 1,$$

and the solutions are

$$p = p(0) + \theta = p^0 + \theta,$$

$$q = \tfrac{1}{2}\theta^2 + p^0\theta + q^0.$$

The reader can easily verify that this is a canonical transformation for all θ.

Remark. In our general discussion we have considered G to be a function of θ as well as of ξ and t. However in most applications of interest G will not in fact depend on θ, as is the case in the above example.

VII-2 SYMMETRIES AND CONSERVATION LAWS

In Chapter III we saw that the invariance of the Lagrangian under a continuous family of point transformations (that is, a symmetry of the Lagrangian) was related to a conservation law. A similiar relation holds for the invariance of the Hamiltonian under a continuous family of canonical transformations. To exhibit this relation we study how the Hamiltonian, or more generally how any dynamical variable $R(\xi, t)$, varies along a θ-orbit.

As before ξ depends on ξ^0, t, and θ, but along a θ-orbit ξ^0 and t are held fixed, so we have

$$\begin{aligned} \frac{dR}{d\theta} &= \frac{\partial R}{\partial \xi_\nu}\frac{d\xi_\nu}{d\theta} \\ &= \frac{\partial R}{\partial \xi_\nu}\gamma_{\nu\mu}\frac{\partial G}{\partial \xi_\mu} \\ &= [R, G], \end{aligned} \tag{10}$$

where we have used Eq. (8). In particular if R is the Hamiltonian of the system we have

$$\frac{dH}{d\theta} = [H, G]. \tag{11}$$

From this we see that the Hamiltonian is invariant (has a symmetry) under the transformation generated by G, that is, H does not vary along a θ-orbit, if and only if $[H, G] = 0$.

This result is interesting because we have seen at Eq. (15) of Chapter VI that the time variation of G, namely,

$$\frac{dG}{dt} = [G, H] + \frac{\partial G}{\partial t},$$

also depends on the Poisson bracket with H. Therefore if $\partial G/\partial t = 0$, it follows that G is a conserved quantity ($dG/dt = 0$) if and only if $dH/d\theta = 0$. To put if differently, H does not vary along the θ-orbits generated by G if and only if G does not vary along the t-orbits (the motions) generated by H. This symmetrical statement depends on the symmetrical assumption that G does not depend on time and H does not depend on θ. The more general statement is that if G is independent of θ, then $dH/d\theta = 0$ if and only if $dG/dt = \partial G/\partial t$.

This is the connection between symmetry and conservation which we wanted to obtain: conserved quantities are the infinitesimal generators of transformations under which the Hamiltonian is invariant.

Let us investigate this connection in more detail. It was shown in Chapter III that cónservation of any component of angular or linear momentum is related to an invariance of the Lagrangian under rotation or translation. A similar result is more easily obtained for the Hamiltonian. Conservation of any generalized momentum p_α is related to an invariance of the Hamiltonian under translations in the canonically conjugate coordinate q_α. This follows at once from the equation

$$\dot{p}_\alpha = -\frac{\partial H}{\partial q_\alpha},$$

for it implies that $\dot{p}_\alpha = 0$ if and only if H does not depend on q_α. We can now state this relation between conservation and the invariance (symmetry) of H in terms of continuous families of canonical transformations and their infinitesimal generators G. Note first that a generalized momentum p_α *is the infinitesimal generator of translations in* q_α. Indeed, if we put $G = p_\alpha$, α fixed, in Eq. (8), we have

$$\frac{dq_\beta}{d\theta} = \delta_{\alpha\beta}, \qquad \frac{dp_\beta}{d\theta} = 0,$$

whose solution is the family of canonical transformations

$$q_\beta = q_\beta^0 + \theta\delta_{\alpha\beta}, \qquad p_\beta = p_\beta^0$$

from q^0, p^0 to q, p. Thus as asserted, $G = p_\alpha$ generates the transformation to new variables which differ from the old only by translation in q_α for that fixed α: the θ-orbits are lines along which only that q_α changes. Finally, then, according to our previous discussion $G = p_\alpha$ is a constant of the motion if and only if $dH/d\theta = 0$ along the θ-orbits generated by p_α, or in other words if H is invariant under translation in q_α.

Remark. Let $R(q, p)$ be any time-independent dynamical variable and $S(q, p)$ be another, such that

$$[S, R] = 1.$$

Then under the family of canonical transformations generated by R we have $dS/d\theta = 1$, and along any θ-orbit $S = S^0 + \theta$, where $S^0 = S(q^0, p^0)$. Thus R is the infinitesimal generator of translations in S in almost the same sense that p_α is the infinitesimal generator of translations in q_α. All that is needed are $2(n - 1)$ additional dynamical variables which, together with S and R, form a coordinate system on phase space. These can in general always be found. In this sense, given a dynamical variable S, we may almost define its canonical conjugate R as the dynamical variable which generates translations in S. We say "*almost* define" because this does not define R uniquely, for one can add to R any dynamical variable T such that $[S, T] = 0$. In addition, there is an ambiguity in sign, for as p_α generates translations in the positive q_α direction, q_α generates translations in the negative p_α direction.

We shall now turn to the family of canonical transformations generated by the Hamiltonian, for they are of special interest. Then Eqs. (8) become

$$\frac{d\xi_\nu}{d\theta} = \gamma_{\nu\mu}\frac{\partial H}{\partial \xi_\mu}, \tag{12}$$

which look like Hamilton's canonical equations, but with θ replacing t. If H is time independent they are in fact identical: the θ-orbits given by Eq. (12) are the same as the t-orbits or motions given by the canonical equations. If H is time dependent they are not identical, since in the canonical equations the independent variable t appears in H whereas in Eq. (12) the independent variable θ does not. We will discuss this more general case later.

First let us consider the case in which H is time independent. Then H itself is a constant of the motion and so it must be invariant under the transformation it generates. This transformation is[3]

$$\xi_\nu(\theta) = Z_\nu(\xi^0; \theta). \tag{13}$$

[3] The t in Eq. (1) reflects the time dependence of the infinitesimal generator. In the present case this generator is H, which is assumed to be time independent. Hence we have dropped t in Eq. (13).

But from what we have just said, $\xi_\nu(\theta)$ is identical to the motion $\xi_\nu(t - t_0)$ which passes through the phase point ξ^0 at time t^0: we only need to equate θ to $t - t_0$. Thus the variables at two different times, namely, $\xi^0 = \xi(0)$ at time t_0 and $\xi = \xi(t - t_0)$ at time t (corresponding to $\theta = 0$ and $\theta = t - t_0$, respectively) are connected by the canonical transformation whose infinitesimal generator is H. One speaks of H, which is usually the energy, as the *generator of translations in time.* Time translation is something like translation in some dynamical variable, say q_α, but it is also somewhat different. Whereas invariance of H under translation in q_α implies conservation of the conjugate momentum p_α, invariance of H under translation in t implies conservation of H itself. This is obvious. In fact, invariance of anything under translation in t implies that that thing is conserved.

Now we turn to time-dependent Hamiltonians. Are the phase points at two different times then also connected by a canonical transformation? We must be careful here because the equations of motion (the t-orbits) are given by

$$\frac{d\xi_\nu}{dt} = \gamma_{\nu\mu}\frac{\partial H(\xi, t)}{\partial \xi_\mu},$$

whereas the canonical transformation we have been talking about is given by

$$\frac{d\xi_\nu}{d\theta} = \gamma_{\nu\mu}\frac{\partial H(\xi, t)}{\partial \xi_\mu}.$$

But we could just as well identify θ with t, that is we can write $H(\xi, \theta)$ in this last equation. Then the two become identical. Thus there is in fact a canonical transformation connecting the points along the actual motion, and this transformation is (infinitesimally) generated by H in this sense.

The fact that the time evolution of a point in phase space is itself a continuous family of canonical transformations will become the basis for a new method of solving problems in mechanics. This method is discussed in the next section.

VII-3 THE HAMILTON-JACOBI EQUATION

We have seen in the last section that the motion in phase space is a continuous family of canonical transformations; obtaining this family is equivalent to solving the equations of motion. In fact in their usual form the equations of motion are differential equations for the family of transformations, equations which are stated in terms of its infinitesimal generator H. But we can go about the problem in a different way: we may attempt to find the (global) generating function F of the transformation and then to use it in one of its four types to obtain algebraic equations for the transformation directly from the initial phase point to the point at time t. In this section we develop a method for finding F. Later we shall see that this procedure has certain important advantages.

Let ξ^0 be the phase point at time t^0. We shall now write ξ as a function of t rather than θ, so that at time $t = t^0 + \theta$ we have $\xi = \xi(t) = \xi(t^0 + \theta) = Z(\xi^0; t^0 + \theta) = Z(\xi^0; t)$, and the time t itself, rather than the time difference $\theta = t - t^0$ between two points, becomes the parameter of the transformation. In an attempt to find the global generator we use Eq. (3) at the beginning of this chapter. Since $H^0(\xi^0, t)$ is G^0, the infinitesimal generator, F^0 satisfies the equation

$$\frac{\partial F^0}{\partial t} = H^0(\xi^0, t) - \lambda_{\mu\nu}\frac{\partial Z_\mu}{\partial t} Z_\nu = H(\xi, t) - \lambda_{\mu\nu}\dot{\xi}_\mu \xi_\nu. \tag{14}$$

Here, having replaced θ by t, we used the fact that the time derivative is taken with ξ^0 held fixed, so that

$$\frac{\partial}{\partial t} Z_\mu(\xi^0; t) = \frac{d\xi_\mu}{dt} = \dot{\xi}_\mu.$$

Equation (14) can be used to find F^0 only if we already know $Z(\xi^0, t)$, that is, only if we already know the motion. Since the motion is what we ultimately wish to find, this approach does not seem too promising. However, we will show shortly how to get around this difficulty. For the present let us suppose the motion is known. Then the right-hand side of (14) is known explicitly as a function of ξ^0 and t, and we can then integrate it to obtain

$$F^0(\xi^0, t, t^0) = \int_{t^0}^{t} [H(Z(\xi^0, t'), t') - \lambda_{\mu\nu}\dot{Z}_\mu(\xi^0, t')Z_\nu(\xi\ , t')]\, dt'. \tag{15}$$

This representation of the global generating function in the form of an integral is still not suitable for obtaining the canonical transformation. We want to put the function into one of the four types, for which we must write it as a function not only of the old variables ξ^0_μ, but of a mixture of some old and some new variables. This we can easily do by specifying the motion itself by the mixed variables. In our discussion of the variational principle in Chapter II, Section 3, we saw that the configuration-space points q^0 and $q(t)$ at the two times completely specify the motion. Let us therefore write the motion in terms of q^0 and $q = q(t)$:

$$\xi_\alpha(t') = Z_\alpha(\xi^0, t') = \bar{Z}(q^0, q, t').$$

Then the generating function will depend on q^0 and q, and we shall write $S(q, t; q^0, t^0) = -F^0(\xi^0, t, t^0)$. Thus we write

$$S(q, t; q^0, t^0) = \int_{t^0}^{t} [\lambda_{\mu\nu}\dot{\bar{Z}}_\mu(q^0, q, t')\bar{Z}_\nu(q^0, q, t') - H(\bar{Z}(q^0, q, t'), t')]\, dt'. \tag{16}$$

Note that S is the negative of F^0, so that it generates the transformation backwards in time from $q(t)$ to q^0. Finally, in the integrand of (16) we shall replace the p_α by their expressions in terms of the q_β and $\dot{q}_\beta$. Then the integrand becomes simply the Lagrangian $L(q, \dot{q}, t)$ [see Eq. (6) of Chapter VI], and we have

$$S(q, t; q^0, t^0) = \int_{t^0}^{t} L(q', \dot{q}', t')\, dt', \tag{17}$$

where q'_α is $\bar{Z}_\alpha(q^0, q, t')$ for $\alpha = 1, \ldots, n$. S is now clearly of type 1. Note the interesting result that the global generator is just what is minimized by the physical path.

Let us take an example: Consider the unit-mass free particle in one dimension: $H = \frac{1}{2}p^2$. The motion which passes through q^0 at time t^0 and q at time t is

$$q\,(t') = q^0 + (q - q^0)\frac{t' - t^0}{t - t^0}\,. \tag{18}$$

Since the Lagrangian is $L = \frac{1}{2}\dot{q}^2$ we see from Eq. (17) that

$$S = \int_{t^0}^{t} \frac{1}{2}\left[\frac{q - q^0}{t - t^0}\right]^2 dt' = \frac{(q - q^0)^2}{2(t - t^0)}\,. \tag{19}$$

From this we can reobtain the motion by writing out the equations of type 1. We have

$$p^0 = \frac{\partial F^1}{\partial q^0} = -\frac{\partial S}{\partial q^0} = \frac{q - q^0}{t - t^0},$$

$$p = -\frac{\partial F^1}{\partial q} = \frac{\partial S}{\partial q} = \frac{q - q^0}{t - t^0} = p^0.$$

Solving the first of these equations for q we get the motion in terms of q^0 and p^0,

$$q = q^0 + p^0\,(t - t^0),$$

which is clearly equivalent to (18).

This example illustrates the case in which we already know the motion. But what we want is a way to obtain S without first knowing the motion, for once we know S we can obtain the motion from the equations

$$p_\alpha = \frac{\partial S}{\partial q_\alpha}, \qquad p_\alpha^0 = -\frac{\partial S}{\partial q_\alpha^0}, \tag{20}$$

which give the time-dependent canonical transformation between q^0, p^0 and q, p. The trick is to note that the first set of equations in (20) expresses the p_α directly as functions of the q_α and q_α^0. We differentiate (17) with respect to time along a motion and obtain

$$\frac{dS}{dt} = \frac{\partial S}{\partial q_\alpha}\dot{q}_\alpha + \frac{\partial S}{\partial t} = L(q, \dot{q}, t), \tag{21}$$

or

$$\frac{\partial S}{\partial t} = L - \frac{\partial S}{\partial q_\alpha}\dot{q}_\alpha = L - p_\alpha \dot{q}_\alpha = -H(q, p, t).$$

Then by inserting the first set of (20) we arrive at

$$\frac{\partial S}{\partial t} = -H\left(q, \frac{\partial S}{\partial q}, t\right). \tag{22}$$

This partial differential equation for S is the celebrated *Hamilton-Jacobi* (HJ) *Equation*, whose solution, the desired generating function S, can be obtained without first knowing the motion.

We will discuss solutions of (22) a little later. So far we have shown that Eq. (17) is a solution of (22) and that this solution gives the canonical transformation according to (20). Also we see that the S given by (17) depends on the n constants $q_1^0, \ldots, q_n^0$. In general, however, the solution we obtain to (22) will not be the same as that given by (17) and so in order to proceed we need to prove that we can work with other types of solutions. To this end we prove the following:

THEOREM. Let $S(q, Q, t)$ be any solution of the Hamilton-Jacobi equation depending on $n + 1$ constants $Q_1, \ldots, Q_n$, C so that $\det(\partial^2 S/\partial q_\alpha\, \partial Q_\beta) \neq 0$. Then the equations

$$p_\alpha = \frac{\partial S}{\partial q_\alpha}, \qquad P_\alpha = -\frac{\partial S}{\partial Q_\alpha}, \tag{23}$$

where $P_1, \ldots, P_n$ are constants, define the canonical transformation from the $2n$ constant of the motion Q_α, P_α to the q_α, p_α which are the motions of the system (satisfy the canonical equations).

Remarks. *1.* We call one of the constants C because it plays a special role: since only derivatives of S appear in (22), if S is a solution, then so is $S + C$, where C is any constant. Most of the time we shall simply leave the constant C out of the function, adding it at the end. Note that C plays no role in the theorem, in particular in Eqs. (23). *2.* The condition on the determinant ensures that the second set of (23) can be solved for the $q_\alpha(Q, P, t)$. These functions can then be inserted into the first set to yield $p_\alpha(Q, P, t)$. Likewise this condition makes it possible to write the Q_α, P_α as functions of q, p, t, so that the Q_α, P_α can constitute $2n$ independent constants of the motion in the sense of Chapter III.

PROOF. We first differentiate the second set of equations in (23) along the alleged motion:

$$\dot{P}_\alpha = 0 = -\frac{d}{dt}\frac{\partial S}{\partial Q_\alpha} = -\frac{\partial^2 S}{\partial Q_\alpha \partial q_\beta}\dot{q}_\beta - \frac{\partial^2 S}{\partial Q_\alpha \partial t}.$$

But from (22) we have

$$\frac{\partial^2 S}{\partial Q_\alpha\, \partial t} = -\frac{\partial \hat{H}}{\partial Q_\alpha} = -\frac{\partial H}{\partial p_\beta}\frac{\partial^2 S}{\partial Q_\alpha\, \partial q_\beta},$$

where we have set

$$H\left(q, \frac{\partial S}{\partial q}, t\right) = \hat{H}(q, Q, t)$$

by writing the $\partial S/\partial q_\alpha$ as functions of q, Q, t. Thus

$$0 = -\frac{\partial^2 S}{\partial Q_\alpha\, \partial q_\beta}\left[\dot{q}_\beta - \frac{\partial H}{\partial p_\beta}\right].$$

Then det $(\partial^2 S/\partial Q_\alpha\, \partial q_\beta) \neq 0$ implies that

$$\dot{q}_\beta = \frac{\partial H}{\partial p_\alpha}.$$

This proves that the alleged motion satisfies the first set of the canonical equations.

To prove that the alleged motion also satisfies the other set of canonical equations we differentiate the first set of (23) along this motion:

$$\dot{p}_\alpha = \frac{d}{dt}\frac{\partial S}{\partial q_\alpha} = \frac{\partial^2 S}{\partial q_\alpha\, \partial q_\beta}\dot{q}_\beta + \frac{\partial^2 S}{\partial q_\alpha\, \partial t}.$$

Again we use (22) to get

$$\frac{\partial^2 S}{\partial q_\alpha\, \partial t} = -\frac{\partial \hat{H}}{\partial q_\alpha} = -\frac{\partial H}{\partial q_\alpha} - \frac{\partial H}{\partial p_\beta}\frac{\partial^2 S}{\partial q_\alpha\, \partial q_\beta},$$

so that

$$\dot{p}_\alpha = -\frac{\partial H}{\partial q_\alpha} + \frac{\partial^2 S}{\partial q_\alpha\, \partial q_\beta}\left(\dot{q}_\beta - \frac{\partial H}{\partial p_\beta}\right) = -\frac{\partial H}{\partial q_\alpha}.$$

This proves that the alleged motion satisfies the canonical equations of motion, which is what was required.

Thus we see that any solution of the HJ equation which depends as in the theorem on $n + 1$ constants allows us to find the motion in terms of $2n$ independent constants of the motion. These $2n$ constants can then be related to the initial values of the variables if that is what we want to do.

We now turn to a brief discussion of the solutions of the HJ equation. The HJ equation is a first-order, generally nonlinear, partial differential equation for S in the $n + 1$ variables $q_1, \ldots, q_n, t$. In general the solution of a partial differential equation, and hence the solution of the HJ equation, depends not on arbitrary constants of integration, but on an arbitrary function. But a fundamental role is played by the so-called *complete solutions*, which depend on constants. General solutions can be obtained by manipulating complete solutions, but for us any complete solution, called *Hamilton's principal function*, is all that is needed. The Q_α which actually arise in any particular complete solution will seldom be the q_α^0 of (20), but the transformation from the q_α^0, p_α^0 to the Q_α, P_α is easily obtained by setting $t = t^0$ in Eqs. (23).

As a result we have arrived at the following HJ prescription for solving the dynamical problem: Obtain a complete solution S of Eq. (22), depending on n arbitrary constants Q_α (we are leaving out the additive constant C) so that $\det(\partial^2 S/\partial q_\alpha \partial Q_\beta) \neq 0$. Insert this solution into (23) and solve algebraically for $q_\alpha(Q, P, t)$, $p_\alpha(Q, P, t)$, where the constants P_α are also arbitrary.

To illustrate these ideas we return to the free particle in one dimension. The Hamiltonian, as before, is $H = \frac{1}{2}p^2$, so that the HJ equation is

$$\frac{\partial S}{\partial t} = -\frac{1}{2}\left(\frac{\partial S}{\partial q}\right)^2.$$

This will be satisfied by a function S such that $-\partial S/\partial t$ and $\frac{1}{2}(\partial S/\partial q)^2$ are the same constant, say Q. Then

$$S = q\sqrt{2Q} - Qt$$

is a solution depending on one constant (two if we add C), and $\partial^2 S/\partial q\, \partial Q = (2Q)^{-1/2} \neq 0$. Inserting this S into (22) yields

$$p = \sqrt{2Q}, \qquad P = t - \frac{q}{\sqrt{2Q}}.$$

Then by solving algebraically we have

$$q = -P\sqrt{2Q} + \sqrt{2Q}t, \qquad p = \sqrt{2Q}$$

which shows that p is a constant. On comparing this with the previous form of the solution following from Eq. (17), or simply by setting $q(t^0) = q^0$, $p(t^0) = p^0$, we see that $Q = \frac{1}{2}(p^0)^2$ and $P = -(q^0/p^0) + t^0$. Note that the present S and that of Eq. (19) are very different indeed. No adjustment of Q or P will bring them to the same form, for this one is linear in q and t, and Eq. (19) is quadratic in q and depends on $1/t$.

To summarize, the HJ equation provides a method for finding a canonical transformation from the ξ_α to a set of $2n$ (independent) constants η_α. As pointed out in Chapter III this is equivalent to solving the dynamical problem.

VII-4 SEPARATION OF VARIABLES; ACTION-ANGLE VARIABLES

(In this section we do *not* use the summation convention)

The only general method we know for solving the HJ equation is called separation of variables, a method we have already used in the last example.

It is a method which works only for certain Hamiltonians and in certain generalized coordinates, but when it works it is very easy to solve the HJ equation. For most other Hamiltonians and in other generalized coordinates we know of no general method for solving the HJ equation.

In the cases to which we shall apply the method of separation of variables, the Hamiltonian will be time independent. If we then restrict our considerations to such Hamiltonians, the HJ equation is

$$\frac{\partial S(q, t; Q)}{\partial t} = -H\left(q, \frac{\partial S}{\partial q}\right), \tag{24}$$

and we shall first try to find a solution that can be written in the separated form

$$S = W(q, Q) + T(t).$$

Substituting this in (24), we get

$$\frac{dT}{dt} = -H\left(q, \frac{\partial W}{\partial q}\right).$$

The left-hand side depends only on t whereas the right-hand side depends only on the q_α; therefore both sides must equal a constant independent of both the q_α and t. Let this constant be $-Q_1$, obtaining

$$\begin{aligned} T &= -Q_1 t, \\ S &= W - Q_1 t, \end{aligned} \tag{25}$$

and the following equation for W:

$$H\left(q, \frac{\partial W}{\partial q}\right) = Q_1. \tag{26}$$

Equation (26) is a partial differential equation in n variables for W, which is called *Hamilton's characteristic function.*

This shows that for time-independent Hamiltonians we can always separate out the time. We can proceed further by separation of variables only if (26) is similarly separable in each of the q_α, that is, only if a solution can be written in the form

$$W = \sum_{\alpha=1}^{n} W_\alpha(q_\alpha, Q). \tag{27}$$

Let us therefore assume that such a solution exists and let us try to find the W_α.

We insert (27) into (26) and try to multiply H by a function $f(q_1, \ldots, q_{n-1})$, of at most the first $n - 1$ variables, so that all the dependence on q_n is contained in one term which does not depend on the other q_α. If such a function f exists, then (26) will become

$$fH\left(q, \frac{\partial W}{\partial q}\right) = H'_{n-1}\left(q, \frac{\partial W}{\partial q}\right) + H_n\left(q_n, \frac{\partial W_n}{\partial q_n}\right) = f(q_1, \ldots, q_{n-1})Q_1,$$

where H'_{n-1} does not depend on q_n or $\partial W_n/\partial q_n$. Then having isolated the q_n-dependence we can split off H_n and write

$$H_n\left(q_n, \frac{\partial W_n}{\partial q_n}\right) = Q_n.$$

This is a first-order ordinary differential equation for W_n and as such, it can always be solved (or at least reduced to an integral). We write $\partial W_n/\partial q_n$ rather than dW_n/dq_n to emphasize that W_n depends on the Q_α also.

Now we are left with a partial differential equation in the $n - 1$ variables $q_1, \ldots, q_{n-1}$:

$$H'_{n-1} - fQ_1 = -Q_n.$$

Repeating the process, we try to separate out another variable, say q_{n-1}, and having done that, to separate still another. In this way, if all the variables can be separated out, we reduce the solution of (26) to the solution of n ordinary first-order differential equations, each one of the form (no summation convention)

$$H_\alpha\left(q_\alpha, \frac{\partial W_\alpha}{\partial q_\alpha}, Q\right) = Q_\alpha. \tag{28}$$

Then according to (25) and (27) the resulting solution of the HJ equation is

$$S(q, t; Q) = \sum_{\alpha=1}^{n} W_\alpha(q_\alpha, Q) - Q_1 t. \tag{29}$$

Equations (23), the transformation equations which give the motion, now become

$$p_\alpha = \frac{\partial W_\alpha}{\partial q_\alpha}, \tag{30}$$

$$P_\alpha = -\sum_\beta \frac{\partial W_\beta}{\partial Q_\alpha} + \delta_{1\alpha} t. \tag{31}$$

This procedure is what we mean by separation of variables, and when it is possible we say that the HJ equation is separable. There are a number of important systems whose HJ equations are separable. The most important of these, a particle moving in a central force, is discussed in detail in Section 6(a). As might be supposed, separable systems have a number of other special properties, which we now explore.

From Eqs. (30) we see that each p_α can be expressed as a function only of its conjugate coordinate q_α and the constants Q_β. Thus we can plot a graph of p_α as a function of q_α for each set of values of the n constants Q_β. Since in any motion these Q_β (as well as the P_β) are fixed, this graph is a projection of the motion on the (q_α, p_α)-plane (we will call this the *α-plane*) of phase space.

For example, a particle subjected to a constant force in the 3-direction has the Hamiltonian

$$H = \frac{1}{2m}(p_1^2 + p_2^2 + p_3^2) - Fq_3,$$

and separation yields

$$p_3 = \pm\sqrt{2mQ_3 + 2mFq_3}.$$

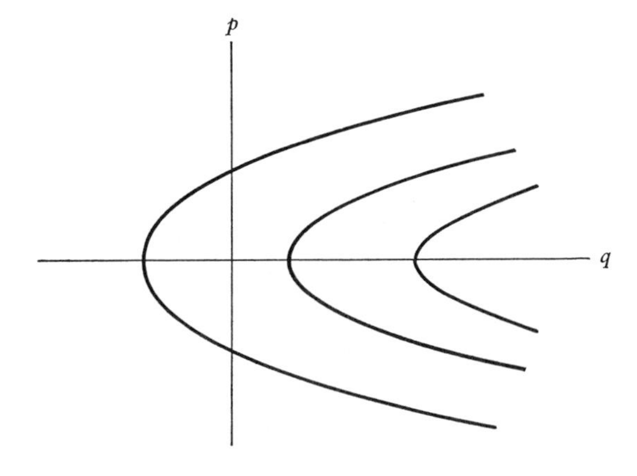

Figure 32. The 3-plane of phase space for a particle subject to a constant force in the 3-direction.

The graph of p_3 as a function of q_3 is a family of parabolas, as shown in Fig. 32, each parabola corresponding to a different value of Q_3.

Remark. In this section when we refer to orbits we mean these graphs in the α-planes of phase space. They should not be confused either with orbits in configuration space (e.g., graphs of r *vs.* θ) or with motions (graphs of q *vs.* t).

The system illustrated in Fig. 32 is unbounded, by which we mean that each orbit includes arbitrarily large values of q. If a system is bounded, on the other hand, separability has further consequences because essentially only the following two types of orbits are possible.

1. *Vibration.* In this case the graph of p_α *vs.* q_α is a closed curve, as shown in Fig. 33. As the value of q_α oscillates back and forth between the turning points, a point on the phase-space orbit moves once around the curve, completing one *cycle.* While this orbit does not depend on the P_β,

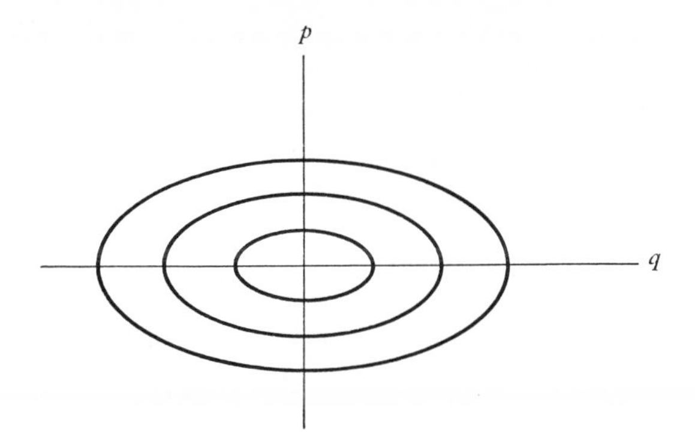

Figure 33. The phase space for vibration.

the time it takes q_α to execute a complete cycle often does. In fact, there may be no well-defined period for the execution of all the cycles, the time changing from one execution of the cycle to the next, depending on the motion of the other coordinates. (For a system with only one degree of freedom, there being no other coordinates, a period does exist.) Since the Hamiltonian is ordinarily quadratic in p_α, the expression for p_α in terms of q_α usually involves a square root. The turning points are then characterized as the points at which p_α becomes complex, and the orbit will be symmetric about a line connecting the turning points, that is, about the q axis (see Figs . 32, 33, and 34). The coordinate q_α is said to be *single valued* in this case because each configuration of the system corresponds to just one value of q_α.

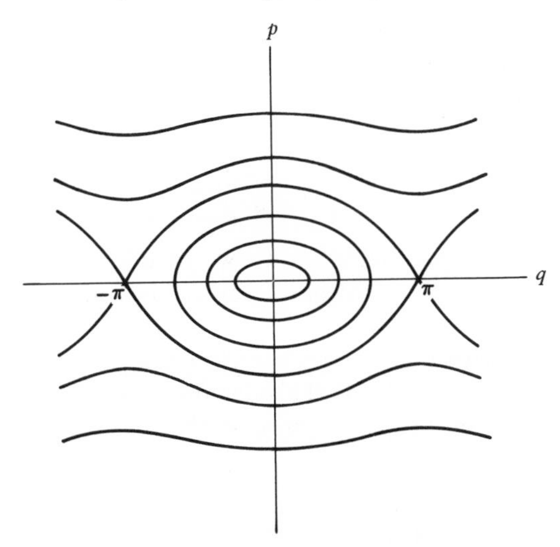

Figure 34. The phase space for a simple pendulum, illustrating vibration (closed curves) and rotation (the other curves).

2. *Rotation.* In this case there exists an interval q_α^0 such that q_α and $q_\alpha + mq_\alpha^0$ correspond to the same configuration if m is an integer. Then p_α is a periodic function of q_α with period q_α^0. This usually occurs for a system which is rotating in some way, and then q_α can sometimes be the angle of rotation, so that $q_\alpha^0 = 2\pi$. (If the angle q_α is a cyclic or ignorable coordinate, the phase-space orbit is $p_\alpha = \text{const}$, which we may also take to be a periodic function with period 2π.) Again, the time it takes the system to go from q_α to $q_\alpha + q_\alpha^0$ is not in general independent of q_α, and therefore the q_α motion need not have a well-defined time period. For rotation, the coordinate q_α is said to be *multiple valued*, because each configuration corresponds to many different values of q_α.

The reader should be able to convince himself that situations other than the two described cannot occur for bounded separable systems. Essentially this is because an orbit cannot cross itself and it cannot have a beginning or ending point.

The variable q_α can change character from multiple to single valued and vice versa depending on the values of the Q_β, or in other words depending on the particular motion. For example, consider the simple pendulum, whose Hamiltonian is

$$H = \frac{p^2}{2Ml^2} - Mgl\cos\theta = Q,$$

where l is the length and M the mass. If $Q < Mgl$ the motion is restricted and θ is single valued. If $Q > Mgl$, the motion passes through $\theta = -\pi$ and $\theta = \pi$, as shown in Fig. 34 so that any given value of θ will describe the same position as $\theta + 2\pi m$. Therefore θ becomes multiple valued.

The special properties of bounded separable systems allow one to perform a canonical transformation to a particularly useful set of variables on phase space, the so-called *action-angle variables.*

The *action variables* J_α are defined by the integrals

$$J_\alpha = \oint p_\alpha\, dq_\alpha = \oint \frac{\partial W_\alpha}{\partial q_\alpha}\, dq_\alpha \tag{32}$$

for each α-plane of phase space. (Remember: we are not using the summation convention in this section.) The integral is taken over a complete cycle of variation of q_α and, as is seen from Figs. 33 and 34, it represents either the area inside a closed orbit (if q_α is single valued) or the area under one cycle of an orbit (if q_α is multiple valued). This demonstrates that J_α will depend only on the Q_β, for the orbits depend only on the Q_β. Analytically this can be seen from the expressions

$$J_\alpha = \oint \frac{\partial W_\alpha(q_\alpha, Q)}{\partial q_\alpha}\, dq_\alpha \equiv \hat{J}_\alpha(Q). \tag{33}$$

Since the integration is over a complete cycle, the q_α dependence integrates out. But the Q_α, being constants of the motion, can be written as functions of q, p. Thus we can write

$$J_\alpha = \hat{J}_\alpha(Q) = J_\alpha(q, p). \tag{34}$$

This is half of the canonical transformation we want to define.

Remarks. 1. In general any constant of the motion can be expanded as a function of the $2n$ independent constants Q_β, P_β, but each J_α depends on the Q_β alone. Moreover, the Q_β determine the orbits uniquely, and the orbits determine their areas, namely the J_α, uniquely. Thus the relation between the Q_β and J_α is invertible: the J_α, like the Q_β, are independent. Analytically, one can prove that det $(\partial \hat{J}_\alpha / \partial Q_\beta) \neq 0$. We leave this to the Problems. *2.* If H is the energy or at any rate if it has the dimensions of energy, the J_α have the dimensions of angular momentum. This is because $\dot{q}_\alpha = \partial H/\partial p_\alpha$ implies that $p_\alpha \dot{q}_\alpha$ has the dimensions of energy, or $\oint p_\alpha \, dq_\alpha$ has the dimensions of energy $\times$ time $=$ angular momentum. In fact if H has the dimensions of energy, the product of any canonically conjugate pair of variables has the dimensions of angular momentum. This means, for instance, that the variables conjugate to the J_α, whatever other properties they may have, will be dimensionless.

Let us now write

$$W_\alpha(q_\alpha, Q(J)) = \hat{W}_\alpha(q_\alpha, J)$$

and $\hat{W} = \Sigma \hat{W}_\alpha$. Note that

$$\frac{\partial W_\alpha}{\partial q_\alpha} = \frac{\partial \hat{W}_\alpha}{\partial q_\alpha}. \tag{35}$$

To complete the canonical transformation started by Eq. (34) we choose the J_α to be generalized momenta, since they have the dimensions of angular momentum. Then the equations [see Eqs. (30) and (35)]

$$p_\alpha = \frac{\partial \hat{W}}{\partial q_\alpha} = \frac{\partial \hat{W}_\alpha}{\partial q_\alpha}$$

can be thought of as the first half of the type-2 equations defining the transformation [see Eq. (57) of Chapter VI], and $\hat{W}(q, J)$ can be thought of as the type-2 generator. If we write w_α for the position variables conjugate to the J_α, the complete set of type-2 equations becomes

$$w_\alpha = \frac{\partial \hat{W}}{\partial J_\alpha}, \qquad p_\alpha = \frac{\partial \hat{W}_\alpha}{\partial q_\alpha}, \tag{36}$$

the first of which may be taken as defining the *angle variables* w_α. Finally, the new Hamiltonian is

$$K = H + \frac{\partial \hat{W}}{\partial t} = H,$$

where H must be written as a function of the J_α and w_α. But the w_α do not appear in H, for $H = Q_1$ is a constant of the motion that can be written in terms of the J_α alone.

The new Hamiltonian immediately gives the equations of motion for the new variables (we already know that the J_α are constants of the motion):

$$\dot{J}_\alpha = -\frac{\partial Q_1}{\partial w_\alpha} = 0; \qquad \dot{w} = \frac{\partial Q_1}{\partial J_\alpha} \equiv \nu_\alpha, \tag{37}$$

where the ν_α are constants because $Q_1(J)$ is a function only of constants (the ν_α depend only on the J_β). The solutions of (37) are

$$J_\alpha = \text{constant},$$
$$w_\alpha = \nu_\alpha t + \phi_\alpha, \tag{38}$$

where the ϕ_α are new constants. The w_α increase without limit and are therefore multiple-valued variables. Among other things, the transformation to action-angle variables thus transforms all generalized coordinates, both single and multiple valued, into multiple-valued coordinates.

According to (36) each of the w_α depends in general on all of the q_β. But this dependence is of a special kind: w_α is periodic in the q_β for $\beta \neq \alpha$ and increases monotonically with q_α, changing by 1 in each α-cycle. We prove this by calculating $\Delta_\beta w_\alpha$, the change in w_α in going once around an orbit in the β-plane of phase space (i.e., when q_β goes through one cycle and all the other q_γ and all the J_γ or Q_γ are held fixed). We have (remember: no summation convention)

$$\Delta_\beta w_\alpha = \oint \frac{\partial w_\alpha}{\partial q_\beta}\, dq_\beta = \oint \frac{\partial^2 \hat{W}}{\partial q_\beta\, \partial J_\alpha}\, dq_\beta$$
$$= \frac{\partial}{\partial J_\alpha} \oint \frac{\partial \hat{W}_\beta}{\partial q_\beta}\, dq_\beta = \frac{\partial J_\beta}{\partial J_\alpha} = \delta_{\alpha\beta},$$

as asserted.

Remark. This calculation does not show that w_α changes by 1 in an α-cycle during an actual motion. During the motion all the q_γ are changing, while in the above calculation for $\beta = \alpha$ all but q_α were held fixed.

Our canonical transformation may now be written in the form

$$w_\alpha = w_\alpha(q, p), \qquad J_\alpha = J_\alpha(q, p), \tag{39}$$

and its inverse in the form

$$q_\alpha = q_\alpha(w, J), \qquad p_\alpha = p_\alpha(w, J). \tag{40}$$

We know something about how the w_β vary as functions of the q_α with the J_α held fixed, namely, about the first set of Eqs. (36). But the first set of

Eqs. (40) are just the solutions of those equations for the q_α as functions of the w_β. Therefore we know something about the way the q_α depend on the w_β when the J_α are held fixed: when w_α changes by 1, q_α has gone through one cycle and all the other q_β, $\beta \neq \alpha$, have returned to their original values. In one cycle a single-valued q_α has then returned to its original value. Thus for such single-valued q_α we have

$$q_\alpha(w + m, J) = q_\alpha(w, J), \tag{41}$$

where $w + m$ is the collection $(w_1 + m_1, w_2 + m_2, \ldots, w_n + m_n)$, in which the m_γ are arbitrary integers. If q_α is multiple valued, we have

$$q_\alpha(w + m, J) = q_\alpha(w, J) + q_\alpha^0 m_\alpha,$$

and if we write $q'_\alpha = q_\alpha - q_\alpha^0 w_\alpha$, we find that q'_α satisfies (41). Similarly, of course, the sine and cosine of q_α will be periodic.

Equation (41) tells us that q_α (and in the multiple-valued case q'_α) can be expanded in a multiple Fourier series:

$$q_\alpha(w, J) = \sum A_{m_1 \cdots m_n, \alpha} \exp [2\pi i(m_1 w_1 + \cdots + m_n w_n)], \tag{42}$$

where the A's are functions of the J_β and the multiple sum is over all the m_γ from $-\infty$ to $+\infty$. This is convenient because Eqs. (38) tell us how the w_α and J_α depend on the time. From them we obtain

$$q_\alpha(t, J, \phi) = \sum B_{m_1 \cdots m_n, \alpha} \exp [2\pi i t(m_1 \nu_1 + \cdots + m_n \nu_n)], \tag{43}$$

where

$$B_{m_1 \cdots m_n, \alpha} = A_{m_1 \cdots m_n, \alpha} \exp [2\pi i(m_1 \phi_1 + \cdots + m_n \phi_n)].$$

Note again that in the case of multiple-valued variables it is the q'_α which can be so expanded.

These results extend immediately to dynamical variables in general. Every dynamical variable that can be written as a function of the q_α and p_α can also be written as a function of the q_α and $\dot{q}_\alpha$. According to Eq. (43) the $\dot{q}_\alpha$ contain the same *frequencies* ν_β as do the q_α, and therefore every single-valued dynamical variable $F(q, p)$ can be written in the form

$$F(q, p) = \sum F_{m_1 \cdots m_n} \exp [2\pi i t(m_1 \nu_1 + \cdots + m_n \nu_n)].$$

The frequencies ν_β do not in general correspond to frequencies of actual motion. That is, neither the system nor even q_β will in general return to its initial state after a time $1/\nu_\beta$. In fact although p_α is a periodic function of q_α in the sense already described, the q_α (and hence also the p_α) need not be periodic functions of the time. Because the q_α have finite ranges, each q_α will in general return repeatedly to any given value, but the time it takes to return is not the same for every repetition.

Suppose, however, that all the ν_α are commensurable, that is that there exist integers k_α such that (remember: no sum) $\nu_\alpha/k_\alpha = \nu_\beta/k_\beta \equiv \mu$ for all α, β. Then after a time $T = 1/\mu$ each exponential in Eq. (43) becomes of the form

$$\exp\,[2\pi i(m_1k_1 + m_2k_2 + \cdots + m_nk_n)] = 1,$$

where the m_α and k_α are integers, and thus each q_α takes on the same value as it had at $t = 0$. The system, said to be *completely degenerate*, is then truly periodic. Another way to see this, not using the Fourier series of (43), is to note that after a time T each w_α has changed by

$$\Delta_T w_\alpha = \nu_\alpha T = k_\alpha,$$

which is an integer. We have seen that everytime a w_α changes by one unit, all the q_β return to their initial values, and now we see that in the time T all of the w_α have changed by an integral number. Thus all of the q_β are back to their initial values.

Conversely, assume that the system has a period T, or that for each α

$$q_\alpha(t + T) = q_\alpha(t).$$

Then since all the q_α return to their initial values after a time T, all the w_α have changed by an integer: there exist integers k_α such that for each α

$$\Delta_T w_\alpha = \nu_\alpha T = k_\alpha,$$

or[4]

$$\frac{\nu_\alpha}{k_\alpha} = \frac{\nu_\beta}{k_\beta} = \frac{1}{T},$$

so that the ν_α are commensurable. Even in this periodic or completely degenerate case, the ν_α are not individually frequencies of motion in general. See Problem 8.

A system in which no two ν_β are commensurable is called *conditionally periodic*. Section 6(b) demonstrates that for a conditionally periodic system the w_β do not change by 1 in each β-cycle of an actual motion. Such a system does not in general return to its initial state, and in configuration space its orbit may never close smoothly, but may be *ergodic* in some region. This means that in some region the orbit will eventually come arbitrarily close to any point.

Whether a system is truly periodic can, as a result, be found by finding whether it is completely degenerate, and hence by studying the ratios of the ν_α. Consider for instance a system with two degrees of freedom, so that

[4] We are assuming that in general all degrees of freedom are excited so that $\Delta_T w_\alpha \neq 0$ for any α. If some degrees of freedom are constrained or happen not to be excited in a given motion, the discussion applies only to those which are excited. In that case we can say only that some of the ν_α are commensurable, and the system is called simply *degenerate*.

$H = Q_1 = Q_1(J_1, J_2)$. This can be solved for $J_1 = \bar{J}_1(Q_1, J_2)$, and then some simple partial differentiation yields

$$\frac{\nu_2}{\nu_1} = \frac{\partial Q_1/\partial J_2}{\partial Q_1/\partial J_1} = -\frac{\partial \bar{J}_1}{\partial J_2}. \tag{44}$$

Then the system may be completely degenerate only if ν_2/ν_1 is rational, that is, only if $\partial \bar{J}_1/\partial J_2$ is a rational number. However $\partial \bar{J}_1/\partial J_2$ can be a rational number and yet the system may not be completely degenerate. For instance, the system could have more than two degrees of freedom, but with only ν_1 and ν_2 commensurable, as discussed in footnote 4.

VII-5 CANONICAL PERTURBATION THEORY

Many important problems in classical mechanics, like satellite motion in the combined field of the Earth and the Moon or in the field of the slightly nonspherical Earth, can not be solved exactly. Therefore it is useful to develop methods for obtaining approximate solutions. *Canonical perturbation theory* is such a method. It is of particular interest here because it makes use of canonical transformations and action-angle variables and because it is analogous to quantum perturbation theory.

The method applies to separable bounded systems of the kind we have been discussing, and involves approximating the generating function of the canonical transformation from q, p to the action-angle variables. It is called *perturbation theory* because it is most useful for a system which differs in some sense only slightly from one whose motion is known exactly. For simplicity, the method will be discussed in detail for a system with one degree of freedom, and the extension to systems with more degrees of freedom will be briefly outlined at the end of the section.

Let $H(q, p, \lambda)$ be a Hamiltonian which gives rise to bounded motion for all values of the parameter λ in some interval that includes $\lambda = 0$. Furthermore let $H(q, p, \lambda)$ be such that for $\lambda = 0$ the motion given by the Hamiltonian

$$H(q, p, 0) \equiv H_0(q, p)$$

is known exactly in terms of action-angle variables. H_0 is called the *unperturbed* Hamiltonian.

Thus we know the canonical transformation from q, p to the action-angle variables J^0, w^0 for H_0. In terms of these new variables the new unperturbed Hamiltonian K_0 is

$$K_0(J^0) = H_0(q, p),$$

and the corresponding equations of the *unperturbed* motion are

$$\dot{w}^0 = \partial K_0/\partial J^0 = \nu^0(J^0), \qquad \dot{J}^0 = 0. \tag{45}$$

Further, let

$$K(w^0, J^0, \lambda) = H(q, p, \lambda)$$

be the new *perturbed* Hamiltonian expressed in terms of these new variables.

By assumption the motion given by the actual perturbed Hamiltonian H is also bounded and so action-angle variables J, w exist in terms of which

$$K(w^0, J^0, \lambda) = H(q, p, \lambda) = E(J, \lambda). \tag{46}$$

Since both w^0, J^0 and w, J are obtained from q, p by canonical transformations there must exist a canonical transformation from one set of these variables to the other. We proceed to find the generator of this transformation.

The transformation is time independent and we know that for sufficiently small λ it must be generated by a type-2 generator, which we shall call $S(w^0, J, \lambda)$. It is of type 2 because at $\lambda = 0$ the transformation is the identity transformation, which is of type 2, and we assume that the generator varies continuously with λ. In particular we assume that S is an analytic function of λ in some interval which includes $\lambda = 0$, so that we can expand it in a power series in λ:

$$S(w^0, J, \lambda) = S_0(w^0, J) + \lambda S_1(w^0, J) + \lambda^2 S_2(w^0, J) + \cdots.$$

Clearly $S_0(w^0, J) = w^0 J$, for it generates the identity transformation. Our problem will be to find S (actually the S_k) and from it the new action-angle variables.

Remark. The generating functions S and $S' = S + aw^0 + bJ$, where a and b are constants, give canonical transformations whose equations for w and J differ only by constants. For this reason we will simplify expressions we may obtain for S by dropping linear terms in w^0 and J which may occur.

The transformation generated by our S function can be written in the form

$$J^0 = \frac{\partial S}{\partial w^0} = J + \lambda \frac{\partial S_1}{\partial w^0} + \lambda^2 \frac{\partial S_2}{\partial w^0} + \cdots, \qquad w = \frac{\partial S}{\partial J} = w^0 + \lambda \frac{\partial S_1}{\partial J} + \lambda^2 \frac{\partial S_2}{\partial J} + \cdots. \tag{47}$$

The Hamiltonian $K(w^0, J^0, \lambda)$ of Eq. (46) can also be expanded in powers of λ:

$$K(w^0, J^0, \lambda) = K_0(J^0) + \lambda K_1(w^0, J^0) + \lambda^2 K_2(w^0, J^0) + \cdots. \tag{48a}$$

Here we have used the fact that w^0, J^0 are the action-angle variables for $K(w^0, J^0, 0) = K_0(J^0)$. Since K is a known function of w^0, J^0, the functions $K_0, K_1, K_2, \ldots$ are also known.

When w^0, J^0 are expressed in terms of w, J, the Hamiltonian becomes

$$K(w^0(w, J), J^0(w, J), \lambda) = E(J, \lambda)$$

and is independent of w. Let us expand $E(J, \lambda)$ in powers of λ. We write

$$E(J, \lambda) = E_0(J) + \lambda E_1(J) + \lambda^2 E_2(J) + \cdots. \tag{48b}$$

To find the Hamiltonian $E(J, \lambda)$ we must find the E_k. This can not be done simply by equating the coefficients of λ in (48*b*) to those in (48*a*), for (48*a*) has addition λ-dependence through the relationship between J and J^0. In order then to find the E_k in terms of the K_k we must first express K in terms of J rather than J^0. This we can do by expanding $K(w^0, J^0, \lambda)$ about $K(w^0, J, \lambda)$

$$K(w^0, J^0, \lambda) = K(w^0, J, \lambda) + \frac{\partial K}{\partial J}(J^0 - J) + \frac{1}{2}\frac{\partial^2 K}{\partial J^2}(J^0 - J)^2 + \cdots,$$

and using (47) to get

$$\begin{aligned} K(w^0, J^0, \lambda) &= K(w^0, J, \lambda) + \frac{\partial K}{\partial J}\left(\lambda \frac{\partial S_1}{\partial w^0} + \lambda^2 \frac{\partial S_2}{\partial w^0} + \cdots\right) \\ &\quad + \frac{1}{2}\frac{\partial^2 K}{\partial J^2}\left(\lambda \frac{\partial S_1}{\partial w^0} + \lambda^2 \frac{\partial S_2}{\partial w^0} + \cdots\right)^2 \\ &= K(w^0, J, \lambda) + \lambda \frac{\partial K}{\partial J}\frac{\partial S_1}{\partial w^0} + \lambda^2\left[\frac{\partial K}{\partial J}\frac{\partial S_2}{\partial w^0} + \frac{1}{2}\frac{\partial^2 K}{\partial J^2}\left(\frac{\partial S_1}{\partial w^0}\right)^2\right] + \cdots, \end{aligned}$$

where we have kept only terms to order λ^2. Next we replace $K(w^0, J, \lambda)$ on the right-hand side by (48*a*) evaluated at $J^0 = J$. Again keeping only terms to order λ^2 we get

$$\begin{aligned} K(w^0, J^0, \lambda) &= K_0(J) + \lambda K_1(w^0, J) + \lambda^2 K_2(w^0, J) + \cdots \\ &\quad + \lambda \frac{\partial S_1}{\partial w^0}\left(\frac{\partial K_0}{\partial J} + \lambda \frac{\partial K_1}{\partial J} + \cdots\right) \\ &\quad + \lambda^2\left[\frac{\partial K_0}{\partial J}\frac{\partial S_2}{\partial w^0} + \frac{1}{2}\frac{\partial^2 K_0}{\partial J^2}\left(\frac{\partial S_1}{\partial w^0}\right)^2 + \cdots\right] + \cdots, \end{aligned}$$

which is equal to Eq. (48*b*). Both series are now functions of λ and J, and so we can equate the coefficients of equal powers of λ to obtain the following important relations for the S_k and E_k:

$$E_0(J) = K_0(J), \tag{49a}$$

$$E_1(J) = K_1(w^0, J) + \nu^0(J)\frac{\partial S_1}{\partial w^0}, \tag{49b}$$

$$E_2(J) = K_2(w^0, J) + \frac{\partial K_1}{\partial J}\frac{\partial S_1}{\partial w^0} + \nu^0(J)\frac{\partial S_2}{\partial w^0} + \frac{1}{2}\frac{\partial \nu^0}{\partial J}\left(\frac{\partial S_1}{\partial w^0}\right)^2. \tag{49c}$$

Here we have used (45) to express $\partial K_0/\partial J$ in terms of the unperturbed frequency ν^0.

Remark. Recall that $K_0, K_1, K_2, \ldots$ are known functions obtained from the expansion of the Hamiltonian expressed as a function of w^0 and J^0. However, since we have expanded K about J, these functions and their derivatives are all evaluated at J rather than J^0. In particular ν^0 is to be evaluated at J and not J^0.

Before preceding, let us stop and see where we are going. We want to determine the S_k ($k = 1, 2 \cdots$) so that the transformation in (47) will be known. We already know the transformation from the q, p to the w^0, J^0:

$$q = q(w^0, J^0), \qquad p = p(w^0, J^0). \tag{50}$$

This a canonical transformation, and as such it is independent of any Hamiltonian. Its special significance is that for the unperturbed Hamiltonian H_0 the (unperturbed) motion is given by $J^0 = \text{const}$ and $w^0 = \nu^0 t + \text{const}$.

Once we find the S_k we will be able to express q, p in terms of w, J by the relations

$$q = q(w^0(w, J), J^0(w, J)), \qquad p = p(w^0(w, J), J^0(w, J)). \tag{51}$$

The special significance of this transformation is that for the perturbed Hamiltonian H the (perturbed) motion is given (approximately) by $J = \text{const}$ and $w = \nu t + \text{const}$. The new frequency ν is obtained from the E_k, as we will soon show.

Because both w, J and w^0, J^0 are action-angle variables, q and p are periodic functions of both w and w^0 with period one. If w^0 changes by one, q and p return to their original values, and thus w changes by some integer. But when we assume that the system is amenable to perturbation theory, we assume that w and w^0 do not differ greatly. Hence the integer by which w changes is also one, and thus $w - w^0$, as well, of course, as J and J^0, are periodic functions of w^0 with period one. This must be true for all values of λ, and hence it follows from the second of Eqs. (47) that each $\partial S_k/\partial J$ is similarly periodic in w_0. Then each S_k is the sum of such a periodic function and an arbitrary function $f_k(w^0)$. According to the first of Eqs. (47) the derivatives $f_k'(w^0)$ must also be periodic, so that each S_k is the sum of a function periodic with period one in w^0 plus a linear function $c_k w^0$. These linear functions introduce a λ-dependent linear function into S, and in the Remark just before Eqs. (47) we have agreed to drop such terms from S. Thus we set $c_k = 0$ and then each S_k can be expressed as a Fourier series of the form

$$S_k(w^0, J) = \sum_{m=-\infty}^{\infty} S_k(J, m) \exp(2\pi i m w^0), \qquad k \neq 0.$$

The $\partial S_k/\partial w^0$ can be expanded similarly, except that these expansions will contain no constant ($m = 0$) term. As a result we obtain the important consequence that

$$\int_0^1 \frac{\partial S_k}{\partial w^0} dw^0 = 0. \tag{52}$$

The importance of this is that it will allow us to calculate the S_k by performing certain integrals.

Let us denote the integral of a function F over w^0 from 0 to 1 by

$$\int_0^1 F\, dw^0 = \langle F \rangle.$$

Then if we integrate (49*b*) in this way and use (52) we arrive at

$$E_1(J) = \langle K_1 \rangle, \tag{53}$$

so that

$$\frac{\partial S_1}{\partial w^0} = \frac{\langle K_1 \rangle - K_1}{\nu^0}. \tag{54}$$

All we need do to find S_1 is to integrate this differential equation. The solution is $S_1(w^0, J) + \eta(J)$, where η is an arbitrary function of J. But Eqs. (47) show that η does not affect the relation between J and J^0 and that it merely adds the function $\lambda\eta'(J)$ to $w - w^0$. Since J is a constant of the motion, we have then simply added a constant to $w - w_0$, which is of no consequence: both w and w^0 are angle variables and the difference $w - w^0$ at, say, $w^0 = 0$ is completely arbitrary. This freedom in the phase difference between w and w^0, reflected in the freedom to add an arbitrary function of J to S_1, allows us to set $\eta(J)$ equal to zero in all subsequent discussions.

We integrate (49*c*) the same way to obtain

$$\begin{aligned} E_2(J) &= \langle K_2 \rangle + \left\langle \frac{\partial K_1}{\partial J} \frac{\partial S_1}{\partial w^0} \right\rangle + \frac{1}{2} \frac{\partial \nu^0}{\partial J} \left\langle \left(\frac{\partial S_1}{\partial w^0} \right)^2 \right\rangle \\ &= \langle K_2 \rangle + \frac{1}{\nu^0} \left[\left\langle \frac{\partial K_1}{\partial J} \right\rangle \langle K_1 \rangle - \left\langle \frac{\partial K_1}{\partial J} K_1 \right\rangle \right] \\ &\quad + \frac{1}{2(\nu^0)^2} \frac{\partial \nu^0}{\partial J} [\langle K_1^2 \rangle - \langle K_1 \rangle^2], \end{aligned} \tag{55}$$

so that

$$\frac{\partial S_2}{\partial w^0} = \frac{1}{\nu^0}\left[E_2 - K_2 - \frac{\partial K_1}{\partial J}\frac{\partial S_1}{\partial w^0} - \frac{1}{2}\frac{\partial \nu^0}{\partial J}\left(\frac{\partial S_1}{\partial w^0}\right)^2\right]$$

$$= \frac{1}{\nu^0}(\langle K_2\rangle - K_2) + \frac{1}{(\nu^0)^2}\left(\left\langle\frac{\partial K_1}{\partial J}\right\rangle\langle K_1\rangle - \left\langle\frac{\partial K_1}{\partial J}K_1\right\rangle\right.$$

$$\left. - \frac{\partial K_1}{\partial J}\langle K_1\rangle + \frac{\partial K_1}{\partial J}K_1\right)$$

$$+ \frac{1}{2(\nu^0)^3}\frac{\partial \nu^0}{\partial J}(\langle K_1^2\rangle - 2\langle K_1\rangle^2 + 2\langle K_1\rangle K_1 - K_1^2). \tag{56}$$

Again, this can be integrated to yield S_2. Then the new variables w, J are found in terms of the old variables w^0, J^0 through Eqs. (47), (54) and (56).

The new perturbed energy can be written in terms of J without first calculating S. Equations (48*b*), (49*a*), (53), and (55) yield

$$E(J) = K_0(J) + \lambda\langle K_1\rangle + \lambda^2\left[\langle K_2\rangle + \frac{1}{\nu^0}\left(\left\langle\frac{\partial K_1}{\partial J}\right\rangle\langle K_1\rangle - \left\langle\frac{\partial K_1}{\partial J}K_1\right\rangle\right)\right.$$

$$\left. + \frac{1}{2(\nu^0)^2}\frac{\partial \nu^0}{\partial J}(\langle K_1^2\rangle - \langle K_1\rangle^2)\right] + \cdots, \tag{57}$$

and from this the new frequency $\nu = \partial E/\partial J$ is readily calculated. This one-dimensional perturbation theory is applied to the simple harmonic oscillator in Section 6(c).

With a little modification this procedure can be applied to a conditionally periodic system with n degrees of freedom. Let $H(q, p, \lambda)$ be a Hamiltonian which gives rise to conditionally periodic motion for all values of λ, such that for $H_0(q, p) = H(q, p, 0)$ the Hamilton-Jacobi equation can be solved and the action-angle variables found. Now q and p stand for the set of coordinates and momenta $q_1, \ldots, q_n$ and $p_1, \ldots, p_n$. For H_0 we designate the action-angle variables by

$$w^0 = (w_1^0, \ldots, w_n^0), \qquad J^0 = (J_1^0, \ldots, J_n^0).$$

The reader can repeat the previous development, extending it from one to n dimensions. Instead of Eqs. (49), we now have (we are still not using the summation convention)

$$E_0(J) = K_0(J), \tag{58a}$$

$$E_1(J) = K_1(w^0, J) + \sum_\alpha \nu_\alpha^0 \frac{\partial S_1}{\partial w_\alpha^0}, \tag{58b}$$

$$E_2(J) = K_2(w^0, J) + \sum_\alpha\left(\frac{\partial K_1}{\partial J_\alpha}\frac{\partial S_1}{\partial w_\alpha^0} + \nu_\alpha^0\frac{\partial S_2}{\partial w_\alpha^0}\right) + \tfrac{1}{2}\sum_{\alpha\beta}\frac{\partial \nu_\alpha^0}{\partial J_\beta}\frac{\partial S_1}{\partial w_\alpha^0}\frac{\partial S_2}{\partial w_\beta^0}. \tag{58c}$$

If we write $\langle F \rangle$ now for the n-fold integral of F over $w_1^0, w_2^0, \ldots, w_n^0$, each from 0 to 1, then from (58b) we obtain

$$E_1(J) = \langle K_1(w^0, J) \rangle, \tag{59}$$

so that

$$\sum_\alpha \nu_\alpha^0 \frac{\partial S_1}{\partial w_\alpha^0} = \langle K_1(w^0, J) \rangle - K_1(w^0, J). \tag{60}$$

The right-hand side of (60) is a known function whose form is

$$\langle K_1 \rangle - K_1 = \sum_{m_1=-\infty}^{\infty} \cdots \sum_{m_n=-\infty}^{\infty} W_{m_1 \cdots m_n}(J) \exp\left(2\pi i \sum m_\beta w_\beta^0\right). \tag{61}$$

Similarly S_1 has the expansion

$$S_1 = \sum_{m_1=-\infty}^{\infty} \cdots \sum_{m_n=-\infty}^{\infty} S_{m_1 \cdots m_n}^{(1)}(J) \exp\left(2\pi i \sum m_\beta w_\beta^0\right),$$

so that

$$\sum_\alpha \nu_\alpha^0 \frac{\partial S_1}{\partial w_\alpha^0} = 2\pi i \sum_{m_1=-\infty}^{\infty} \cdots \sum_{m_n=-\infty}^{\infty} \sum_\alpha \nu_\alpha^0 m_\alpha S_{m_1 \cdots m_n}^{(1)}(J) \exp\left(2\pi i \sum m_\beta w_\beta^0\right). \tag{62}$$

By comparing (61) and (62) we see that the expansion coefficients of S_1 are given by

$$S_{m_1 \cdots m_n}^{(1)} = \left[2\pi i \sum_\alpha \nu_\alpha^0 m_\alpha\right]^{-1} W_{m_1 \cdots m_n}, \tag{63}$$

provided that $\sum \nu_\alpha^0 m_\alpha \neq 0$. For a conditionally periodic system the ν_α^0 are not commensurable, so this last condition is always valid and (63) determines all the expansion coefficients except the constant term $S_{00\cdots 0}^{(1)}$. But as in the one-dimensional case, this term plays no essential role and can be dropped.

This method breaks down for a degenerate system, in which there exist numbers $k_1, \ldots, k_n$ such that $\sum \nu_\alpha^0 k_\alpha = 0$, so that the coefficient $S_{k_1 \cdots k_n}^{(1)}$ becomes undeterminable. However one can still proceed if for all such sets $k_1, \ldots, k_n$ the corresponding $W_{k_1 \cdots k_n}$ are also zero. An example is given in Section 6(d).

Once S_1 is determined, (58c) can be integrated over $w_1^0, \ldots, w_0^n$ to determine $E_2(J)$ and $\partial S_2/\partial w_\alpha^0$. But this is now straightforward, though complicated, and we leave the details to the reader. See Section 6(d).

VII-6 EXAMPLES, APPLICATIONS, AND EXTENSIONS

(a) The Hamilton-Jacobi Equation for the Central Force Problem

The Hamilton-Jacobi equation for the central force problem can be solved by the method of separation of variables. The Hamiltonian for the two-dimensional problem, namely,

$$H = \frac{1}{2m}\left(p_r^2 + \frac{1}{r^2} p_\theta^2\right) + V(r),$$

is time independent so that, according to (25), we may switch from $S(q, Q, t)$ to $W(q, Q)$ by

$$S = W - Q_1 t. \tag{64}$$

Then writing

$$W = W_r(r, Q) + W_\theta(\theta, Q),$$

we find that Eq. (26), the equation for W, becomes

$$\frac{1}{2m}\left[\left(\frac{\partial W_r}{\partial r}\right)^2 + \frac{1}{r^2}\left(\frac{\partial W_\theta}{\partial \theta}\right)^2\right] + V(r) = Q_1. \tag{65}$$

As in the description of separability following Eq. (27), we multiply (65) by $2mr^2$, obtaining

$$r^2\left(\frac{\partial W_r}{\partial r}\right)^2 + \left(\frac{\partial W_\theta}{\partial \theta}\right)^2 + 2mr^2V(r) = 2mr^2Q_1.$$

Then we can split off $(\partial W_\theta/\partial\theta)^2$, and we arrive at

$$W_\theta' = Q_2,$$

$$r^2W_r'^2 + 2mr^2V(r) - 2mr^2Q_1 = -Q_2^2,$$

where we have written Q_2 for what we called $Q_2^{1/2}$ in Section 4. Actually we could have arrived at this result more quickly, for the only θ dependence in (65) is in the W_θ' term, which must therefore be a constant.

The equations we have obtained are of the form of (28) and can be immediately integrated to yield

$$W_\theta = Q_2\theta,$$

$$W_r = \sqrt{2m}\int\sqrt{Q_1 - V(r) - (2mr^2)^{-1}\,Q_2^2}\,dr. \tag{66}$$

From (64) and (66) we have

$$P_1 = -\frac{\partial S}{\partial Q_1} = -\frac{\partial W_r}{\partial Q_1} + t,$$

or

$$t - P_1 = \sqrt{m/2}\int[Q_1 - V(r) - (2mr^2)^{-1}Q_2^2]^{-1/2}\,dr, \tag{67}$$

which is almost identical to Eq. (50) of Chapter II. We see that Q_1 and Q_2 are the energy E and angular momentum l of the system, respectively, and that P_1, the momentum conjugate to $Q_1 = E$, is the initial time.

Similarly we have

$$P_2 = -\frac{\partial S}{\partial Q_2} = -\frac{\partial W_r}{\partial Q_2} - \theta,$$

or

$$\theta + P_2 = \frac{Q_2}{\sqrt{2m}} \int [Q_1 - V(r) - (2mr^2)^{-1}Q_2^2]^{-\frac{1}{2}} \frac{dr}{r^2}, \tag{68}$$

which is almost identical to Eq. (51) of Chapter II. The constant P_2, the momentum conjugate to $Q_2 = l$, is the negative of the initial angle. Equation (67) can be integrated to give r as a function of t (the motion), and (68) can be integrated to give r as a function of θ (the orbit).

(b) Action-Angle Variables in the Central Force Problem

For bounded motion in a central field we can use action-angle variables to describe the motion. According to Eqs. (66) of the previous example we have

$$p_\theta = \frac{\partial W_\theta}{\partial \theta} = Q_2,$$

$$p_r = \frac{\partial W_r}{\partial r} = \sqrt{2m}\sqrt{Q_1 - V(r) - Q_2^2/2mr^2}.$$

Then for bounded motion Eq. (32), defining the action variables, gives

$$J_\theta = \int Q_2 \, d\theta = 2\pi Q_2,$$

$$J_r = \sqrt{2m} \oint \sqrt{Q_1 - V(r) - Q_2/2mr^2} \, dr. \tag{69}$$

The equation for J_r holds only for bounded motion, and then a complete cycle consists of r going from the inner turning point r_1 to the outer one r_2 and back again. Thus the integral for J_r is

$$J_r = 2\sqrt{2m} \int_{r_1}^{r_2} \sqrt{Q_1 - V(r) - J_\theta^2/8\pi^2 mr^2} \, dr, \tag{70}$$

where r_1 and r_2 are the zeros of the integrand.

We consider as a specific example the potential

$$V(r) = -\frac{k}{r} + \frac{\beta}{r^2},$$

which is the Kepler potential with the perturbation β/r^2 (see Problem 15 of Chapter II where we wrote x instead of r). With this potential Eq. (70) can be integrated (see Appendix B) to yield

$$J_r = -\sqrt{J_\theta^2 + 8\pi^2 m\beta} + \sqrt{-\frac{2m}{E}} \pi k.$$

Solving this for Q_1, we have

$$Q_1 = E = H(J_r, J_\theta) = \frac{-2m\pi^2k^2}{(J_r + \sqrt{J_\theta^2 + \alpha})^2}, \tag{71}$$

where $\alpha = 8\pi^2 m\beta$.

Equation (37) for the frequencies then gives

$$\nu_r = \frac{\partial Q_1}{\partial J_r} = \frac{4m\pi^2k^2}{(J_r + \sqrt{J_\theta^2 + \alpha})^3} = \sqrt{\frac{2}{m}}\frac{1}{\pi k}(-Q_1)^{-3/2} \tag{72}$$

and

$$\nu_\theta = \frac{\partial Q_1}{\partial J_\theta} = \nu_r \frac{J_\theta}{\sqrt{J_\theta^2 + \alpha}}. \tag{73}$$

The expression for ν_r in terms of Q_1 is the same as in the pure Kepler problem. For the pure Kepler problem (with $\alpha = 0$) we have $\nu_\theta = \nu_r$ and the motion is degenerate; the orbit closes on itself after one oscillation. But in general the orbit does not close.

In order to see this we go on to the angle variables, which we find by first expressing W_θ and W_r as functions of J_θ and J_r. Equations (69) and (71), when inserted into (66), yield

$$\hat{W}_\theta = \frac{1}{2\pi} J_\theta \theta,$$

$$\hat{W}_r = \sqrt{2m}\int\left[\frac{-2m\pi^2k^2}{(J_r + \sqrt{J_\theta^2 + \alpha})^2} + \frac{k}{r} - \frac{\beta}{r^2} - \frac{J_\theta^2}{8\pi^2 m r^2}\right]^{1/2} dr. \tag{74}$$

Then from the definition (36) of the angle variables we obtain

$$w_\theta = \frac{\partial \hat{W}_\theta}{\partial J_\theta} + \frac{\partial \hat{W}_r}{\partial J_\theta} = \frac{\theta}{2\pi} + \frac{\partial \hat{W}_r}{\partial J_\theta}, \tag{75}$$

$$w_r = \frac{\partial \hat{W}_\theta}{\partial J_r} + \frac{\partial \hat{W}_r}{\partial J_r} = \frac{\partial \hat{W}_r}{\partial J_r}. \tag{76}$$

The importance of these equations lies in the simple time dependence of w_θ and w_r:

$$\begin{aligned} w_\theta &= \nu_\theta t + \phi_\theta, \\ w_r &= \nu_r t + \phi_r, \end{aligned} \tag{77}$$

where ϕ_θ and ϕ_r are new constants and ν_θ and ν_r are given by (72) and (73).

On inserting (74) and (76) into (77) we get

$$w_r(t) = \nu_r t + \phi_r = \sqrt{\frac{m}{2}}\,\nu_r \int\left[Q_1 + \frac{k}{r} - \frac{Q_2^2 + 2m\beta}{2mr^2}\right]^{-1/2} dr, \tag{78}$$

which, when integrated, gives the time dependence of r. If we integrate over one cycle in r, as we know, w_r changes by one [as is easily verified by performing the integration in (78)]. Call the time of one such r-cycle τ_r. It follows then, that $\tau_r = 1/\nu_r$, and ν_r is thus the frequency in time.

On the other hand, the same procedure will not produce a similar result for w_θ. On inserting (74) and (75) into (77) we get

$$w_\theta(t) = \nu_\theta t + \phi_\theta$$
$$= \frac{\theta}{2\pi} + \frac{Q_2}{\pi}\int\left[Q_1 + \frac{k}{r} - \frac{Q_2^2 + 2m\beta}{2mr^2}\right]^{-1/2}\left(\nu_\theta - \frac{Q_2}{2\pi m r^2}\right) dr, \quad (79)$$

where ν_θ on the right-hand side can be expressed in terms of Q_1, Q_2 through Eqs. (73), (72), and (69). In a motion, then, w_θ depends both on θ and r. If θ changes through one cycle (from 0 to 2π), w_θ changes by one only if r does not change. But in a motion r does change. Thus in the time τ_θ of one θ-cycle w_θ does not change by one, for there is a contribution to w_θ from the integral in (79).

The integral can actually be evaluated and it is then found that

$$\theta + P_2 = \frac{\nu_\theta}{\nu_r}\sin^{-1} F(r) = \frac{\nu_\theta}{\nu_r}\sin^{-1} G(t),$$

where P_2 is a constant,

$$F(r) = \frac{kr - (Q_2^2 + 2m\beta)/m}{r\sqrt{k^2 + 2Q_1(Q_2^2 + 2m\beta)/m}},$$

and $G(t)$ is obtained from $F(r)$ by solving (78) for $r(t)$. As r varies between r_1 and r_2, $F(r)$ varies between $+1$ and -1. Since $r(t)$ is periodic with period τ_r, so is $G(t)$, and hence $\sin^{-1} G(t)$ changes by $2\pi n$ when t changes by $n\tau_r$, where n is an integer. Thus $\sin(\theta + P_2)$ can be periodic with some period T if and only if $n\nu_\theta/\nu_r$ is an integer. When $\beta = \alpha = 0$, then $\nu_\theta/\nu_r = 1$ and θ is periodic with the same period as r. For nonzero values of β, the ratio ν_θ/ν_r will almost always be irrational, depending, of course, on the actual value of J_θ. In general then, θ is not periodic and the orbit does not close. We thus see that one must be very careful in speaking of ν_θ as a frequency: it is the frequency for the angle variable w_θ, but it is not the frequency of any actual motion.

Let us again (see Problem 15, Chapter II) consider the case in which α/J_θ^2 is small but nonzero. Then the orbit will be nearly the closed elliptical orbit of the pure Kepler problem. If $\alpha < 0$ we have $\nu_\theta < \nu_r$ or $\tau_\theta > \tau_r$. In the time it takes to complete one cycle in r the particle completes only the fraction τ_r/τ_θ of a cycle in θ. Thus the point where r takes on any given value on the nearly elliptical orbit, say the perihelion of the orbit, moves by an angle

$\phi = 2\pi(\tau_r/\tau_\theta - 1)$ in one r-period τ_r. Therefore the rate of precession of the perihelion is

$$\dot{\phi} = \phi/\tau_r = \frac{2\pi}{\tau_r}\left(\frac{\nu_\theta}{\nu_r} - 1\right)$$
$$= \frac{2\pi}{\tau_r}([1 + \alpha/J_\theta^2]^{-1/2} - 1).$$

Since we are considering $|\alpha/J_\theta^2| \ll 1$ we shall calculate $\dot{\phi}$ only to the lowest order in this quantity. The result is

$$\dot{\phi} = -\frac{\pi\alpha}{\tau_r J_\theta^2} = -\frac{\alpha}{4\pi\tau_r l^2}.$$

When $\alpha = 0$ the angular momentum l and the eccentricity e of the elliptical orbit are related by (Goldstein, 1950)

$$l^2 = \frac{mk^2(1 - e^2)}{-2E} = mak(1 - e^2),$$

where a is the semimajor axis of the unperturbed ellipse. Then

$$\dot{\phi} = -\frac{\alpha}{4\pi mak\tau_r}(1 - e^2)^{-1}. \tag{80}$$

This calculation is used in a test of general relativity, which predicts (Fock, 1964) that in the first approximation the gravitational potential actually deviates from the pure Newtonian $1/r$ law by a term of the form β/r^2. Thus it predicts that the perihelion of a planetary orbit precesses in accordance with our present result. This precession is largest for orbits with short periods (small radii) and large eccentricity. In fact it has been observed only for the planet Mercury, where it amounts to only 42 seconds of arc per century.

(c) Canonical Perturbation Theory Applied to the Simple Harmonic Oscillator

To illustrate the use of perturbation theory we will apply it first to the one-dimensional simple harmonic oscillator. Let the Hamiltonian be

$$H = \frac{1}{2m}p^2 + \tfrac{1}{2}(k + \lambda)q^2$$
$$= H_0 + \lambda H_1 + \lambda^2 H_2, \tag{81}$$

where

$$H_0 = \frac{1}{2m}p^2 + \tfrac{1}{2}kq^2,$$
$$H_1 = \tfrac{1}{2}q^2, \tag{82}$$
$$H_2 = 0.$$

Of course the motion can be found exactly for the Hamiltonian of Eq. (81) but we want to show how perturbation theory approximates the exact solution.

Let us first solve the problem exactly in terms of action-angle variables. Equation (26) for Hamilton's characteristic function W is

$$\frac{1}{2m}\left(\frac{\partial W}{\partial q}\right)^2 + \tfrac{1}{2}(k+\lambda)q^2 = Q,$$

so that Eq. (32) for the action yields

$$J = \oint \frac{\partial W}{\partial q}\,dq = \oint \sqrt{2mQ - m(k+\lambda)q^2}\,dq = Q/\nu,$$

where $\nu = (2\pi)^{-1}\sqrt{(k+\lambda)/m}$. The angle variable is given by (36):

$$\begin{aligned} w &= \frac{\partial \hat{W}}{\partial J} = \frac{\partial}{\partial J}\int \frac{\partial \hat{W}}{\partial q}\,dq \\ &= m\nu \int [2m\nu J - m(k+\lambda)q^2]^{-1/2}\,dq \\ &= (2\pi)^{-1}\sin^{-1}\left\{\left[\frac{2\nu J}{(k+\lambda)}\right]^{-1/2} q\right\}. \end{aligned}$$

Thus we have

$$q = \sqrt{\frac{2\nu J}{k+\lambda}}\sin 2\pi w \tag{83a}$$

and

$$p = 2\pi m\nu\sqrt{\frac{2\nu J}{k+\lambda}}\cos 2\pi w. \tag{83b}$$

For the unperturbed case ($\lambda = 0$) we have

$$E_0 = Q = \nu^0 J^0,$$

$$q = \sqrt{\frac{2\nu^0 J^0}{k}}\sin 2\pi w^0, \tag{84a}$$

and

$$p = 2\pi m\nu^0\sqrt{\frac{2\nu^0 J^0}{k}}\cos 2\pi w^0, \tag{84b}$$

where

$$\nu^0 = (2\pi)^{-1}\sqrt{k/m} = \nu(1+\lambda/k)^{-1/2}. \tag{85}$$

Equations (83) and (84) are canonical transformations connecting J, w and

J^0, w^0 respectively, to q, p. By eliminating q, p, we can obtain the transformations connecting J, w directly to J^0, w^0. From (83) we have

$$q^2 + \frac{p^2}{(2\pi m\nu)^2} = \frac{2\nu J}{k + \lambda},$$

and then by inserting the expressions for p and q in (84) we obtain

$$J = J^0\left(1 + \frac{\lambda}{k}\right)^{-1/2}\left(1 + \frac{\lambda}{k}\sin^2 2\pi w^0\right). \tag{86a}$$

Either by inserting (84*a*) for q and (86*a*) for J into Eq. (83*a*) or by equating p/q obtained from (83) to p/q obtained from (84), we get

$$2\pi w = \tan^{-1}\left\{\left(1 + \frac{\lambda}{k}\right)^{1/2}\tan 2\pi w^0\right\}. \tag{86b}$$

Equations (86) give the canonical transformation connecting J, w and J^0, w^0.

In order to compare exact results with perturbation theory, we will want our relations expanded in power series in λ. Thus we expand Eqs. (86) in Taylor's series about $\lambda = 0$. We start with (86*a*), expanding J^0 as a function of J and w^0, and it then becomes

$$J^0 = J\left[1 + \frac{1}{2}\frac{\lambda}{k}\cos 4\pi w^0 - \frac{1}{4}\left(\frac{\lambda}{k}\right)^2(\cos 4\pi w^0 - \tfrac{1}{2}\cos 8\pi w^0) + \cdots\right]. \tag{87a}$$

Equation (86*b*) becomes

$$w = w^0 + \frac{1}{8\pi}\frac{\lambda}{k}\sin 4\pi w^0 - \frac{1}{16\pi}\left(\frac{\lambda}{k}\right)^2 \times (\sin 4\pi w^0 - \tfrac{1}{4}\sin 8\pi w^0) + \cdots. \tag{87b}$$

Finally, by expanding (85) in a similar series, we obtain

$$\nu = \nu^0\left[1 + \frac{1}{2}\frac{\lambda}{k} - \frac{1}{8}\left(\frac{\lambda}{k}\right)^2 + \cdots\right]. \tag{88}$$

We now apply canonical perturbation theory to the same problem, and as we shall see, the results will be the same. From the Hamiltonian (81) we can construct the function $K(w^0, J^0, \lambda)$ of (46) and obtain the K_i in accordance with (48*a*). This is easily done by using H_0, H_1 and H_2 of Eq. (82) and the transformation (84). We obtain

$$K_0(w^0, J^0) = \nu^0 J^0,$$

$$K_1(w^0, J^0) = \frac{1}{k}\nu^0 J^0 \sin^2 2\pi w^0 = \frac{1}{2k}\nu^0 J^0(1 - \cos 4\pi w^0),$$

$$K_2 = 0.$$

From these we can calculate $E_1(J)$ and $\partial S_1/\partial w^0$ by (53) and (54). (Remember that J^0 gets replaced by J in these equations.) Then we have

$$E_1(J) = \langle K_1 \rangle = \frac{\nu^0 J}{2k}, \tag{89}$$

and

$$\frac{\partial S_1}{\partial w^0} = \frac{[\langle K_1 \rangle - K_1]}{\nu_0} = \frac{J}{2k} \cos 4\pi w^0.$$

Integrating, we obtain

$$S_1 = \frac{J}{8\pi k} \sin 4\pi w^0. \tag{90}$$

Next, we calculate $E_2(J)$ and $\partial S_2/\partial w^0$ by (55) and (56). For this purpose note that

$$\frac{\partial \nu^0}{\partial J} = \frac{\partial^2 E_0(J)}{\partial J^2} = \frac{\partial^2(\nu^0 J)}{\partial J^2} = 0.$$

This simplifies both (55) and (56), which now become

$$\begin{aligned} E_2(J) &= \frac{1}{\nu^0}\left[\left\langle \frac{\partial K_1}{\partial J} \right\rangle \frac{\nu^0 J}{2k} - \left\langle \frac{\partial K_1}{\partial J} K_1 \right\rangle\right] \\ &= \frac{\nu^0 J}{k^2} [\tfrac{1}{4} - \langle \sin^4 2\pi w^0 \rangle] \\ &= -\frac{\nu^0 J}{8k^2}, \end{aligned} \tag{91}$$

and

$$\begin{aligned} \frac{\partial S_2}{\partial w^0} &= \frac{1}{\nu^0}\left[-\frac{\nu^0 J}{8k^2} - \frac{J}{2k}\frac{\partial K_1}{\partial J} + \frac{1}{\nu^0}\frac{\partial K_1}{\partial J} K_1\right] \\ &= -\frac{J}{4k^2}[\cos 4\pi w^0 - \tfrac{1}{2}\cos 8\pi w^0]. \end{aligned}$$

Integrating, we obtain

$$S_2 = -\frac{J}{16\pi k^2}[\sin 4\pi w^0 - \tfrac{1}{4}\sin 8\pi w^0]. \tag{92}$$

Now we can construct $E(J)$ to second order in λ by using E_1 and E_2 from Eqs. (89) and (91). We obtain

$$E(J) = \nu^0 J\left[1 + \frac{1}{2}\frac{\lambda}{k} - \frac{1}{8}\left(\frac{\lambda}{k}\right)^2 + \cdots\right],$$

where we have used $E_0(J) = \nu^0 J$. According to Eq. (88), this expression agrees with $E = \nu J$ to second order in λ.

To check whether we get the correct expressions for w and J^0 in terms of w^0 and J, we use (47) with Eqs. (90) and (92) for S_1 and S_2. Then we have

$$J^0 = J + \frac{\lambda J}{2k} \cos 4\pi w^0 - \frac{\lambda^2 J}{4k^2} (\cos 4\pi w^0 - \tfrac{1}{2} \cos 8\pi w^0) + \cdots$$

and

$$w = w^0 + \frac{\lambda}{8\pi k} \sin 4\pi w^0 - \frac{\lambda^2}{16\pi k^2} (\sin 4\pi w^0 - \tfrac{1}{4} \sin 8\pi w^0) + \cdots,$$

which agree with Eqs. (87) and hence with the exact solution to second order in λ.

This example shows in detail how perturbation theory works in a case for which the exact solution is known. Section 6(d) is an example in two degrees of freedom, for which the exact solution is not known. The Problems give the reader an opportunity to apply the method to some other examples.

(d) Canonical Perturbation Theory in Two Degrees of Freedom

Here we shall outline the use of canonical perturbation theory for a system with two degrees of freedom. Most of the details will be left to the reader.

Consider the Hamiltonian

$$H(q, p, \lambda) = \tfrac{1}{2}(p_1^2 + p_2^2) + \tfrac{1}{2}(a_1^2 q_1^2 + a_2^2 q_2^2) + \lambda a_1^2 a_2^2 q_1^2 q_2^2.$$

The unperturbed Hamiltonian

$$H_0 = \tfrac{1}{2}(p_1^2 + p_2^2) + \tfrac{1}{2}(a_1^2 q_1^2 + a_2^2 q_2^2)$$

represents a simple harmonic oscillator in two (uncoupled) dimensions. The transformation to the corresponding action-angle variables, as found in Section 6(c), is (recall: no summation convention)

$$\begin{aligned} q_\alpha &= \sqrt{\frac{J_\alpha^0}{\pi a_\alpha}} \sin 2\pi w_\alpha^0, \\ p_\alpha &= \sqrt{\frac{a_\alpha J_\alpha^0}{\pi}} \cos 2\pi w_\alpha^0, \end{aligned} \tag{93}$$

$\alpha = 1, 2$. As before, we write the Hamiltonian in terms of w^0, J^0, obtaining

$$H(q, p, \lambda) = K(w^0, J^0, \lambda) = K_0 + \lambda K_1 + \lambda^2 K_2 + \cdots,$$

where

$$\begin{aligned} K_0 &= \frac{1}{2\pi} \sum a_\alpha J_\alpha^0, \\ K_1 &= \frac{a_1 a_2}{\pi^2} J_1^0 J_2^0 (\sin^2 2\pi w_1^0)(\sin^2 2\pi w_2^0), \\ K_i &= 0, \qquad i \geq 2. \end{aligned}$$

We now take the double integral of K_1 over the w_α and replace J_α^0 by J_α to obtain

$$E_1 = \langle K_1 \rangle = \frac{a_1 a_2 J_1 J_2}{4\pi^2}.$$

Hence to first order in λ the perturbed Hamiltonian in terms of its own action-angle variables is

$$\begin{aligned} E(J) &= K_0(J) + \lambda E_1 \\ &= \nu_1^0 J_1 + \nu_2^0 J_2 + \lambda \nu_1^0 \nu_2^0 J_1 J_2, \end{aligned}$$

where the unperturbed frequencies are

$$\nu_\alpha^0 = \frac{a_\alpha}{2\pi}.$$

The perturbed frequencies to first order in λ are

$$\begin{aligned} \nu_1 &= \frac{\partial E}{\partial J_1} = \nu_1^0 + \lambda \nu_1^0 \nu_2^0 J_2, \\ \nu_2 &= \frac{\partial E}{\partial J_2} = \nu_2^0 + \lambda \nu_1^0 \nu_2^0 J_1. \end{aligned} \tag{94}$$

Remark. These frequencies can be calculated to first order in λ whether or not the system is degenerate.

In order to write w^0, J^0 in terms of w, J to first order in λ, we use (60) and (61):

$$\begin{aligned} \sum_\alpha \nu_\alpha \frac{\partial S_1}{\partial w_\alpha^0} &= \langle K_1 \rangle - K_1 \\ &= \nu_1^0 \nu_2^0 J_1 J_2 [1 - 4(\sin^2 2\pi w_1^0)(\sin^2 2\pi w_2^0)] \\ &= \tfrac{1}{4} \nu_1^0 \nu_2^0 J_1 J_2 \sum_{m_1, m_2} \tilde{W}_{m_1 m_2} \exp\left(2\pi i \sum_\alpha m_\alpha w^0{}_\alpha\right), \end{aligned}$$

where we have factored $\xi = \frac{1}{4}\nu_1^0 \nu_2^0 J_1 J_2$ out of the $W_{m_1 m_2}$. The Fourier coefficients are easily calculated. One obtains

$$\begin{aligned} \tilde{W}_{22} = \tilde{W}_{-2,-2} = \tilde{W}_{-2,2} = \tilde{W}_{2,-2} &= -1, \\ \tilde{W}_{02} = \tilde{W}_{0,-2} = \tilde{W}_{20} = \tilde{W}_{-2,0} &= 2, \end{aligned}$$

and all other $\tilde{W}_{ij}$ vanish. Then if we are to use the Fourier expression (62) for S_1 together with (63), we need

$$S_{02}^{(1)} = -S_{0,-2}^{(1)} = (2\pi i\nu_2^0)^{-1}\xi,$$
$$S_{20}^{(1)} = -S_{-2,0}^{(1)} = (2\pi i\nu_1^0)^{-1}\xi,$$
$$S_{22}^{(1)} = -S_{-2,-2}^{(1)} = -[4\pi i(\nu_1^0 + \nu_2^0)]^{-1}\xi,$$
$$S_{2,-2}^{(1)} = -S_{-2,2}^{(1)} = -[4\pi i(\nu_1^0 - \nu_2^0)]^{-1}\xi.$$

On writing out the Fourier series, we obtain the following expression for S to first order in λ:

$$S = \sum_\alpha w_\alpha J_\alpha^0 + \lambda \frac{J_1 J_2}{4\pi}\left[\nu_2^0 \sin 4\pi w_1^0 + \nu_1^0 \sin 4\pi w_2^0 \right.$$
$$- 2\frac{\nu_1^0\nu_2^0}{\nu_1^0 + \nu_2^0}\sin 4\pi(w_1^0 + w_2^0)$$
$$\left. - 2\frac{\nu_1^0\nu_2^0}{\nu_1^0 - \nu_2^0}\sin 4\pi(w_1^0 - w_2^0)\right].$$

Finally, by taking derivatives of this with respect to the w_α^0 and J_α, we can obtain expressions for the w_α and J_α^0 of the form

$$\begin{aligned} w_\alpha &= w_\alpha^0 + \lambda f_\alpha(w^0, J), \\ J_\alpha^0 &= J_\alpha + \lambda g_\alpha(w^0, J). \end{aligned} \tag{95}$$

If we want to obtain the q_α and p_α as functions of time, we solve Eqs. (95) for the w_α^0, J_α^0 to first order in λ, obtaining

$$\begin{aligned} w_\alpha^0 &= w_\alpha - \lambda f_\alpha(w, J), \\ J_\alpha^0 &= J_\alpha + \lambda g_\alpha(w, J), \end{aligned} \tag{96}$$

for when we replace w^0 by w in the terms of order λ we make an error only of order λ^2. Now we know that the J_α are constants of the motion and that the w_α vary linearly with time according to

$$w_\alpha = \nu_\alpha t + \phi_\alpha,$$

where the ν_α are given (to first order in λ) by (94). If we insert this time dependence into (96) and the result into (93), we obtain $q_\alpha(t)$ and $p_\alpha(t)$ to first order in λ. See Problem 15.

This procedure fails to work only if $\nu_1^0 = \nu_2^0$. Other degenerate cases are correctly given by our expression for S because the dangerous $W_{m_1m_2}$ terms are zero. For instance, if $\nu_1^0 = 2\nu_2^0$ we would get into trouble if and only if $W_{n,-2n}$ were not zero for some n. Of course, as we proceed to calculate higher order terms, for example S_2, we may still run into trouble for this case. But to first order we only have to develop special techniques if $\nu_1^0 = \nu_2^0$.

(e) Adiabatic Invariance

Consider a system whose Hamiltonian $H(q, p, \lambda)$ can be treated by action-angle variables for all values of the constant parameter λ in some interval. Now suppose that the same parameter λ is not a constant, but is given an explicit time dependence; then the action-angle variables found assuming λ were constant are no longer appropriate. In particular, the action is no longer a constant. However, if λ changes very slowly with time, the action-angle variables remain useful because it turns out that the action is very nearly constant. More precisely, we shall show that if λ changes by an amount $\Delta\lambda$ in a time T, the change in J is proportional to $\Delta\lambda/T$. This means that for a given $\Delta\lambda$, no matter how large, the change in J can be made as small as we like by causing the change in λ to take place over a long enough time T.

To prove this assertion we assume that we have found the action-angle variables for constant λ. Thus we have (remember, no sum)

$$J_\alpha(Q, \lambda) = \oint p_\alpha \, dq_\alpha,$$

where

$$p_\alpha(q_\alpha, Q, \lambda) = \frac{\partial W_\alpha}{\partial q_\alpha}.$$

Now these equations are inverted to obtain $Q_\alpha(J, \lambda)$, and $\hat{W}(q, J, \lambda)$ is constructed in the usual way. The equations

$$p_\alpha = \frac{\partial \hat{W}}{\partial q_\alpha}, \qquad w_\alpha = \frac{\partial \hat{W}}{\partial J_\alpha}, \tag{97}$$

as usual, then define the canonical transformation from q, p to J, w, and if λ were in fact a constant, J would also be constant. In any case, however, whether λ is constant or not, Eqs. (97) define a proper canonical transformation, and we may ask how the w_α and J_α vary in time.

For this purpose we find the new Hamiltonian

$$K = H + \frac{\partial \hat{W}}{\partial t} = H(J, \lambda) + \frac{\partial \hat{W}}{\partial \lambda}\dot{\lambda}.$$

Note that H is independent of w. This is because J, w would be the action-angle variables for the Hamiltonian H if λ were constant, and we know, of course, that that Hamiltonian is then independent of w. Now we write $\partial\hat{W}/\partial\lambda$ as a function of w, J rather than q, J. Call this function $W_\lambda(w, J, \lambda)$. Thus

$$K = H(J, \lambda) + W_\lambda(w, J, \lambda)\dot{\lambda}. \tag{98}$$

Then the equations of motion for the new variables are

$$\dot{w}_\alpha = \frac{\partial K}{\partial J_\alpha} = \frac{\partial H}{\partial J_\alpha} + \frac{\partial W_\lambda}{\partial J_\alpha}\dot{\lambda}$$

$$= \nu_\alpha(J, \lambda) + \frac{\partial W_\lambda}{\partial J_\alpha}\dot{\lambda}, \tag{99a}$$

and

$$\dot{J}_\alpha = -\frac{\partial K}{\partial w_\alpha} = -\frac{\partial W_\lambda}{\partial w_\alpha}\dot{\lambda}. \tag{99b}$$

The $\nu_\alpha(J, \lambda)$ are called the *local frequencies*. They are the frequencies the system would have if λ were fixed.

It is clear that the J_α are not constants of the motion, as they would be if λ were fixed. But what we wish to show is that under certain conditions the J_α vary extremely slowly, and that it is a good approximation to take them as constants. For this purpose, as is seen from Eq. (99*b*), we must study the behavior of $\partial W_\lambda/\partial w_\alpha$. Now we have seen that every single-valued dynamical variable can be written as a multiple Fourier series,[5] and it can be shown that W_λ is single valued. Indeed, in one α-cycle W_λ changes by

$$\Delta_\alpha W_\lambda = \oint \frac{\partial W_\lambda}{\partial q_\alpha}\, dq_\alpha = \frac{\partial}{\partial \lambda}\oint \frac{\partial \hat{W}}{\partial q_\alpha}\, dq_\alpha = \frac{\partial J_\alpha}{\partial \lambda}.$$

Recall that the partial derivative with respect to λ in the definition of W_λ is taken holding the q_α and J_α fixed (since the λ-dependence comes from the explicit time dependence of W). Therefore $\partial J_\alpha/\partial\lambda = 0$ and $\Delta_\alpha W_\lambda = 0$, so that W_λ is single valued. If we then calculate $\partial W_\lambda/\partial w_\alpha$ by taking the derivative of the Fourier series expression for W_λ, no terms with $m_\alpha = 0$ will be present. Hence

$$\int_0^1 \frac{\partial W_\lambda}{\partial w_\alpha}\, dw_\alpha = 0, \tag{100}$$

for all terms with $m_\alpha \neq 0$ average out to zero in one cycle of w_α.

Now let us assume that $\lambda(t)$ is a slowly varying function of the time. Specifically, let us assume that we can adjust λ so as to have

$$|\dot{\lambda}| < \epsilon, \tag{101}$$

where ϵ is an arbitrary small positive number. We want to show that the change in J_α in some time T is, to first order in ϵ, independent of T and can be made arbitrarily small. This means, for instance, that we can make the total change in λ arbitrarily large (by making T long enough) while keeping

[5] The Fourier series is still valid when written in terms of $\exp\,[2\pi i(m_1 w_1 + \cdots + m_n w_n)]$. Now, however $w_\alpha \neq \nu_\alpha t + \phi_\alpha$, so that the Fourier series in time is no longer valid.

the changes in the J_α arbitrarily small (by making ϵ small enough). Let the initial time be $t = 0$ and choose $w_\alpha(0) = 0$ for convenience. Then the change in J_α in the time T satisfies the equation (no summation on α)

$$|\Delta_T J_\alpha| = \left|\int_0^T \dot{\lambda}\frac{\partial W_\lambda}{\partial w_\alpha}\,dt\right| = \left|\int_0^{w_\alpha(T)} \dot{\lambda}\frac{\partial W_\lambda}{\partial w_\alpha}\frac{dw_\alpha}{\dot{w}_\alpha}\right|,$$

where we assume that we can express t as a function of w_α, that is that each equation $w_\alpha = w_\alpha(t)$ can be inverted. Now $\dot{w}_\alpha$ as given by (99a) can be inserted into this expression. Using (101), we have

$$|\dot{w}_\alpha| \leq |\nu_\alpha| + \left|\frac{\partial W_\lambda}{\partial J_\alpha}\right|\epsilon,$$

or

$$|\dot{w}_\alpha|^{-1} \leq |\nu_\alpha^{-1}| + 0(\epsilon),$$

where ν_α is the value of the local frequency at $t = 0$, and $0(\epsilon)$ contains terms of order ϵ and higher. Thus, again using (101), we have

$$|\Delta_T J_\alpha| < \epsilon\left|\frac{1}{\nu_\alpha}\int_0^{w_\alpha(T)} \frac{\partial W_\lambda}{\partial w_\alpha}\,dw_\alpha + 0(\epsilon)\right|.$$

Since, according to (100), $\partial W_\lambda/\partial w_\alpha$ averages to zero over one cycle, the integral is bounded:

$$\left|\int_0^{w_\alpha(T)} \frac{\partial W_\lambda}{\partial w_\alpha}\,dw_\alpha\right| \leq B,$$

and B does not depend on T. Therefore we obtain the result

$$|\Delta_T J_\alpha| < \epsilon B/|\nu_\alpha| + 0(\epsilon^2).$$

This is what we set out to prove: by making ϵ small enough the change in J_α can be made arbitrarily small independent of T. A dynamical variable of this kind is called an *adiabatic invariant*. Thus in the *adiabatic approximation*, that is, for extremely slow variation of the Hamiltonian, the J_α are invariants.

Actually this derivation is not completely rigorous, for the result depends on $\dot{w}_\alpha$ being constant enough and on inverting $w_\alpha(t)$, which in turn depend on $\lambda(t)$. In order partly to see how this works and partly to see how adiabatic invariance is used, we present the following simple example. Consider the slightly perturbed simple harmonic oscillator,

$$H = \tfrac{1}{2}(p^2 + \lambda^2 q^2) \tag{102}$$

where $\lambda = \lambda(t)$. The HJ equation is easily solved for W, and J is easily calculated. One obtains

$$W = \int \sqrt{2Q - \lambda^2 q^2}\, dq,$$

$$J = \frac{2\pi Q}{\lambda}, \qquad Q = \frac{\lambda J}{2\pi}, \qquad \nu = \frac{\partial Q}{\partial J} = \frac{\lambda}{2\pi}.$$

Then

$$\hat{W} = \lambda \int \sqrt{\frac{J}{\pi\lambda} - q^2}\, dq,$$

from which we obtain

$$w = \frac{1}{2\pi} \sin^{-1} \frac{q}{\sqrt{J/\pi\lambda}}$$

and

$$q = \sqrt{\frac{J}{\pi\lambda}} \sin 2\pi w.$$

Now we must calculate $\partial \hat{W}/\partial\lambda$ and write it in terms of w, J to get W_λ. Straightforward calculation yields

$$W_\lambda = \frac{J}{4\pi\lambda} \sin 4\pi w,$$

so that finally, using Eqs. (99), we arrive at

$$\dot{w} = \frac{\lambda}{2\pi} + \dot{\lambda} \frac{\sin 4\pi w}{4\pi\lambda}, \tag{103a}$$

$$\dot{J} = -\frac{J}{\lambda} \dot{\lambda} \cos 4\pi w. \tag{103b}$$

To proceed, we pick a simple form for $\lambda(t)$. Suppose we start at $t = 0$ with some initial positive value λ_i and allow λ to vary linearly in time:

$$\lambda(t) = \lambda_i + \epsilon t.$$

Then

$$\dot{w} = \frac{\lambda_i}{2\pi} + \epsilon \frac{t}{2\pi} + \epsilon \frac{\sin 4\pi w}{4\pi(\lambda_i + \epsilon t)}.$$

This shows that for small enough ϵ [e.g., such that $\epsilon/\lambda_i < 2\lambda_i$] $\dot{w}$ is always positive, and it is therefore possible to invert the relationship and find $t(w)$.

Then from (103) we obtain

$$\int_0^T \frac{\dot{J}}{J}\,dt = \left|\ln\frac{J(T)}{J(0)}\right| = \left|\int_0^T \frac{\dot{\lambda}}{\lambda}\cos 4\pi w\,dt\right|$$

$$= \epsilon\left|\int_0^{w(T)} \frac{\cos 4\pi w}{\dot{w}\lambda}\,dw\right|$$

$$= \frac{\epsilon}{2}\left|\int_0^{w(T)} \frac{\cos 4\pi w}{\lambda^2 + \frac{\epsilon}{2}\sin 4\pi w}\,4\pi\,dw\right|.$$

For small enough ϵ the denominator is always positive. Let its minimum value be $1/B$ (in fact this is greater than $\lambda_i^2 - \frac{1}{2}\epsilon$). Then

$$0 \leq \left|\ln\frac{J(T)}{J(0)}\right| \leq \tfrac{1}{2}\epsilon B\left|4\pi\int_0^{w(T)}\cos 4\pi w\,dw\right| \leq \tfrac{1}{2}\epsilon B.$$

Therefore by making ϵ small enough the ratio of $J(T)$ to $J(0)$ can be brought arbitrarily close to 1, or the difference between $J(T)$ and $J(0)$ can be made arbitrarily small.

How is this used in a specific problem? Suppose we have the Hamiltonian of (102) and the motion is given initially as

$$q = A_i \cos \lambda_i t, \qquad p = -\lambda_i A_i \sin \lambda_i t,$$

where A_i is the initial amplitude and λ_i is the initial (circular) frequency. Then initially

$$J_i = \oint p\,dq = \lambda_i^2 A_i^2 \oint \sin^2 \lambda_i t\,dt = \pi\lambda_i A_i^2,$$

and finally, at some later time

$$J_f = \oint p\,dq = \pi\lambda_f A_f^2,$$

where A_f and λ_f are the final amplitude and frequency. But if, as is true to arbitrary accuracy, we can vary λ so slowly that $J_f = J_i$, then $\lambda_f A_f^2 = \lambda_i A_i^2$. Let us assume that λ has changed by some total factor a, so that $\lambda_f = a\lambda_i$. Then $A_f = a^{-1/2}A_i$. Thus as the frequency increases, the amplitude decreases. On the other hand the energy is in general $E = \frac{1}{2}\lambda^2 A^2$, so that

$$E_f = aE_i.$$

Thus the energy can be changed by any finite factor a by choosing T long enough and ϵ small enough, while at the same time holding J essentially fixed.

Problems

1. Let

$$\phi = \arc\tan\,(x_1/x_2)$$

be the angle of rotation about the 3-axis. Let $G(\mathbf{x}, \mathbf{p})$ be a canonical variable conjugate to ϕ defined by $[\phi, G] = 1$. Discuss any difference between G and l_3, the component of angular momentum in the 3-direction. Find the θ-orbits generated by G.

2. The HJ prescription yields a canonical transformation from the ξ_α to constant η_α. Find the new Hamiltonian function K.

3. From Eq. (33) prove that $\det\,(\partial \hat{J}_\alpha/\partial Q_\beta) \neq 0$. [*Hint:* Show, in the course of the proof, that (no sum)

$$\det\left(\oint \frac{\partial W}{\partial q_\alpha\,\partial Q_\beta}\,dq_\alpha\right) = \oint\cdots\oint \det\left(\frac{\partial W}{\partial q_\alpha\,\partial Q_\beta}\right)dq_1\cdots dq_n.$$

This follows from separability.]

4. Consider the Hamiltonian

$$H = \frac{1}{2m}p_1^2 + \frac{1}{2m}(p_2 - kq_1)^2.$$

Solve for the motion by the Hamilton-Jacobi prescription. Find the general orbit from the solution of the HJ equation. What physical system might this correspond to? Solve this problem in three other ways: (a) by solving the canonical equations; (b) by making a canonical transformation with $Q_1 = Ap_1$, $P_1 = B(p_2 - kq_1)$, choosing Q_2 and P_2 conveniently (choose also the constants A and B), solving for Q_α, P_α, and transforming back; (c) by going from the solution of the HJ equation to action-angle variables.

5. In calculating $\Delta_\beta w_\alpha$ we wrote (no sum)

$$\oint \frac{\partial \hat{W}_\beta}{\partial q_\beta\,\partial J_\alpha}\,dq_\beta = \frac{\partial}{\partial J_\alpha}\oint \frac{\partial \hat{W}_\beta}{\partial q_\beta}\,dq_\beta.$$

Moving the derivative outside the integral in this way is legitimate in general only if the limits of integration do not depend on J_α. But the orbit over which we integrate does in fact depend on J_α. Prove that this step is nevertheless legitimate in this case.

6. (a) Consider the two-dimensional simple harmonic oscillator, whose Hamiltonian is $H = \frac{1}{2}\xi_\alpha\xi_\alpha$, where α is summed from $\alpha = 1$ to $\alpha = 4$. Let $G = \omega_{\mu\nu}\xi_\mu\xi_\nu$, with $\omega_{\mu\nu} = \omega_{\nu\mu}$, be an infinitesimal generator of a continuous family of canonical transformations under which H is invariant. Find the general form of the matrix Ω of the $\omega_{\mu\nu}$. Show that every such G can be written as a linear combination of four infinitesimal generators $G^\alpha = \omega^\alpha_{\mu\nu}\xi_\mu\xi_\nu$ each of which separately leaves H invariant. In other words, each Ω can be written as a sum of four Ω^α, or the possible Ω's form a four-dimensional vector space. Find the canonical transformations

corresponding to these four G^α or Ω^α. (b) Since $\xi_\alpha \xi_\alpha$ is the "magnitude" of the position vector ξ in phase space, the Hamiltonian $H = \frac{1}{2}\xi_\alpha \xi_\alpha$ is left invariant by the rotations in four dimensions (phase space here has four dimensions). There are *six* elementary infinitesimal rotations in four dimensions, one in each plane. Why are there only *four* in part (a)? Explain, and prove your explanation by demonstration.

7. Prove that every canonical transformation of types 2 and 3 lies on a continuous 1-parameter family including the identity transformation. What can you say about types 1 and 4?
8. Assume that a dynamical system is periodic, that is, that there exists a time T such that $q_\alpha(t + T) = q_\alpha(t)$ for each α and for any t. Let τ_α be the α-period, that is, the *shortest* time such that $q_\alpha(t + \tau_\alpha) = q_\alpha(t)$ for each α and for any t. Show that there exist integers k_α such that (no sum) $k_\alpha \tau_\alpha = k_\beta \tau_\beta$ for all α, β, that is, that the α-periods are commensurable. Show that in this case each $\nu_\alpha \tau_\alpha$ is an integer.
9. A particle of mass m is constrained to move on the x axis subject to the potential
$$V = a \sec^2 \left(\frac{x}{l}\right).$$
Solve the HJ equation, obtaining an integral expression for S. (Actually, it is not difficult to find the motion $x(t)$ by the HJ prescription.) Go to action-angle variables. Find J and ν. In this case ν is the actual frequency. How does it depend on amplitude? For small amplitude it should correspond to the frequency for small vibrations. Check this.
10. Suppose l in the previous problem varies slowly as a function of the time. When l has changed to zl, where z is some number, how much will the energy have changed? Where does this change in energy come from?
11. Let $H(q, p)$ be time independent and let q_n be cyclic. Prove that Hamilton's principal function can be written in the form
$$S = -Q_1 t + Q_n q_n + W_{n-1}(q, Q),$$
where $\partial W_{n-1} / \partial q_n = 0$.
12. The string of a simple pendulum is slowly shortened by being pinched at the point of suspension and by having the pinch move downward. Use adiabatic invariance to find the amplitude and energy as a function of the length of the string. Assume initially a small amplitude. Then at what length will the motion deviate measurably from simple harmonic? Where does the energy come from that is fed to the string?
13. In kinetic theory a model of a gas is sometimes taken as a collection of hard spheres confined to a box. Consider a single free particle of mass m confined to a finite interval of the x axis (a one-dimensional box), moving very rapidly and bouncing elastically back and forth between the two opposite ends of the interval (the walls). Suppose the walls are now slowly brought together. Use adiabatic invariance to find how the *pressure* varies as a function of the length of the box. How does the energy (temperature)

vary? In an adiabatic process PV^γ is constant. From your result calculate γ for such a *one-dimensional gas*. Compare with the result from the usual kinetic theory. Now suppose the length of the interval changes suddenly. Which thermodynamic variables (P, V, T) remain constant? Explain.

14. Use canonical perturbation theory to find the motion of a simple pendulum for large oscillations. There are two limits in which the action-angle variables can be found exactly: (1) $Q/mgl \ll 1$. Here H_0 is the harmonic oscillator Hamiltonian and $H_1 = -mgl(\frac{1}{24}\theta^4 - \cdots)$. (2) $Q/mgl \gg 1$. Here H_0 is the free particle Hamiltonian and $H_1 = mgl(1 - \cos\theta)$. Find the motion as an expansion about each of these limits. Compare and discuss the results.
15. Complete the example of Section 6(d) to first order in λ. That is, find the $q_\alpha(t)$ and $p_\alpha(t)$ to this order.
16. If we follow the procedure outlined in Eqs. (27) to (29) for separation of variables in the HJ equation, it seems that in addition to Q_1 there are n constants Q_α that will appear, making a total of $n + 1$. Actually there are only n. Explain why.
17. Write down the HJ equation for a charged particle in the field of an electric dipole. Separate the equation and obtain (at least) integral expressions for the motion and the orbit.

CHAPTER VIII

The Theory of Fields

The Lagrangian and Hamiltonian formulations of classical particle mechanics can be extended to describe continuous systems. In a continuous system, such as a vibrating rod, each point $\mathbf{x}$ in the system can move independently of every other point, subject to certain continuity restrictions. Thus the system must be described by a continuous function (or set of functions) $\psi(\mathbf{x}, t)$, which gives the state of each point at time t. Such a function is called a *field.*

In this chapter the formulation of classical (nonquantum) field theory is given. We also examine the electromagnetic field, which requires the introduction of the special theory of relativity. It is questionable whether relativity can be studied usefully in the framework of classical particle mechanics, but it fits in naturally in a discussion of field theory.

VIII-1 THE LAGRANGIAN FORMULATION OF FIELD THEORY

In this section we show how the Lagrangian formulation of particle dynamics can be extended to describe fields. We start with an example of a system with an infinite number of discrete particles labeled by an index $i = \ldots, -2, -1, 0, 1, 2, \ldots$, and then we will let i become a continuous variable. In the process the field theory equations emerge from the equations of motion of the particles.

Consider then a one-dimensional array of mass points, each of mass m, connected by a set of springs of length a and with spring constant k, as shown in Fig. 35. The equilibrium position of the ith mass is $x = ia$ for all i.If the generalized coordinate q_i of the ith mass is taken to be its displacement from its equilibrium position, the kinetic energy of the system can be written

$$T = \sum_{i=-\infty}^{\infty} \tfrac{1}{2} m \dot{q}_i^2,$$

and the potential energy stored in the springs is

$$V = \sum_{i=-\infty}^{\infty} \tfrac{1}{2} k (q_i - q_{i+1})^2.$$

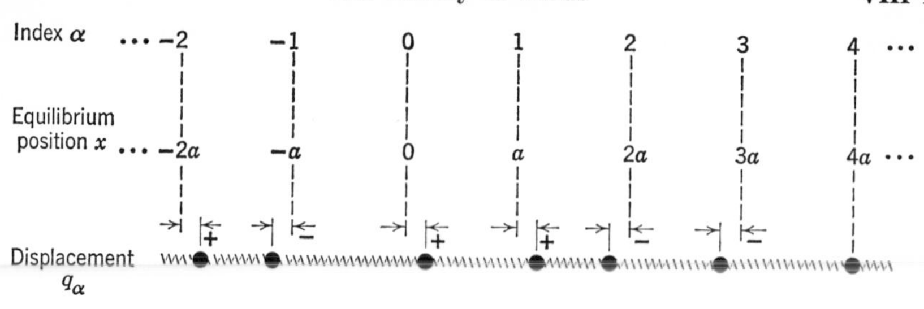

Figure 35. A one-dimensional array of mass points connected by springs.

From the Lagrangian

$$L = \sum_{i=-\infty}^{\infty} [\tfrac{1}{2}m\dot{q}_i^2 - \tfrac{1}{2}k(q_i - q_{i+1})^2] \tag{1}$$

we easily obtain the Euler-Lagrange equations of motion

$$\frac{d}{dt}\frac{\partial L}{\partial \dot{q}_j} - \frac{\partial L}{\partial q_j} = m\ddot{q}_j + k(q_j - q_{j+1}) - k(q_{j-1} - q_j) = 0. \tag{2}$$

(To obtain this result remember that in the expression for L contributions to $\partial L/\partial q_j$ come from both the $j = i$ and the $j = i + 1$ terms in the sum.) These equations are similar to the coupled differential equations discussed in Chapter IV, Section 5(b), except that here we have an infinite number of them. They can be solved by methods similar to those in Chapter IV. We do not go into this here, but see Brioullin (1953).

Instead, we prepare to make the system continuous rather than discrete by subdividing each particle into M equal subparticles of mass m/M and distributing them uniformly along the springs. We label each subparticle by its equilibrium position, writing $q(x)$ instead of q_i. That is, $q(x)$ is the displacement from equilibrium of the subparticle whose equilibrium position is x. Letting $\Delta x = a/M$ be the separation between the subparticles, $\mu = m/a$ be the linear mass density, and $Y = ak$ be Young's modulus, the Lagrangian can be written in the form[1]

$$L = \sum \Delta x\{\tfrac{1}{2}\mu\dot{q}^2(x) - \tfrac{1}{2}Y([q(x + \Delta x) - q(x)]/\Delta x)^2\}, \tag{3}$$

where the sum is over the discrete equilibrium positions. The Euler-Lagrange equations obtained from this Lagrangian are

$$\frac{d}{dt}\frac{\partial L}{\partial \dot{q}(x)} - \frac{\partial L}{\partial q(x)} = \mu\,\Delta x\ddot{q}(x) - \frac{Y}{\Delta x}[q(x + \Delta x) - 2q(x) + q(x - \Delta x)] = 0. \tag{4}$$

[1] The reader should show that if the spring constant is k for the spring of length a, it is Mk for a piece of the same spring of length a/M.

We can now make the system continuous by going to the limit $M \to \infty$ or $\Delta x \to 0$ in Eqs. (3) and (4). In this limit (3) becomes

$$L = \int dx \{\tfrac{1}{2}\mu \dot{q}^2(x) - \tfrac{1}{2}Y(\partial q/\partial x)^2\}, \tag{5}$$

where partial derivatives are used because q is a function of both x and t. (That is, for each equilibrium position x the displacement q is a function of t.) To obtain a nonzero limit for Eq. (4) we must first divide it by Δx. Then in the limit we get

$$\mu \frac{\partial^2 q}{\partial t^2} - Y \frac{\partial^2 q}{\partial x^2} = 0, \tag{6}$$

which is the wave equation for longitudinal waves in an elastic rod. In going from the discrete to the continuous system the equations of motion have changed from a set of ordinary differential equations to a single partial differential equation.

It is possible to obtain the wave equation directly from the continuum Lagrangian (5). Let us write $L = \int \mathscr{L}\, dx$ where $\mathscr{L}$ is called the *Lagrangian density*. The wave equation is obviously obtained from the operation

$$\frac{\partial}{\partial t} \frac{\partial \mathscr{L}}{\partial(\partial q/\partial t)} + \frac{\partial}{\partial x} \frac{\partial \mathscr{L}}{\partial(\partial q/\partial x)} = 0, \tag{7}$$

which is the generalization of the Euler-Lagrange equation to this case.

From this example we see that the Lagrangian formalism is preserved in a natural way when we go from the discrete to the continuous case. It is a simple matter to generalize from one spatial dimension to three and from one field variable q to a set of N field variables ψ_A, $A = 1, \ldots, N$. (For $N = 3$, for example, the ψ_A might be the components of the three-dimensional displacement of a mass point from its equilibrium position, and so could describe elastic waves in a solid medium.) We use superscripts to designate the three spatial coordinates of a point: x^1, x^2, x^3. Furthermore, by setting $t = x^0$ we can conveniently let x represent the set of four components x^0, x^1, x^2, x^3 of a point or *event* in four-dimensional space-time or in *four-space*. We will use Greek indices to run from 0 to 3 and Latin indices for the three-space components alone; that is, the components of x are x^α, $\alpha = 0, 1, 2, 3$, and those of $\mathbf{x}$ are x^k, $k = 1, 2, 3$. Then with the notation

$$\frac{\partial \psi_A}{\partial x^\alpha} = \psi_{A\alpha},$$

the generalization of (7) to three dimensions and N field variables can be written

$$\frac{\partial}{\partial x^\alpha} \frac{\partial \mathscr{L}}{\partial \psi_{A\alpha}} = 0,$$

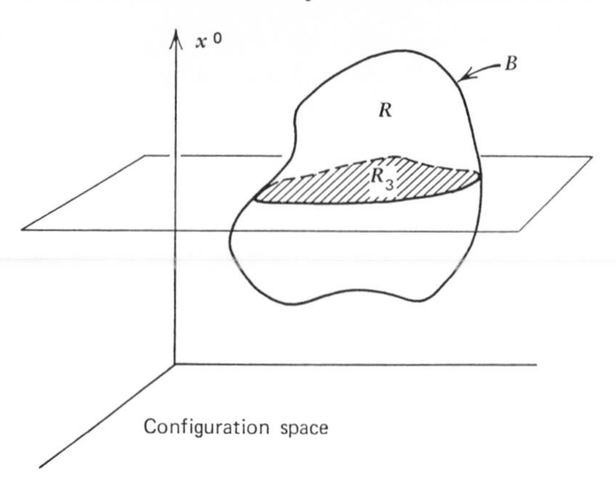

Figure 36. An arbitrary region of four-space. R_3 is a three-dimensional subregion at some fixed time.

where the summation convention is used. With $\mathscr{L}$ properly chosen, for instance, this would be the wave equation for three-dimensional elastic waves.

We now want to establish a formulation of continuum mechanics which is independent of reference to discrete systems. For this purpose we will use a variational principle very similar to the one we used in Section 3 of Chapter II. The problem in field theory is to find the field functions in some arbitrary bounded region R of four-space (see Fig. 36), given their values on the boundary B of R. The equations satisfied by these field functions are to be obtained from the Lagrangian density $\mathscr{L}$ of the system according to a variational prescription which we want to develop. In general $\mathscr{L} = \mathscr{L}(\psi, \psi', x)$ will be a function of ψ, the set of field functions $\psi_A(x)$; of ψ' the set of first derivatives $\psi_{A\alpha}(x)$ of the field functions; and of x, the point in four-space. If an arbitrary set of functions $\psi_A(x)$ is inserted into the Lagrangian density, $\mathscr{L}$ becomes a function of the x^α, and the integral

$$S = \int L\,dt = \int_R \mathscr{L}(\psi(x), \psi'(x), x)\,d^4x \tag{8}$$

is just a number, where $d^4x = dx^0\,dx^1\,dx^2\,dx^3$. This number will depend on the set of functions ψ_A used. The variational principle of field theory then says that the *true physical field functions* $\psi_A(x)$ *are those which have the required value on B and for which S is an extremum.* [See the analogous discussion at Eq. (31) in Chapter II.]

In order to find the field equations from this principle, it must be quantified and made more precise. Instead of all possible functions $\psi_A(x)$ which take on the required values on B, we consider a one-parameter family of them. We write $\psi_A(x, \varepsilon)$ for the ε-family and require that it be a differentiable function of ε. Let this family include the actual physical field functions, and choose the parameterization so that for these functions $\varepsilon = 0$. Since all members of this family take on the same values on B we have

$$\psi_A(x, \varepsilon) = \psi_A(x, 0) = \psi_A(x) \qquad \text{for } x \text{ on } B,$$

or

$$\frac{\partial \psi_A}{\partial \varepsilon} = 0 \qquad \text{on } B. \tag{9}$$

When this ε-family of functions is inserted into $\mathscr{L}$, the integral S in Eq. (8) becomes a function of ε. Then an accurate statement of the variational principle is that for every ε-family of functions satisfying the conditions we have named above,

$$\left.\frac{dS}{d\varepsilon}\right|_{\varepsilon=0} = \left[\frac{d}{d\varepsilon}\int_R \mathscr{L}\, d^4x\right]_{\varepsilon=0} = 0. \tag{10}$$

The field equations can easily be derived from this condition. Since R is independent of ε, we may take the derivative with respect to ε inside the integral, to obtain

$$\frac{\partial \mathscr{L}}{\partial \varepsilon} = \frac{\partial \mathscr{L}}{\partial \psi_A}\frac{\partial \psi_A}{\partial \varepsilon} + \frac{\partial \mathscr{L}}{\partial \psi_{A\alpha}}\frac{\partial \psi_{A\alpha}}{\partial \varepsilon},$$

where the repeated index A is summed from 1 to N and the repeated index α is summed, as always, from 0 to 3. Now we have

$$\frac{\partial \psi_{A\alpha}}{\partial \varepsilon} = \frac{\partial}{\partial \varepsilon}\frac{\partial \psi_A}{\partial x^\alpha} = \frac{\partial}{\partial x^\alpha}\frac{\partial \psi_A}{\partial \varepsilon},$$

so that some simple manipulation yields

$$\frac{\partial \mathscr{L}}{\partial \varepsilon} = \left[\frac{\partial \mathscr{L}}{\partial \psi_A} - \frac{\partial}{\partial x^\alpha}\frac{\partial \mathscr{L}}{\partial \psi_{A\alpha}}\right]\frac{\partial \psi_A}{\partial \varepsilon} + \frac{\partial}{\partial x^\alpha}\left[\frac{\partial \mathscr{L}}{\partial \psi_{A\alpha}}\frac{\partial \psi_A}{\partial \varepsilon}\right]. \tag{11}$$

So far we have not used Eq. (9). We insert (11) into (10) and get

$$0 = \int_R \left[\frac{\partial \mathscr{L}}{\partial \psi_A} - \frac{\partial}{\partial x^\alpha}\frac{\partial \mathscr{L}}{\partial \psi_{A\alpha}}\right]\frac{\partial \psi_A}{\partial \varepsilon}\, d^4x + \int_R \frac{\partial}{\partial x^\alpha}\left[\frac{\partial \mathscr{L}}{\partial \psi_{A\alpha}}\frac{\partial \psi_A}{\partial \varepsilon}\right] d^4x \tag{12}$$

at $\varepsilon = 0$. The second integral can be converted to an integral over B by the divergence theorem, whose four-dimensional form is obtained by an obvious generalization of the well-known three-dimensional form (see Appendix A). We have

$$\int_R \frac{\partial}{\partial x^\alpha}\left[\frac{\partial \mathscr{L}}{\partial \psi_{A\alpha}}\frac{\partial \psi_A}{\partial \varepsilon}\right] d^4x = \oint_B \frac{\partial \mathscr{L}}{\partial \psi_{A\alpha}}\frac{\partial \psi_A}{\partial \varepsilon}\, dB_\alpha,$$

where dB_α, the (four-dimensional) *outward normal surface element* on B, is the differential expression that enters into the divergence theorem. Then from Eq. (9) we see that this last integral vanishes as a consequence of the boundary condition on the functions $\psi_A(x, \varepsilon)$ and so the first integral in (12) must alone equal zero. Furthermore it must be zero for arbitrary ε-family of functions $\psi_A(x, \varepsilon)$ which satisfy (9) and the conditions named above (9); this implies that the expression in square brackets is zero separately for each value of A at $\varepsilon = 0$. Thus the field equations derived from the variational principle are

$$\frac{\partial}{\partial x^\alpha}\frac{\partial \mathscr{L}}{\partial \psi_{A\alpha}} - \frac{\partial \mathscr{L}}{\partial \psi_A} = 0. \tag{13}$$

It is obvious that the generalization we obtained of Eq. (7) is a special case of this, valid when $\mathscr{L}$ is a function of $\psi_{A\alpha}$ only. We see then that a variational principle closely analogous to Hamilton's principle for particle motion leads to the field equations without previous reference to discrete systems. Now the problem is to write the Lagrangian density $\mathscr{L}$ for an arbitrary system.

Consider, for instance, a system of two fields ψ_A and χ_A, and assume we know the *free-field* Lagrangian densities $\mathscr{L}_1(\psi)$ and $\mathscr{L}_2(\chi)$ for each in the absence of the other. The new density $\mathscr{L}_0 = \mathscr{L}_1 + \mathscr{L}_2$ will give, by (13), the same free-field equations for ψ_A and χ_A as we would get from $\mathscr{L}_1$ and $\mathscr{L}_2$ separately. To give any interaction between the fields, an interaction term $\mathscr{L}_{\mathrm{I}}(\psi, \chi)$ must be added to $\mathscr{L}_0$, and the *coupled* field equations are then obtained by (13) from the total Lagrangian density

$$\mathscr{L} = \mathscr{L}_1 + \mathscr{L}_2 + \mathscr{L}_{\mathrm{I}}.$$

It is not in general easy to find such $\mathscr{L}_{\mathrm{I}}$'s for the systems that occur in nature. In fact to find them has been among the principal tasks of theoretical physics. An example is given in Section 4(c).

Expressions for conserved quantities, and hence for the energy and momentum carried by a field, can be derived from $\mathscr{L}$. To do this we make use of a result similar to the one discussed in Chapter III, namely that conserved quantities are related to symmetries of the Lagrangian density.

We start with an ε-family $\psi_A(x, \varepsilon)$ of transformations of the fields, no longer restricted to some region R of four-space, but still with the property that at $\varepsilon = 0$ the fields satisfy the field equations (13). From Eq. (11) we can calculate $\partial\mathscr{L}/\partial\varepsilon$ at $\varepsilon = 0$. The first term of (11) vanishes since the $\psi_A(x, 0)$ satisfy (13), and so we have

$$\frac{\partial \mathscr{L}}{\partial \varepsilon} = \frac{\partial}{\partial x^\alpha}\left[\frac{\partial \mathscr{L}}{\partial \psi_{A\alpha}}\frac{\partial \psi_A}{\partial \varepsilon}\right], \qquad \varepsilon = 0. \tag{14}$$

This is the field-equivalent of Eq. (21) of Chapter III. That equation led to conservation when the left-hand side was the total time derivative of some function G. Equation (14) leads to conservation when the left-hand side can be written in the form

$$\frac{\partial \mathscr{L}}{\partial \varepsilon} = \frac{\partial G^\alpha}{\partial x^\alpha}, \qquad \varepsilon = 0, \tag{15}$$

where the G^α are a set of four functions of the x^μ. If (15) is satisfied, we have

$$\frac{\partial}{\partial x^\alpha}\left[\frac{\partial \mathscr{L}}{\partial \psi_{A\alpha}}\frac{\partial \psi_A}{\partial \varepsilon} - G^\alpha\right] = 0, \qquad \varepsilon = 0, \tag{16}$$

and we will show that this implies that the quantity

$$\int\left[\frac{\partial \mathscr{L}}{\partial \psi_{A0}}\frac{\partial \psi_A}{\partial \varepsilon} - G^0\right] d^3x, \qquad \varepsilon = 0, \tag{17}$$

is a constant, where $d^3x = dx^1\, dx^2\, dx^3$ is the three-space volume element.

Whenever a set of four functions S^μ of the field variables satisfies the relation

$$\frac{\partial S^\mu}{\partial x^\mu} = 0, \tag{18a}$$

we say that the S^μ form a *conserved current*. The quantity actually conserved is $\int S^0\, d^3x$, where the integral is over all three-space. To prove this, we write (18a) in the form

$$\frac{\partial S^\mu}{\partial x^\mu} = \frac{\partial S^0}{\partial t} + \nabla \cdot \mathbf{S} = 0, \tag{18b}$$

where $\mathbf{S} = (S^1, S^2, S^3)$ as usual. Integrating this equation over a fixed volume V and using the divergence theorem (Appendix A), we obtain

$$\frac{d}{dt}\int_V S^0\, d^3x = -\oint_{\partial V} \mathbf{S} \cdot d\mathbf{\Sigma},$$

where $d\mathbf{\Sigma}$ is the outward normal surface element on the surface ∂V of V. If $\mathbf{S}$ goes to zero with increasing r more rapidly than r^{-2}, where r is the distance to the origin, the integral on the right-hand side approaches zero as V is chosen larger and larger, and in the limit when V is all of three-space we obtain

$$\frac{d}{dt}\int S^0\, d^3x = 0. \tag{19a}$$

Thus for such sufficiently rapidly decreasing $\mathbf{S}$, the three-space integral of S^0 is a constant.

This proof may be illustrated by a diagram in four-space. We represent three-space in the diagram by a two-dimensional plane, as in Fig. 36, then if we choose V to be a sphere it will be represented in the diagram by a circular

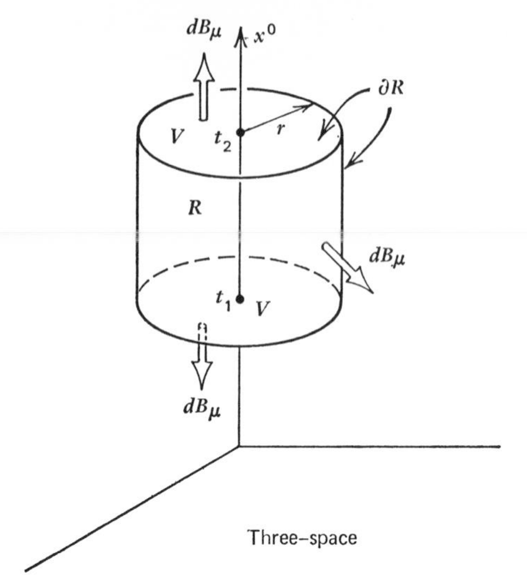

Figure 37. A bounded region in four-space consisting of a three-dimensional sphere for a finite time.

disc. The result is Fig. 37. Now consider the integral of $\partial S^\mu/\partial x^\mu = 0$ over the four-space region R of the figure. By the four-dimensional divergence theorem [see the discussion following Eq. (12)], this may be written

$$0 = \int_R \frac{\partial S^\mu}{\partial x^\mu}\, d^4x = \oint_{\partial R} S^\mu \, dB_\mu.$$

The components of dB_μ are easily read off the figure for the upper and lower spheres (discs): for the upper one $dB_\mu = (d^3x, 0, 0, 0)$, and for the lower one $dB_\mu = (-d^3x, 0, 0, 0)$. On the other surface, dB_μ has no zeroth (time) component and its magnitude is $dx^0\, d\Sigma$, where $d\mathbf{\Sigma}$ is the surface element of the three-space sphere: for example, $d\Sigma = r^2 \sin\theta\, d\varphi\, d\theta$. Thus $dB_\mu = (0, dx^0\, d\Sigma_1, dx^0\, d\Sigma_2, dx^0\, d\Sigma_3) = (0, dx^0\, d\mathbf{\Sigma})$. We have, then,

$$\int_V S^0(t_2)\, d^3x - \int_V S^0(t_1)\, d^3x + \int_{t_1}^{t_2} dt \oint_{\partial V} \mathbf{S} \cdot d\mathbf{\Sigma} = 0. \tag{19b}$$

As before, the third term is assumed to vanish for large enough V, and the three-space integral of S^0 is thus seen to be constant, for it is the same at time t_2 as at t_1.

Returning then to (16), we see that if the G^μ exist, the quantity in square brackets is a conserved current and (17) is the conserved quantity.

We can now investigate the energy and momentum in the field. Recall that in Chapter III it was found that energy and momentum were the conserved quantities associated with invariance of the Lagrangian under time and space displacements, respectively. Thus we may expect to learn something about energy and momentum by considering the ε-family of four-space displacements of the form

$$\psi_A(x, \varepsilon) = \psi_A(x + \varepsilon h),$$

where the h^μ are a set of four arbitrary constants. At $\varepsilon = 0$ the derivative with respect to ε is

$$\frac{\partial \psi_A}{\partial \varepsilon} = \frac{\partial \psi_A}{\partial x^\mu} h^\mu = \psi_{A\mu} h^\mu,$$

which we need to insert into Eq. (16). A conserved current will be obtained only if we can find functions G^μ which satisfy (15). Recall that $\partial \mathscr{L}/\partial \varepsilon$ is calculated by first inserting the functions $\psi_A(x, \varepsilon)$ and $\psi_{A\mu}(x, \varepsilon)$ into $\mathscr{L}$, so that $\mathscr{L}$ itself becomes a function of the x^μ and ε, and by then taking the ε-derivative. If $\mathscr{L}$ does not depend on x^μ explicitly, its dependence on both x^μ and ε comes entirely from its dependence on the $\psi_A(x + \varepsilon h)$ and $\psi_{A\mu}(x + \varepsilon h)$. For such $\mathscr{L}$ we have, at $\varepsilon = 0$,

$$\frac{\partial \mathscr{L}}{\partial \varepsilon} = \frac{\partial \mathscr{L}}{\partial x^\mu} h^\mu = \frac{\partial}{\partial x^\mu} (h^\mu \mathscr{L}).$$

Here $\partial/\partial x^\mu$ is the *formal derivative.* This is an extension of the definition (Chapter II, footnote 2) from one independent variable (the time t) to several (the x^μ). Then we may write

$$G^\mu = h^\mu \mathscr{L}.$$

Since the G^μ exist, we get the conserved currents by inserting the expressions for $\partial \psi_A/\partial \varepsilon$ and G^μ into (16):

$$\frac{\partial}{\partial x^\alpha} \left[\frac{\partial \mathscr{L}}{\partial \psi_{A\alpha}} \psi_{A\mu} h^\mu - h^\alpha \mathscr{L} \right] = h^\mu \frac{\partial T_\mu{}^\alpha}{\partial x^\alpha} = 0,$$

where

$$T_\mu{}^\alpha = \frac{\partial \mathscr{L}}{\partial \psi_{A\alpha}} \psi_{A\mu} - \delta_\mu^\alpha \mathscr{L}. \tag{20}$$

Here δ_μ^α is the usual Kronecker δ symbol with one of the indices written as a superscript. These $T_\mu{}^\alpha$ are known as the elements of the *stress tensor* of the field. Since the h^μ are arbitrary, it follows that

$$\frac{\partial T_\mu{}^\alpha}{\partial x^\alpha} = 0.$$

Thus there are four conserved currents, one for each value of μ, and hence also four conserved constants:

$$P_\mu = \int T_\mu{}^0 \, d^3x.$$

Of the four conserved P_μ, the zeroth, namely P_0, is associated with time translations (i.e., with h^0). Thus we may take it to be the energy of the system. Each of the other P_k is associated with translation in one of the three-space directions. Thus we take them to be the components of the momentum carried by the field. Then the integrals defining the P_μ allow us to interpret the $T_\mu{}^0$ as energy ($\mu = 0$) and momentum ($\mu = k$) densities.

Remarks. *1.* If $\mathscr{L}$ has the dimensions of energy density (energy per unit volume) so has $T_\mu{}^\nu$, and thus the P_μ all have the dimensions of energy. The P_k can then not quite be the momenta, but it turns out that they are constant multiples of them. This is true in general: what we obtain are not necessarily the energy and momentum, but multiples of them. *2.* The other elements $T_\mu{}^k$ of the stress tensor also have physical meaning: they give the local momentum-energy flux in the field, as shown in Problem 1.

In this section we have treated the time and space variables in a symmetric way, which leads, among other things, to a unified derivation of energy and momentum conservation. Such considerations lead in a natural way to a discussion of special relativity, to which we turn in the next section.

VIII-2 RELATIVISTIC FIELD THEORY

In this section we will describe fields which are consistent with the special theory of relativity. To do so we must review briefly some aspects of relativity and must adapt our notation to its requirements. For more details the reader is referred to the literature, for example, Bergman (1948).

In prerelativity physics the Galilei transformations, which we discussed in Chapter III, play a fundamental role. By using these transformations two observers who are moving at constant velocity relative to each other should be able to agree about certain properties of events which they both observe. For instance they should both observe the same time interval between any two such events. Experimentally, however, it turns out that they will not observe the same time interval, and that the difference in their observations becomes greater as their relative velocity approaches the velocity of light. With his special theory of relativity Einstein incorporated this and similar experimental results into a consistent framework.

Consider two events; let $\Delta\mathbf{x}$ be the spatial separation between them and Δt the time interval between them as viewed by one observer, and $\Delta\mathbf{x}'$ and $\Delta t'$ be the separation and time interval as viewed by another. Experiment indicates that

$$(c\,\Delta t)^2 - \Delta\mathbf{x} \, . \, \Delta\mathbf{x} = (c\,\Delta t')^2 - \Delta\mathbf{x}' \cdot \Delta\mathbf{x}' \tag{21}$$

(here c is the velocity of light), a relation which is very easily shown not to hold if the primed and unprimed quantities are related by the Galilei transformation. Einstein showed that Eq. (21) will hold, however, if the Galilei transformations are replaced by the *Lorentz transformations*. The most general *homogeneous* Lorentz transformation is of the form $R'\Lambda_3 R$ where R and R' are rotations in three-space leaving the time invariant, and Λ_3 is given by

$$x'^1 = x^1, \qquad x'^2 = x^2,$$

$$x'^3 = \frac{x^3 - vct/c}{\sqrt{1 - v^2/c^2}}, \tag{22}$$

$$ct' = \frac{-vx^3/c + ct}{\sqrt{1 - v^2/c^2}},$$

where v is the constant relative velocity of the two observers. *Inhomogeneous* Lorentz transformations include also a shift of origins in space and time [see Eq. (27)].

The Lorentz transformations are thus characterized as those which preserve the quadratic form of Eq. (21), just as in Chapter IV, Section 5(c), the rotations were characterized as the transformations which preserve the quadratic form $\Delta\mathbf{x} \cdot \Delta\mathbf{x} = (\Delta x^1)^2 + (\Delta x^2)^2 + (\Delta \mathbf{x}^3)^2$. For rotations in three-space this quadratic form is important not only for displacement vectors $\Delta\mathbf{x}$, but also for other physical vectors like force $\mathbf{F}$ or momentum $\mathbf{p}$, for which it represents the square of the magnitude (e.g., $F^2 = \mathbf{F} \cdot \mathbf{F}$). Similarly, in relativity theory there exist *four-vectors* (w^0, w^1, w^2, w^3), physical four-component objects other than the *four-displacement* $(c\,\Delta t, \Delta\mathbf{x})$, for which the quadratic form $(w^0)^2 - (w^1)^2 - (w^2)^2 - (w^3)^2$ is of fundamental importance. That is, two observers moving at constant relative velocity v will calculate the same value for this quadratic form. For example the relativistic energy E and momentum $\mathbf{p}$ form such a four-vector [see Eq. (32)]. Because this form is preserved only by Lorentz transformations, four-vectors must also transform under Lorentz transformations. Just as in three-space the same rotation is applied to different vectors on going from one coordinate system to a rotated one, so in *four-space* the same Lorentz transformation is applied to different four-vectors on going from one observer to another.

In order to treat this material in a consistent formalism, a somewhat modified notation is ordinarily used. We shall take $c = 1$ for the velocity of light, so that the unit of time is the time it takes light to travel one unit of length (e.g., a *light-meter*), or the unit of length is the distance light travels in one unit of time (e.g., a *light-year*). As in Section 1, we shall use upper indices ranging from 0 to 3 to designate t and $\mathbf{x}$. Thus $x^\alpha = (x^0, x^1, x^2, x^3) = (x^0, \mathbf{x})$ where $x^0 = t$. A four-vector w will have what we shall call *contravariant* components $w^\alpha = (w^0, w^1, w^2, w^3) = (w^0, \mathbf{w})$. In order to write the quadratic form of Eq. (21) in a summation convention, we define the *covariant* components w_α of the four-vector w by

$$w_0 = w^0, \qquad w_i = -w^i,$$

or $w_\alpha = (w^0, -\mathbf{w}) = (w_0, -\mathbf{w})$. (Recall that Greek indices run from 0 to 3, and Latin from 1 to 3.) We do this so the fundamental quadratic form can then be written using the summation convention as

$$(w^0)^2 - \mathbf{w}\cdot\mathbf{w} = w^\alpha w_\alpha = |w|^2, \tag{23}$$

which we will call the *norm* of w. Note that despite the notation this norm is not positive-definite (see Chapter IV).

In this notation sums will almost always take place on one upper and one lower index (except for sums over Latin indices). We introduce the *metric tensor* $g_{\mu\nu}$ by the equation

$$w_\mu = g_{\mu\nu} w^\nu, \tag{24}$$

so that $g_{\mu\nu} = 0$ if $\mu \neq \nu$, $g_{00} = 1$, and $g_{ij} = -\delta_{ij}$. Then (23) can be written in the equivalent form

$$|w|^2 = g_{\mu\nu} w^\mu w^\nu. \tag{25}$$

We also define $g^{\mu\nu}$ (with upper indices) by

$$g^{\mu\nu} g_{\nu\lambda} = g_{\lambda\nu} g^{\nu\mu} = \delta^\mu_\lambda. \tag{26}$$

It then follows that $w^\alpha = g^{\alpha\beta} w_\beta$.

An inhomogeneous Lorentz transformation between two four-space coordinate systems x^μ and x'^μ is of the form

$$x'^\mu = a^\mu{}_\nu x^\nu + b^\mu \tag{27}$$

where the $a^\mu{}_\nu$ are the matrix elements of a homogeneous Lorentz transformation, described at Eq. (22), and the b^μ are constants. Technically a four-vector w is an object whose contravariant components w'^α in the primed system are related to those in the unprimed system by the homogeneous part of (27), namely by

$$w'^\alpha = a^\alpha{}_\beta w^\beta. \tag{28}$$

The statement that (27) preserves the quadratic form (21), or that (28) preserves (23) or (25), may be written as

$$a^\mu{}_\nu a^\alpha{}_\beta g_{\mu\alpha} = g_{\nu\beta}. \tag{29}$$

It follows that

$$w'_\mu = g_{\mu\nu} w'^\nu = g_{\mu\nu} a^\nu{}_\beta g^{\beta\alpha} w_\alpha \equiv a_\mu{}^\alpha w_\alpha, \tag{30}$$

defining the $a_\mu{}^\alpha$ (note the difference between $a_\mu{}^\alpha$ and $a^\alpha{}_\mu$).

We extend the technical definition of a vector to that of a tensor $t^{\alpha_1\cdots\alpha_n}{}_{\beta_1\cdots\beta_m}$ contravariant of *rank* n and covariant of rank m (the whole tensor of rank $n + m$). This is a tensor if and only if its components in the primed system are given in terms of its components in the unprimed system by

$$t^{\alpha_1\cdots\alpha_n}{}_{\beta_1\cdots\beta_m} = a^{\alpha_1}{}_{\mu_1}\cdots a^{\alpha_n}{}_{\mu_n}a_{\beta_1}{}^{\lambda_1}\cdots a_{\beta_m}{}^{\lambda_m}t^{\mu_1\cdots\mu_n}{}_{\lambda_1\cdots\lambda_m}. \tag{31}$$

A tensor of rank zero is called a *scalar*. (See Problem 2.) As with vectors, the $g^{\mu\nu}$ and $g_{\mu\nu}$ are used to raise and lower indices; for example,

$$t^{\alpha\beta} = g^{\mu\beta}t^{\alpha}{}_{\mu}.$$

We have already done this in defining $a_\mu{}^\nu$ at Eq. (30).

The reason these technical definitions of vectors and tensors are important is that virtually all physical statements can be made in terms of such *covariant objects*, that is, scalars, vectors, tensors. (The word "covariant" is used in two senses.) In fact this assertion is essentially the Principle of Relativity, one of the building blocks of the special theory. For instance, as we will see below, the electromagnetic field is a tensor of rank 2.

A vector w is called *timelike* if [see Eq. (23)] $|w|^2 > 0$, *spacelike* if $|w|^2 < 0$, and *lightlike* or *null* if $|w|^2 = 0$. The relativistic three-momentum $\mathbf{p}$ of a particle of (rest) mass m is not $m\mathbf{v}$ but $m\mathbf{v}(1 - v^2)^{-\frac{1}{2}}$, where $\mathbf{v}$ is its velocity (remember: $c = 1$). Its relativistic energy E is $m(1 - v^2)^{-\frac{1}{2}}$. If we write $E = p^0$, then its *four-momentum* p is defined by $p^\mu = (p^0, \mathbf{p})$, which is a timelike (contravariant) four-vector if $m > 0$ (it is lightlike for photons and neutrinos, for which $m = 0$), for a simple calculation will show that

$$|p|^2 = p^\mu p_\mu = m^2. \tag{32}$$

Generally, relativistic field functions are covariant quantities. Both as an example and because it is physically very important, we illustrate some of the ideas by stating some simple properties of the electromagnetic field. In Gaussian units and with $c = 1$ Maxwell's equations are

$$\nabla\cdot\mathbf{E} = \rho, \qquad \nabla\times\mathbf{E} + \frac{\partial\mathbf{B}}{\partial t} = 0,$$

$$\nabla\cdot\mathbf{B} = 0, \qquad \nabla\times\mathbf{B} - \frac{\partial\mathbf{E}}{\partial t} = \mathbf{J},$$

where $\mathbf{E}$ and $\mathbf{B}$ are the electric and magnetic fields, ρ is the electric charge density, and $\mathbf{J}$ is the electric current density. It follows from Maxwell's inhomogeneous equations that charge is conserved [see Eqs. (18) and (19)]:

$$\nabla\cdot\mathbf{J} + \frac{\partial\rho}{\partial t} = 0.$$

To put all this in relativistic notation we identify $\mathbf{E}$ and $\mathbf{B}$ with an antisymmetric covariant tensor $f_{\mu\nu} = -f_{\nu\mu}$ of rank two according to the scheme

$$\begin{aligned}\mathbf{E} &= (f_{10}, f_{20}, f_{30}) = (E_1, E_2, E_3),\\ \mathbf{B} &= (f_{23}, f_{31}, f_{12}) = (B_1, B_2, B_3).\end{aligned} \tag{33}$$

Since $f_{\mu\nu}$ is antisymmetric, it has only six independent components, and (33) defines it completely. Then Maxwell's inhomogeneous equations can be written

$$\frac{\partial f^{\mu\nu}}{\partial x^{\nu}} = j^{\mu}, \tag{34}$$

where $j^0 = \rho$, $j^k = J_k$, and j^{μ} is called the *electric current four-vector*. The antisymmetry of $f^{\mu\nu}$ then implies charge conservation [again, see Eqs. (18) and (19)]:

$$\frac{\partial j^{\mu}}{\partial x^{\mu}} = 0. \tag{35}$$

Maxwell's homogeneous equations are

$$\frac{\partial f_{\mu\nu}}{\partial x^{\lambda}} + \frac{\partial f_{\nu\lambda}}{\partial x^{\mu}} + \frac{\partial f_{\lambda\mu}}{\partial x^{\nu}} = 0, \tag{36}$$

and they imply the existence of a *four-vector potential* A^{μ} such that

$$f_{\mu\nu} = \frac{\partial A_{\mu}}{\partial x^{\nu}} - \frac{\partial A_{\nu}}{\partial x^{\mu}}. \tag{37a}$$

The A^{μ} (or A_{μ}) are not uniquely determined by the $f_{\mu\nu}$. In fact, if A_{μ} is replaced by $\bar{A}_{\mu} = A_{\mu} + \partial\chi/\partial x^{\mu}$, where χ is any scalar function, then

$$f_{\mu\nu} = \frac{\partial \bar{A}_{\mu}}{\partial x^{\nu}} - \frac{\partial \bar{A}_{\nu}}{\partial x^{\mu}}$$

as well. The transformation from A_{μ} to $\bar{A}_{\mu}$ is a *gauge transformation*. In terms of **E** and **B**, the fields and potentials are related by

$$\mathbf{B} = \nabla \times \mathbf{A}, \qquad \mathbf{E} = -\nabla\Phi - \frac{\partial \mathbf{A}}{\partial t}, \tag{37b}$$

where $A^{\mu} = (\Phi, \mathbf{A})$. The gauge transformation is

$$\bar{A}^{\mu} = A^{\mu} + g^{\mu\nu}\,\partial\chi/\partial x^{\nu},$$

so that

$$\bar{\Phi} = \Phi + \frac{\partial\chi}{\partial t}, \qquad \bar{\mathbf{A}} = \mathbf{A} - \nabla\chi.$$

In terms of the potentials, Maxwell's inhomogeneous equations are

$$g^{\mu\nu}\frac{\partial^2 A^{\lambda}}{\partial x^{\mu}\,\partial x^{\nu}} - g^{\lambda\mu}\frac{\partial^2 A^{\mu}}{\partial x^{\mu}\,\partial x^{\nu}} = j^{\lambda}.$$

The second term on the left-hand side can be eliminated by a gauge transformation if χ is chosen (this is called the *Lorentz gauge*) as any solution of

$$\Box\chi = -\frac{\partial A^{\lambda}}{\partial x^{\lambda}},$$

where

$$\Box = g^{\mu\nu} \frac{\partial^2}{\partial x^\mu \, \partial x^\nu} = \frac{\partial^2}{\partial t^2} - \nabla^2 \tag{38}$$

is the *d'Alembertian* operator. Then $\partial \bar{A}^\lambda / \partial x^\lambda = 0$, and Maxwell's equations become

$$\Box \bar{A}^\lambda = j^\lambda. \tag{39}$$

Further gauge transformation by any solution of $\Box\chi = 0$ leaves (39) invariant.

We want now to proceed to a theory of relativistic fields. In order to construct one, we merely restrict the discussion of Section 1 so that the quantities appearing in the equations, particularly the field functions, are all covariant. The fields will still be written ψ_A, but now A must represent in general a collection of covariant indices. That is, a field may have a part which is a scalar, a part which is a vector, etc.,[2] and A runs through the components of all these parts. In order that (13) be covariant, $\mathscr{L}$ must transform properly, namely like a scalar. Since the ψ_A are covariant quantities, scalar combinations can be formed of them (see Problem 2), and this is then the form of $\mathscr{L}$: it is composed of scalar combinations of the ψ_A and $\psi_{A\mu}$. Then $\partial\mathscr{L}/\partial\psi_A$ and $\partial\mathscr{L}/\partial\psi_{A\mu}$ will have the obvious covariant characters indicated by their indices.

Remark. Because $\mathscr{L}$ is a scalar of this form, its change under a Lorentz transformation can be entirely described in terms of the changes of the x^μ. That is, although the components of the covariant fields, the ψ_A and $\psi_{A\mu}$, will also be transformed, their transformations will cancel out because they appear in scalar combinations. The fields themselves contribute to any change in $\mathscr{L}$ only through the x^μ in their arguments. One says that $\mathscr{L}$ is *invariant under Lorentz transformations.* This is used later, at Eq. (55). See Problem 2(b).

For example, consider the free (i.e., empty-space) electromagnetic field. Let the field functions be the four-vector potential A_μ defined through Eqs. (37). We will show that the Lagrangian density

$$\mathscr{L} = -\tfrac{1}{4} f_{\alpha\beta} f^{\alpha\beta} \tag{40}$$

will give the correct field equations, where Eq. (37*a*) is now taken as a definition of the $f_{\alpha\beta}$ in terms of the A_μ. Since the A_μ (or equivalently the A^μ) are the ψ_A, the index A represents the four-vector index μ, and $\psi_{A\alpha}$ becomes

$$A_{\mu\alpha} = \frac{\partial A_\mu}{\partial x^\alpha}. \tag{41}$$

[2] In general this may also include *spinors* but we do not discuss them here. See, however, Chapter IX, Section 4(b).

Then $\partial\mathscr{L}/\partial A_\mu = 0$, and the field equations are

$$\frac{\partial}{\partial x^\nu}\frac{\partial\mathscr{L}}{\partial A_{\mu\nu}} = 0. \tag{42}$$

Using (37*a*) and (40), we have

$$\begin{aligned}\frac{\partial\mathscr{L}}{\partial A_{\mu\nu}} &= -\frac{1}{4}\left(\frac{\partial f_{\alpha\beta}}{\partial A_{\mu\nu}} f^{\alpha\beta} + f_{\alpha\beta}\frac{\partial f^{\alpha\beta}}{\partial A_{\mu\nu}}\right)\\ &= -\tfrac{1}{2}f^{\alpha\beta}\frac{\partial f^{\alpha\beta}}{\partial A_{\mu\nu}} = -\tfrac{1}{2}f^{\alpha\beta}(\delta^\mu_\alpha\delta^\nu_\beta - \delta^\nu_\alpha\delta^\mu_\beta)\\ &= -\tfrac{1}{2}(f^{\mu\nu} - f^{\nu\mu}) = -f^{\mu\nu},\end{aligned}$$

and the field equations can be written

$$-\frac{\partial f^{\mu\nu}}{\partial x^\nu} = 0, \tag{43}$$

which is (34) with $j^\mu = 0$. Of course these equations can also be written in terms of the field functions A^μ. The rest of Maxwell's equations (the ordinarily homogeneous ones) are simply identities in the field functions A^μ.

We have seen in the previous section that if $\mathscr{L}$ does not depend explicitly on the x^α, then

$$T_\mu^{\ \alpha} = \frac{\partial\mathscr{L}}{\partial\psi_{A\alpha}}\psi_{A\mu} - \delta^\alpha_\mu\mathscr{L} \tag{44}$$

forms four conserved currents. If $\mathscr{L}$ is a scalar and if A is a collection of covariant indices $T_\mu^{\ \alpha}$ is a covariant quantity, namely a tensor. This tensor, as we have mentioned, is called the stress tensor, and $P_\mu = \int T_\mu^{\ 0}\, d^3x$ is called the *energy-momentum vector*. We will prove that P_μ is indeed a (covariant) four-vector.

For the proof, consider a conserved current vector S^μ, that is, one for which $\partial S^\mu/\partial x^\mu = 0$. Then as we have seen, $\int S^0\, d^3x$ is time independent for a given observer. We want first to show that this integral is equal to $\int S'^0\, d^3x'$, where S'^0 is the time component of S'^μ, the conserved current in some other Lorentz frame, and the new integral is over the region $x'^0 = 0$ rather than over $x^0 = 0$, as in the original integral. We may assume without loss of generality that the relative velocity is in the three-direction. We shall integrate $\partial S^\mu/\partial x^\mu = 0$ through the entire four-space region shown in Fig. 38. First consider just half the region, namely the wedge-like region labeled W_2. Then we have

$$\begin{aligned}0 &= \int_{W_2}\frac{\partial S^\mu}{\partial x^\mu}\, d^4x = \oint_{\partial W_2} S^\mu\, dB_\mu\\ &= \int_{\text{bottom}} S^\mu\, dB_\mu + \int_{\text{top}} S^\mu\, dB_\mu + \int_\Sigma S^\mu\, dB_\mu.\end{aligned}$$

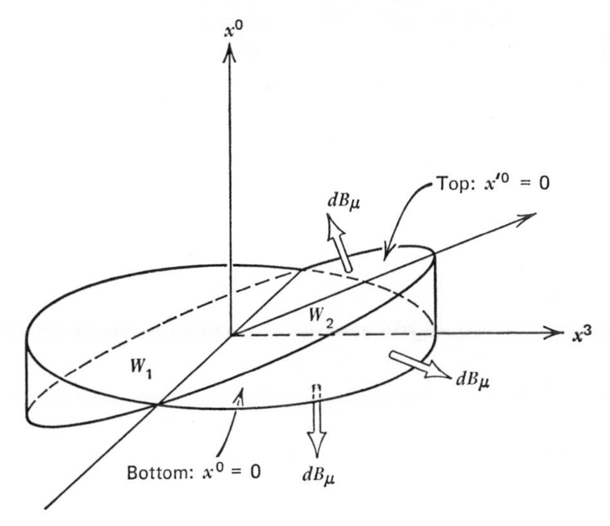

Figure 38. Region of integration used to prove that P_μ is a four-vector.

We proceed as we did in using Fig. 37 to illustrate the proof of Eq. (19). On the bottom, $dB_\mu = (-d^3x, 0, 0, 0)$. On Σ, $dB_\mu = (0, dx^0\, d\mathbf{\Sigma})$. On the top $S^\mu\, dB_\mu = S'^\mu\, dB'_\mu$ by Lorentz invariance, and $dB'_\mu = (d^3x', 0, 0, 0)$, so that

$$\int_{\text{bottom}} S^0\, d^3x = \int_{\text{top}} S'^0\, d^3x' + \int_\Sigma \mathbf{S} \cdot d\mathbf{\Sigma}\, dx_0.$$

Now we require that as r increases the integral over Σ goes to zero. This is a more restrictive condition than the one we needed earlier for Eq. (19), because the time interval in the present integral increases without bound, but nevertheless we accept it. Wedge W_1 gives a similar result. Combining the two, we arrive at

$$\int S^0\, d^3x = \int S'^0\, d^3x', \tag{45}$$

where the integrals are over all three-space. This tells us that the total amount of the conserved quantity is independent of the observer. For instance, Eq. (35), which states that the electric current four-vector is a conserved current, implies not only that charge is conserved, but also that the total amount of charge in the universe (or the charge of the electron) is independent of the observer's state of motion.

We now show that $P_\mu = \int T_\mu{}^0\, d^3x$ is a vector. Let w^μ be an arbitrary constant vector, and write $S^\nu = w^\mu T_\mu{}^\nu$. Then because $T_\mu{}^\nu$ is a tensor forming four conserved currents, S^ν is a vector and $\partial S^\nu/\partial x^\nu = 0$. Therefore $w^\mu P_\mu = \int S^0\, d^3x$ is independent of observer according to (45), and consequently P_μ is a vector (Problem 2).

So far we have discussed only those conservation laws which depend on $\mathscr{L}$ being independent of the x^μ and hence being symmetric under space-time translations. But if $\mathscr{L}$ is a scalar also under homogeneous Lorentz transformations, other conservation laws are obtained, as well. To obtain these we must first discuss the transformation properties of the $\psi_A(x)$.

Consider the homogeneous Lorentz transformation

$$x'^\mu = a^\mu{}_\nu x^\nu. \tag{46}$$

Under such a transformation each of the ψ_A can be written as a function of the x'^μ rather than the x^μ. But in addition, one ordinarily transforms the ψ_A into new linear combinations of themselves. For instance if the index A is a vector index, so that the ψ_A are components of a vector field, the component of the field along, say, the 3′-direction is a linear combination of the components along all four unprimed directions. These new components, or more generally these new linear combinations, we call ψ'_A. Thus

$$\psi'_A(x') = D_A^B \psi_B(x), \tag{47}$$

where the x^μ and x'^μ are related by (46), and the D_A^B, $A, B = 1, \ldots, N$ depend on the particular Lorentz transformation (on the $a^\mu{}_\nu$) and on what the index A actually stands for. If we want to find the functional form of the ψ'_A we actually use (46), or rather its inverse, to obtain (let $x' = y$)

$$\psi'_A(y) = D_A^B \psi_B(a^{-1}y), \tag{48}$$

where, according to Eqs. (26) and (29),

$$(a^{-1}y)^\mu = (a^{-1})^\mu{}_\nu y^\nu = a_\nu{}^\mu y^\nu. \tag{49}$$

In order to use Eqs. (15) and (16) to find conserved quantities, we must consider ε-families of Lorentz transformations. Thus let $a_\nu{}^\mu = a_\nu{}^\mu(\varepsilon)$ depend on a parameter ε so that $a_\nu{}^\mu(0) = \delta_\nu^\mu$, and calculate $\partial\psi_A/\partial\varepsilon$. At $\varepsilon = 0$ we have, according to Eq. (47),

$$\begin{aligned}\frac{\partial}{\partial\varepsilon}\psi_A(x,\varepsilon) &= \frac{\partial}{\partial\varepsilon}\{D_A^B(\varepsilon)\psi_B(a^{-1}[\varepsilon]x)\}\\ &= d_A^B\psi_B(x) + D_A^B(0)\psi_{B\mu}\frac{\partial}{\partial\varepsilon}(a_\nu{}^\mu x^\nu)\\ &= d_A^B\psi_B(x) + \omega_\nu{}^\mu x^\nu \psi_{A\mu}\end{aligned} \tag{50}$$

(we use $D_A^B(0) = \delta_A^B$, for at $\varepsilon = 0$ the transformation is the identity), where

$$\omega_\nu{}^\mu = \frac{da_\nu{}^\mu}{d\varepsilon} \quad \text{at} \quad \varepsilon = 0, \tag{51}$$

and

$$d_A^B = \frac{dD_A^B}{d\varepsilon} \quad \text{at} \quad \varepsilon = 0. \tag{52}$$

Equation (50) can be simplified by using some properties of the $\omega_\mu{}^\nu$ and by writing d_A^B in terms of the $\omega_\nu{}^\mu$. We know that $a^\mu{}_\nu a_\mu{}^\lambda = \delta_\nu^\lambda$ independent of ε, and that the $g^{\mu\nu}$ and $g_{\mu\nu}$, which are used to raise and lower indices, are independent of ε. Thus from (51) we have, at $\varepsilon = 0$,

$$0 = \frac{d}{d\varepsilon}(a^\mu{}_\nu a_\mu{}^\lambda) = \omega^\mu{}_\nu \delta_\mu^\lambda + \delta_\nu^\mu \omega_\mu{}^\lambda = \omega^\lambda{}_\nu + \omega_\nu{}^\lambda,$$

where we have used $a_\mu{}^\lambda(0) = a^\lambda{}_\mu(0) = \delta_\mu^\lambda$. Then by raising the ν or lowering the λ we have

$$\omega^{\nu\lambda} + \omega^{\lambda\nu} = 0 \qquad \text{or} \qquad \omega_{\nu\lambda} + \omega_{\lambda\nu} = 0. \tag{53}$$

As for the d_A^B, we have already pointed out that the D_A^B are functions of the $a_\mu{}^\nu$, so that from (52) we have, at $\varepsilon = 0$,

$$d_A^B = \frac{\partial D_A^B}{\partial a^\mu{}_\nu}\omega^\mu{}_\nu \equiv \tfrac{1}{2}\Delta_{A\mu}^{B\ \nu}\omega^\mu{}_\nu = \tfrac{1}{2}\Delta_A^{B\mu\nu}\omega_{\mu\nu}, \tag{54}$$

defining $\Delta_A^{B\mu\nu}$.[3] We now insert (53) and (54) into (50), obtaining

$$\begin{aligned}\frac{\partial \psi_A}{\partial \varepsilon} &= \tfrac{1}{2}\Delta_A^{B\mu\nu}\omega_{\mu\nu}\psi_B + \omega_{\mu\nu}x^\mu \psi_A^\nu \\ &= \tfrac{1}{2}\omega_{\mu\nu}[\Delta_A^{B\mu\nu}\psi_B + x^\mu\psi_A^\nu - x^\nu\psi_A^\mu],\end{aligned}$$

where $\psi_A^\mu = g^{\mu\nu}\psi_{A\nu}$, as usual.

Now we must calculate $\partial\mathscr{L}/\partial\varepsilon$ and find G^α according to (15). We have, at $\varepsilon = 0$,

$$\begin{aligned}\frac{\partial\mathscr{L}}{\partial\varepsilon} &= \frac{\partial\mathscr{L}}{\partial x^\alpha}\frac{\partial}{\partial\varepsilon}(a_\beta{}^\alpha x^\beta) = \frac{\partial\mathscr{L}}{\partial x^\alpha}\omega_\beta{}^\alpha x^\beta \\ &= \omega_{\beta\mu}x^\beta g^{\alpha\mu}\frac{\partial\mathscr{L}}{\partial x^\alpha} = \tfrac{1}{2}\omega_{\beta\mu}\left[x^\beta\frac{\partial}{\partial x^\alpha}(\mathscr{L}g^{\alpha\mu}) - x^\mu\frac{\partial}{\partial x^\alpha}(\mathscr{L}g^{\alpha\beta})\right] \\ &= \frac{\partial}{\partial x^\alpha}[\tfrac{1}{2}\omega_{\beta\mu}(x^\beta\mathscr{L}g^{\alpha\mu} - x^\mu\mathscr{L}g^{\alpha\beta})] \equiv \frac{\partial G^\alpha}{\partial x^\alpha},\end{aligned} \tag{55}$$

defining the G^α. We used the scalar property of $\mathscr{L}$ here, as mentioned in the Remark preceding Eq. (40). Thus Eq. (16) becomes

$$\begin{aligned}0 = \tfrac{1}{2}\omega_{\mu\nu}\frac{\partial}{\partial x^\alpha}\Bigg[&(\Delta_A^{B\mu\nu}\psi_B + x^\mu\psi_A^\nu - x^\nu\psi_A^\mu)\frac{\partial\mathscr{L}}{\partial\psi_{A\alpha}} \\ &- x^\mu\mathscr{L}g^{\alpha\nu} + x^\nu\mathscr{L}g^{\alpha\mu}\Bigg] \equiv \tfrac{1}{2}\omega_{\mu\nu}\frac{\partial M^{\mu\nu\alpha}}{\partial x^\alpha},\end{aligned}$$

[3] Actually this defines only the antisymmetric part of $\Delta_A^{B\mu\nu}$. But that is all we need. We shall take $\Delta_A^{B\mu\nu}$ to be antisymmetric in μ and ν, that is, we take

$$\Delta_A^{B\mu\nu} = \frac{\partial D_A^B}{\partial a_{\mu\nu}} - \frac{\partial D_A^B}{\partial a_{\nu\mu}}.$$

defining $M^{\mu\nu\alpha}$. This is true for arbitrary antisymmetric $\omega_{\mu\nu}$ and thus since $M^{\mu\nu\alpha}$ was constructed antisymmetric in μ and ν, we have

$$\frac{\partial M^{\mu\nu\alpha}}{\partial x^\alpha} = 0, \tag{56}$$

which means that the antisymmetric tensor

$$m^{\mu\nu} = \int M^{\mu\nu 0}\, d^3x \tag{57}$$

is conserved. Note that $M^{\mu\nu\alpha}$ can be written partly in terms of $T^{\mu\nu}$.

$$M^{\mu\nu\alpha} = \Delta_A^{B\mu\nu}\psi_B \frac{\partial \mathscr{L}}{\partial \psi_{A\alpha}} + (x^\mu T^{\nu\alpha} - x^\nu T^{\mu\alpha}). \tag{58}$$

The space components of $m^{\mu\nu}$, namely, the m^{kl}, correspond to the angular momentum carried by the field. To see this, one needs to *symmetrize* the stress tensor $T^{\mu\nu}$. That is, one finds a tensor $R^{\mu\nu}$ such that if one writes $\bar{T}^{\nu\mu} = T^{\mu\nu} + R^{\mu\nu}$, then $\bar{T}^{\mu\nu}$ is symmetric, and

$$\int \bar{T}^{\mu 0}\, d^3x = \int T^{\mu 0}\, d^3x = P^\mu,$$

or

$$\int R^{\mu 0}\, d^3x = 0.$$

It is shown in Problem 5 that

$$R^{\mu\nu} = \frac{\partial S^{\mu\nu\alpha}}{\partial x^\alpha} \tag{59}$$

will do the trick, where

$$S^{\mu\nu\alpha} = \frac{1}{2}\left(\Delta_A^{B\nu\mu}\psi_B \frac{\partial \mathscr{L}}{\partial \psi_{A\alpha}} + \Delta_A^{B\nu\alpha}\psi_B \frac{\partial \mathscr{L}}{\partial \psi_{A\mu}} - \Delta_A^{B\alpha\mu}\psi_B \frac{\partial \mathscr{L}}{\partial \psi_{A\nu}}\right). \tag{60}$$

For convenience and for reasons of gauge invariance, as we shall see in Section 4(a), it is $\bar{T}^{\mu\nu}$ rather than $T^{\mu\nu}$ which is used in practice. Then it is shown in Problem 5 that if

$$\bar{M}^{\mu\nu\alpha} = x^\mu \bar{T}^{\nu\alpha} - x^\nu \bar{T}^{\mu\alpha} \tag{61}$$

rather than $M^{\mu\nu\alpha}$ is used to define $m^{\mu\nu}$, the result will be the same.

Remark. The symmetry relation $\bar{T}^{k0} = \bar{T}^{0k}$ means that the density of k-component of momentum is equal to the flux of energy in the k-direction. $\bar{T}^{kl} = \bar{T}^{lk}$ implies other such equalities (see Problem 1).

It is Eq. (61) that verifies that m^{kl} is the angular momentum. We have seen that $T^{\mu 0}$, and hence $\bar{T}^{\mu 0}$, is the energy-momentum density. Hence $\bar{M}^{kl0} = x^k \mathscr{P}^l - x^l \mathscr{P}^k$ where $\mathscr{P}$ is the momentum density. This is very much like the particle equation $\mathbf{l} = \mathbf{r} \times \mathbf{p}$, and hence is interpreted as angular momentum density. (It is interesting, incidentally, that unlike particles, fields carry a certain amount of *intrinsic* angular momentum, independent of the point about which it is calculated.) The other components of the form $\bar{M}^{klj}$ of the tensor can be interpreted as the local angular-momentum flux in the field.

Angular-momentum conservation emerges from Lorentz invariance of $\mathscr{L}$ because, as noted just above Eq. (22), the three-space rotations are included among the Lorentz transformations. It is invariance of $\mathscr{L}$ under rotations, as usual, that gives rise to conservation of angular momentum. The other Lorentz transformations, those involving things like the Λ_3 of Eq. (22), lead to conserved currents of the type $\bar{M}^{0\mu\nu}$, which are considerably more difficult to interpret (Møller, 1949; Pryce, 1948).

Section 4 gives some examples of relativistic fields and their properties.

VIII-3 THE HAMILTONIAN FORMALISM

In particle mechanics the momentum conjugate to a coordinate q_α is defined as $p_\alpha = \partial L/\partial \dot{q}_\alpha$, and it has the property that if L is independent of q_α, then $dp_\alpha/dt = 0$. In field theory if one of the field functions, say ψ_B, does not appear in $\mathscr{L}$, then the field equations (13) imply that

$$\frac{\partial}{\partial x^\mu}\frac{\partial \mathscr{L}}{\partial \psi_{B\mu}} = 0.$$

Thus $\partial\mathscr{L}/\partial\psi_{B\mu}$ is a conserved current and, as we have seen in Section 1, this implies that

$$\frac{d}{dt}\int \frac{\partial \mathscr{L}}{\partial \psi_{B0}}\, d^3x = 0. \tag{62}$$

There is a certain similarity here between the field and particles cases, for ψ_{B0} is the time derivative of ψ_B. But there are important differences as well. In particular the exact analog of a cyclic q_α would be a cyclic *value* $\psi_B(x)$ for a fixed $\mathbf{x}$, from which we could then define the momentum conjugate to that $\psi_B(x)$ for the fixed $\mathbf{x}$. Since such a cyclic $\psi_B(x)$ is hardly meaningful, we instead choose to define the momentum canonically conjugate to the *function* ψ_B for all $\mathbf{x}$ as the function given by

$$\pi^B(x) = \frac{\partial \mathscr{L}}{\partial \psi_{B0}}. \tag{63}$$

Then if ψ_B does not appear in $\mathscr{L}$, as is shown by (62), $\int \pi^B\, d^3x$ is a constant.

In this section we start from such considerations and construct for fields the entire canonical formalism we had for particles. Recall that the Hamiltonian is defined in Chapter VI, Section 1, for systems of particles by

$$H = \dot{q}_\alpha p_\alpha - L,$$

involving a sum on α from 1 to n, the number of degrees of freedom. Keeping in mind the definition (63) of canonical momentum for fields, we define H analogously by

$$H = \int \psi_{A0}\pi^A \, d^3x - L$$
$$= \int (\psi_{A0}\pi^A - \mathscr{L}) \, d^3x, \tag{64}$$

or

$$H = \int \mathscr{H} \, d^3x,$$

where

$$\mathscr{H} = \psi_{A0}\pi^A - \mathscr{L} \tag{65}$$

is the *Hamiltonian density*. As in the case of particles, Eq. (63) is solved for the ψ_{A0} as functions of the π^B, the ψ_B, the ψ_{Bk}, and the x^μ, and these solutions are inserted into (65). Then $\mathscr{H}$ becomes a function of the π^B, the ψ_B, the ψ_{Bk}, and the x^μ.

Remarks. 1. Remember that $\psi_{Bk} = \partial\psi_B/\partial x^k$, where k runs from 1 to 3. *2.* Recall also that the 0, 0 component of the stress tensor is the energy density. In general, according to Eq. (20)

$$T_0^0 = \frac{\partial\mathscr{L}}{\partial\psi_{A0}} \psi_{A0} - \mathscr{L},$$

which is the same as (65). Thus $\mathscr{H} = T_0^0$ and, as in the particle case, $H = \int \mathscr{H} \, d^3x$ is the total energy.

Again, it is difficult to formulate the exact analog of the canonical equations, for the derivative of H with respect to q_α, for instance, is analogous to the derivative of the integral in Eq. (64) with respect to a value $\psi_A(x)$ for fixed **x**, whatever that may mean. We shall return to this point soon, but first let us see what we get if we take derivatives of $\mathscr{H}$ with respect to the functions π^B and ψ_B. In $\mathscr{H}$ the functions ψ_{A0} are replaced everywhere by their expressions in terms of the ψ_B, π^B, ψ_{Bk}, and (possibly) x^μ. Furthermore, in accordance with the whole Hamiltonian approach, the π^B and ψ_B are considered to be independent variables. Then we have [see the derivation of Eqs. (7), Chapter VI]

$$\frac{\partial\mathscr{H}}{\partial\pi^B} = \psi_{B0} + \frac{\partial\psi_{A0}}{\partial\pi^B}\pi^A - \frac{\partial\mathscr{L}}{\partial\psi_{A0}}\frac{\partial\psi_{A0}}{\partial\pi^B}.$$

The last two terms cancel in accordance with (63), and we have

$$\psi_{B0} = \frac{\partial\mathscr{H}}{\partial\pi^B}. \tag{66}$$

We now take the derivative with respect to ψ_B:

$$\frac{\partial \mathscr{H}}{\partial \psi_B} = \pi^A \frac{\partial \psi_{A0}}{\partial \psi_B} - \frac{\partial \mathscr{L}}{\partial \psi_{A0}} \frac{\partial \psi_{A0}}{\partial \psi_B} - \frac{\partial \mathscr{L}}{\partial \psi_B}.$$

The first two terms cancel in accordance with (63). The equations of motion tell us that the last term is

$$\frac{\partial \mathscr{L}}{\partial \psi_B} = \frac{\partial}{\partial x^\mu} \frac{\partial \mathscr{L}}{\partial \psi_{B\mu}} = \frac{\partial}{\partial x^k} \frac{\partial \mathscr{L}}{\partial \psi_{Bk}} + \frac{\partial \pi^B}{\partial x^0},$$

where we have used (63) again. But $\partial \mathscr{L} / \partial \psi_{Bk}$ can be written in terms of $\mathscr{H}$, for

$$\frac{\partial \mathscr{H}}{\partial \psi_{Bk}} = \pi^A \frac{\partial \psi_{A0}}{\partial \psi_{Bk}} - \frac{\partial \mathscr{L}}{\partial \psi_{A0}} \frac{\partial \psi_{A0}}{\partial \psi_{Bk}} - \frac{\partial \mathscr{L}}{\partial \psi_{Bk}},$$

and the first two terms on the right-hand side again cancel. Inserting these results into the expression for $\partial \mathscr{H} / \partial \psi_B$, we arrive at

$$\pi_0^B = -\frac{\partial \mathscr{H}}{\partial \psi_B} + \frac{\partial}{\partial x^k} \frac{\partial \mathscr{H}}{\partial \psi_{Bk}}, \tag{67}$$

where we have written $\pi_0^B = \partial \pi^B / \partial x^0$. Equations (66) and (67) are the analogs of Hamilton's canonical equations for a particle system.

These equations can actually be made to look even more like the canonical equations of particle physics. Let us try to calculate something like the derivative of $L = \int \mathscr{L}\, d^3x$ with respect to a value $\psi_A(x)$ for fixed $\mathbf{x}$. To do this, we vary ψ_A as we have so often, writing $\psi_A(x, \varepsilon)$, and then we choose $\psi_A(x, \varepsilon)$ to be the same as $\psi_A(x)$ everywhere except in a very small neighborhood R around our fixed $\mathbf{x}$. Then, as we have seen,

$$\frac{dL}{d\varepsilon} = \int \left[\frac{\partial \mathscr{L}}{\partial \psi_A} - \frac{\partial}{\partial x^\mu} \frac{\partial \mathscr{L}}{\partial \psi_{A\mu}}\right] \frac{\partial \psi_A}{\partial \varepsilon}\, d^3x + \sigma,$$

where σ is a surface integral. Now the integral appearing here is over all three-space, and we shall assume, as we have done all along, that all functions involved vanish so rapidly that their surface integrals vanish as the volumes approach infinity. Thus we set $\sigma = 0$. By assumption $\partial \psi_A / \partial \varepsilon$ is zero everywhere except in R; in the limit as R gets smaller and smaller, closing down on $\mathbf{x}$, we write

$$\lim \frac{dL/d\varepsilon}{\partial \psi_A / \partial \varepsilon} \equiv \frac{\delta L}{\delta \psi_A} = \frac{\partial \mathscr{L}}{\partial \psi_A} - \frac{\partial}{\partial x^\mu} \frac{\partial \mathscr{L}}{\partial \psi_{A\mu}}. \tag{68}$$

This is called the *functional* or *variational derivative*[4] of L with respect to ψ_A. We get a similar result for H, except that $\mathscr{H}$ depends on two sets of

[4] Equation (68) defines $\delta L / \delta \psi_A$ at $\mathbf{x}$, but as usual with definitions of derivatives this defines a function of $\mathbf{x}$ (and of x^0). It is this function which is the functional derivative.

functions ψ_A and π^B and does not depend on the time derivatives ψ_{A0} of the ψ_A or on any of the derivatives $\pi^A{}_\mu$ of the π^A. Thus the functional derivatives of H are

$$\frac{\delta H}{\delta \pi^B} = \frac{\partial \mathscr{H}}{\partial \pi^B},$$

$$\frac{\delta H}{\delta \psi_B} = \frac{\partial \mathscr{H}}{\partial \psi_B} - \frac{\partial}{\partial x^k}\frac{\partial \mathscr{H}}{\partial \psi_{Bk}},$$

and the canonical equations (66) and (67) can be written

$$\psi_{B0} = \frac{\delta H}{\delta \pi^B}, \qquad \pi^B_0 = -\frac{\delta H}{\delta \psi_B}. \tag{69}$$

Incidentally, the field equations (13) can now be written $\delta L/\delta\psi_A = 0$.

Let $\mathscr{F}$ be any function of the ψ_A, ψ_{Ak}, π^A, and x^μ, and let

$$F = \int \mathscr{F}\, d^3x. \tag{70}$$

We want to calculate the time derivative of F when the fields satisfy the field equations (or, equivalently, the canonical equations). We have (recall that $x^0 = t$)

$$\frac{dF}{dt} = \int\left[\frac{\partial \mathscr{F}}{\partial t} + \frac{\partial \mathscr{F}}{\partial \psi_A}\psi_{A0} + \frac{\partial \mathscr{F}}{\partial \psi_{Ak}}\psi_{Ak0} + \frac{\partial \mathscr{F}}{\partial \pi^A}\pi^A_0\right] d^3x,$$

where $\psi_{Ak0} = \partial\psi_{Ak}/\partial x^0 = \partial\psi_{A0}/\partial x^k$. We may write (actually we are integrating this term by parts)

$$\frac{\partial \mathscr{F}}{\partial \psi_{Ak}}\psi_{Ak0} = \frac{\partial}{\partial x^k}\left(\frac{\partial \mathscr{F}}{\partial \psi_{Ak}}\psi_{A0}\right) - \frac{\partial}{\partial x^k}\left(\frac{\partial \mathscr{F}}{\partial \psi_{Ak}}\right)\psi_{A0}.$$

Then if we throw out surface integrals as usual, we have

$$\begin{aligned}\frac{dF}{dt} &= \int \frac{\partial \mathscr{F}}{\partial t}\, d^3x + \int\left[\frac{\delta F}{\delta \psi_A}\psi_{A0} + \frac{\delta F}{\delta \pi^A}\pi^A_0\right] d^3x \\ &= \int \frac{\partial \mathscr{F}}{\partial t}\, d^3x + \int\left[\frac{\delta F}{\delta \psi_A}\frac{\delta H}{\delta \pi^A} - \frac{\delta F}{\delta \pi^A}\frac{\delta H}{\delta \psi_A}\right] d^3x \\ &= \int \frac{\partial \mathscr{F}}{\partial t}\, d^3x + [F, H]^{\mathrm{f}},\end{aligned} \tag{71}$$

defining the *Poisson bracket for fields*. The similarities to the particles case are obvious.

Consider two points in three-space $\mathbf{y}$ and $\mathbf{z}$. We may ask about the Poisson bracket of ψ_A at $\mathbf{y}$ and π^B at $\mathbf{z}$, both of which are functions of $t = x^0$ if $\mathbf{y}$ and $\mathbf{z}$ are held fixed. This Poisson bracket $[\psi_A(x^0, \mathbf{y}), \pi^B(x^0, \mathbf{z})]^{\mathrm{f}}$ could be calculated if we could write ψ_A and π^B as integrals, like F in Eq. (70). In fact this can be done with the aid of Dirac delta functions:

$$\psi_A(x^0, \mathbf{y}) = \int \delta(\mathbf{x} - \mathbf{y})\psi_A(x)\, d^3x,$$

$$\pi^B(x^0, \mathbf{z}) = \int \delta(\mathbf{x} - \mathbf{z})\pi^B(x)\, d^3x.$$

Then we have (we leave the x^0 out of the arguments of ψ_A and π^B)

$$\frac{\delta\psi_A(\mathbf{y})}{\delta\psi_C} = \frac{\partial}{\partial\psi_C}\{\delta(\mathbf{x} - \mathbf{y})\psi_A(x)\} = \delta(\mathbf{x} - \mathbf{y})\delta_A^C,$$

$$\frac{\delta\pi^B(\mathbf{z})}{\delta\psi_C} = \frac{\partial}{\partial\psi_C}\{\delta(\mathbf{x} - \mathbf{z})\pi^B(x)\} = 0,$$

and similar results for $\delta/\delta\pi^C$. The Poisson bracket is then

$$[\psi_A(\mathbf{y}), \pi^B(\mathbf{z})]^{\mathrm{f}} = \int \delta(\mathbf{x} - \mathbf{y})\, \delta_A^C\, \delta(\mathbf{x} - \mathbf{z})\, \delta_C^B\, d^3x = \delta_A^B\, \delta(\mathbf{y} - \mathbf{z}). \tag{72}$$

Again the analogy to systems of particles is obvious: here the delta function plays the role of the Kronecker delta in the particle case. We may proceed to calculate also the Poisson bracket of ψ_{Ak} with π^B and of ψ_A with ψ_{Bk}. We leave it to the reader to show that

$$[\psi_{Ak}(\mathbf{y}), \pi^B(\mathbf{z})]^{\mathrm{f}} = \delta_A^B \frac{\partial}{\partial z^k}\, \delta(\mathbf{z} - \mathbf{y}), \tag{73}$$

$$[\psi_{Ak}(\mathbf{y}), \psi_B(\mathbf{z})]^{\mathrm{f}} = 0.$$

Equations (72) and (73) can be used, with the familiar rules for Poisson brackets of products, etc., to perform calculations.

As a very simple example, we return to the elastic rod, whose Lagrangian density we found to be

$$\mathscr{L} = \frac{\mu}{2}\left(\frac{\partial q}{\partial t}\right)^2 - \frac{Y}{2}\left(\frac{\partial q}{\partial x}\right)^2.$$

The momentum conjugate to q is

$$\pi = \frac{\partial \mathscr{L}}{\partial(\partial q/\partial t)} = \mu\frac{\partial q}{\partial t},$$

so that the Hamiltonian density is

$$\mathscr{H} = \frac{\pi}{\mu}\pi - \frac{\mu}{2}\frac{\pi^2}{\mu^2} + \frac{Y}{2}\left(\frac{\partial q}{\partial x}\right)^2$$
$$= \frac{\pi^2}{2\mu} + \frac{Y}{2}\left(\frac{\partial q}{\partial x}\right)^2.$$

The canonical equations (66) and (67) are

$$\frac{\partial q}{\partial t} = \frac{\pi}{\mu}, \qquad \frac{\partial \pi}{\partial t} = \frac{\partial}{\partial x}\left[Y\frac{\partial q}{\partial x}\right].$$

As usual with the Hamiltonian formalism, one of these equations is merely a repetition of the definition of π. The other is the actual field equation, as can be seen by using the first equation to eliminate $\partial\pi/\partial t$:

$$\mu\frac{\partial^2 q}{\partial t^2} = Y\frac{\partial^2 q}{\partial x^2}.$$

To illustrate the use of Poisson brackets we prove that for this system the integral F of

$$\mathscr{F} = \mu\frac{\partial q}{\partial t}\frac{\partial q}{\partial x} = \pi\frac{\partial q}{\partial x}$$

is a constant of the motion. We have

$$\frac{dF}{dt} = [F, H]^{\text{f}} = \int\left[\frac{\delta F}{\delta q}\frac{\delta H}{\delta \pi} - \frac{\delta F}{\delta \pi}\frac{\delta H}{\delta q}\right]dx,$$

for there is only one space dimension. In the integrand we have

$$\frac{\delta F}{\delta q} = -\frac{\partial}{\partial x}\frac{\partial \mathscr{F}}{\partial(\partial q/\partial x)} = -\frac{\partial \pi}{\partial x},$$

$$\frac{\delta F}{\delta \pi} = \frac{\partial \mathscr{F}}{\partial \pi} = \frac{\partial q}{\partial x},$$

$$\frac{\delta H}{\delta q} = -\frac{\partial}{\partial x}\frac{\partial \mathscr{H}}{\partial(\partial q/\partial x)} = -Y\frac{\partial^2 q}{\partial x^2},$$

$$\frac{\delta H}{\delta \pi} = \frac{\partial \mathscr{H}}{\partial \pi} = \frac{\pi}{\mu}.$$

Thus

$$\frac{dF}{dt} = \int \left(-\frac{1}{\mu} \pi \frac{\partial \pi}{\partial x} + Y \frac{\partial q}{\partial x} \frac{\partial^2 q}{\partial x^2} \right) dx$$

$$= \int \frac{\partial}{\partial x} \left[\frac{\pi^2}{2\mu} - \frac{Y}{2} \left(\frac{\partial q}{\partial x} \right)^2 \right] dx = 0, \qquad (74$$

which vanishes because we assume that all functions go to zero for large enough $|x|$.

The same result can be obtained for this example by using Eqs. (72) and (73), but this is left to the reader (Problem 11).

The Hamiltonian formalism for fields is of importance because it is particularly suitable for making the transition from the classical field theory we have been discussing to the quantum theory of fields. Unfortunately it is not easy to apply to relativistic fields because it treats the space and time coordinates very differently. The Lagrangian formalism is more naturally applicable to classical relativistic fields, largely because it yields the field equations more directly.

VIII-4 EXAMPLES, APPLICATIONS, AND EXTENSIONS

(a) The Stress Tensor of the Electromagnetic Field

From the Lagrangian density

$$\mathscr{L} = -\tfrac{1}{4} f_{\mu\nu} f^{\mu\nu}$$

of the electromagnetic field we can find its symmetrized stress tensor. First we form $T^{\mu\nu}$ and then calculate $R^{\mu\nu}$ according to Eq. (59) to obtain $\bar{T}^{\mu\nu} = T^{\mu\nu} + R^{\mu\nu}$. By Eq. (44) we have

$$T_\mu{}^\nu = A_{\rho\mu} \frac{\partial \mathscr{L}}{\partial A_{\rho\nu}} - \delta_\mu^\nu \mathscr{L}.$$

In obtaining the field equations (43) we already calculated that $\partial \mathscr{L} / \partial A^{\rho\nu} = -f^{\rho\nu}$. Thus

$$T_\mu{}^\nu = -A_{\rho\mu} f^{\rho\nu} + \tfrac{1}{4} \delta_\mu^\nu f_{\rho\sigma} f^{\rho\sigma}.$$

Raising the μ we have

$$T^{\mu\nu} = \tfrac{1}{4} g^{\mu\nu} f_{\rho\sigma} f^{\rho\sigma} + g^{\mu\sigma} A_{\rho\sigma} f^{\rho\nu}.$$

Although the first term is symmetric in μ and ν, the second is not. Furthermore $T^{\mu\nu}$ depends on A_μ and is thus not even gauge invariant. To go on and calculate $R^{\mu\nu}$ we must first find the $\Delta_A^{B\mu\nu}$, and for that we need to find the

D_A^B. But these D_A^B are just the transformation matrices of the A_μ, namely [see Eqs. (30) and (47)],

$$A'_\alpha = a_\alpha{}^\beta A_\beta.$$

Thus D_A^B, or rather D_α^β (since B and A are just vector indices), is $a_\alpha{}^\beta$, and $\Delta_A^{B\mu\nu}$, or rather $\Delta_\alpha^{\beta\mu\nu}$, is given by

$$\Delta_\alpha^{\beta\mu\nu} = \frac{\partial D_\alpha^\beta}{\partial a_{\nu\mu}} - \frac{\partial D_\alpha^\beta}{\partial a_{\mu\nu}} = g^{\beta\lambda}\left(\frac{\partial a_{\alpha\lambda}}{\partial a_{\nu\mu}} - \frac{\partial a_{\alpha\lambda}}{\partial a_{\mu\nu}}\right)$$
$$= g^{\beta\lambda}(\delta_\alpha^\nu\,\delta_\lambda^\mu - \delta_\alpha^\mu\,\delta_\lambda^\nu) = g^{\beta\mu}\,\delta_\alpha^\nu - g^{\beta\nu}\,\delta_\alpha^\mu.$$

Then

$$\Delta_A^{B\mu\nu}\psi_B\frac{\partial\mathscr{L}}{\partial\psi_{A\lambda}} \to \Delta_\alpha^{\beta\mu\nu}A_\beta\frac{\partial\mathscr{L}}{\partial A_{\alpha\lambda}} = (g^{\beta\mu}\,\delta_\alpha^\nu - g^{\beta\nu}\,\delta_\alpha^\mu)A_\beta(-f^{\alpha\lambda}) = -A^\mu f^{\nu\lambda} + A^\nu f^{\mu\lambda}.$$

Inserting this into (60), we obtain

$$S^{\mu\nu\lambda} = \tfrac{1}{2}(-A^\mu f^{\nu\lambda} + A^\nu f^{\mu\lambda} - A^\lambda f^{\nu\mu} + A^\nu f^{\lambda\mu} + A^\mu f^{\lambda\nu} - A^\lambda f^{\mu\nu}) = A^\mu f^{\lambda\nu},$$

and then

$$R^{\mu\nu} = \frac{\partial}{\partial x^\rho}(A^\mu f^{\rho\nu}) = g^{\mu\sigma}A_{\sigma\rho}f^{\rho\nu},$$

where we have used the field equations (43). Finally, we obtain

$$\bar{T}^{\mu\nu} = T^{\mu\nu} + R^{\mu\nu} = (\tfrac{1}{4}g^{\mu\nu}f_{\rho\sigma}f^{\rho\sigma} - g^{\mu\sigma}A_{\rho\sigma}f^{\rho\nu}) + g^{\mu\sigma}A_{\sigma\rho}f^{\rho\nu}$$
$$= \tfrac{1}{4}g^{\mu\nu}f_{\rho\sigma}f^{\rho\sigma} - g^{\mu\sigma}(A_{\rho\sigma} - A_{\sigma\rho})f^{\rho\nu}$$
$$= \tfrac{1}{4}g^{\mu\nu}f_{\rho\sigma}f^{\rho\sigma} - g^{\mu\sigma}f_{\rho\sigma}f^{\rho\nu}.$$

Note that $\bar{T}^{\mu\nu}$ unlike $T^{\mu\nu}$, depends on the A_μ only through the $f_{\mu\nu}$ and is thus gauge invariant.

The above expression for $\bar{T}^{\mu\nu}$ can be written out in terms of the fields **E** and **B**. The energy density is

$$\bar{T}^{00} = \tfrac{1}{4}f_{\rho\sigma}f^{\rho\sigma} - f_{\rho 0}f^{\rho 0}.$$

Now $\mathbf{B} = (f_{23}, f_{31}, f_{12}) = (f^{23}, f^{31}, f^{12})$, and $E_k = f_{k0} = -f^{k0}$ (since $f_{\mu\nu}$ is antisymmetric, this takes care of all of its components). The expression for $f_{\rho\sigma}f^{\rho\sigma}$ squares all the components (with appropriate sign) twice, and $f_{\rho 0}f^{\rho 0} = f_{k0}f^{k0} = -E^2$. Thus

$$\bar{T}^{00} = \tfrac{1}{4}(2B^2 - 2E^2) + E^2 = \tfrac{1}{2}(E^2 + B^2).$$

The momentum density or energy flux has components

$$\bar{T}^{k0} = f_{\rho k}f^{\rho 0}$$

Then $\bar{T}^{10}$, for instance, is

$$\bar{T}^{10} = f_{21}f^{20} + f_{31}f^{30} = B_3E_2 - B_2E_3 = (\mathbf{E} \times \mathbf{B})_1.$$

More generally,

$$\bar{T}^{k0} = (\mathbf{E} \times \mathbf{B})_k.$$

This is the Poynting vector. The three-dimensional stress tensor $\bar{T}^{kl}$ is given by

$$\bar{T}^{kl} = -\tfrac{1}{4}\delta^{kl}f_{\rho\sigma}f^{\rho\sigma} + f_{\rho k}f^{\rho l}.$$

We have already calculated the first term $f_{\rho\sigma}f^{\rho\sigma}$. The 1, 2 component, for instance, of the second term is

$$f_{\rho 1}f^{\rho 2} = f_{01}f^{02} + f_{30}f^{32} = -E_1E_2 - B_2B_1.$$

The 1, 1 component is

$$f_{\rho 1}f^{\rho 1} = f_{01}f^{01} + f_{21}f^{21} + f_{31}f^{31} = -E_1^2 + B_3^2 + B_2^2 = -E_1^2 - B_1^2 + B^2.$$

Thus $f_{\rho k}f^{\rho l}$ may be written

$$f_{\rho k}f^{\rho l} = -E_kE_l - B_kB_l + \delta_{kl}B^2,$$

and the flux of k-momentum in the l-direction is then

$$\begin{aligned}\bar{T}^{kl} &= -\tfrac{1}{4}\delta^{kl}(2B^2 - 2E^2) - E_kE_l - B_kB_l + \delta_{kl}B^2 \\ &= \tfrac{1}{2}\delta^{kl}(E^2 + B^2) - E_kE_l - B_kB_l.\end{aligned}$$

These expressions are the usual ones obtained in the non-Lagrangian formulation of electromagnetic theory. Because we have taken $c = 1$ the dimensions may seem incorrect, but they can be corrected by including appropriate factors of c.

Remark. The electromagnetic field is a relatively complicated one because of its gauge invariance, or because the photon mass is zero. We will not go more deeply, however, into its many unique properties.

(b) Other Relativistic Fields

The electromagnetic field is only one example of relativistic fields. Here we present two other examples.

The scalar meson (uncharged). Let ψ be a scalar, and form the obviously scalar Lagrangian density

$$\mathscr{L} = \frac{1}{2}\left(g^{\alpha\beta}\frac{\partial\psi}{\partial x^\alpha}\frac{\partial\psi}{\partial x^\beta} - \lambda^2\psi^2\right), \tag{75}$$

where λ is a constant. Notice that the first term in $\mathscr{L}$ is very similar to the $\mathscr{L}$ we obtained in Eq. (6) for longitudinal vibrations of an elastic rod. The difference is that here μ and Y both equal 1, and three space coordinates enter rather than just one. We might expect then that the solutions obtained from the $\mathscr{L}$ of (75), if λ were zero, would be waves propagating with velocity $\sqrt{Y/\mu} = 1$ (recall that $c = 1$).

The field equation for ψ is

$$\frac{\partial}{\partial x^\mu}\frac{\partial \mathscr{L}}{\partial \psi_\mu} - \frac{\partial \mathscr{L}}{\partial \psi} = \frac{1}{2}\frac{\partial}{\partial x^\mu}\left[g^{\alpha\beta}\,\delta^\mu_\alpha\,\frac{\partial \psi}{\partial x^\beta} + g^{\alpha\beta}\frac{\partial \psi}{\partial x^\alpha}\,\delta^\mu_\beta\right] + \lambda^2\psi$$

$$= g^{\mu\nu}\frac{\partial^2 \psi}{\partial x^\mu\,\partial x^\nu} + \lambda^2\psi$$

$$= (\Box + \lambda^2)\psi = 0. \tag{76}$$

where $\psi_\mu = \partial\psi/\partial x^\mu$. This is known as the Klein-Gordon equation. It is the relativistic quantum-mechanical equation for an uncharged spinless free particle of mass λ (with not only $c = 1$, but also Planck's constant $\hbar = 1$). For $\lambda = 0$, moreover, this does indeed become the simple wave equation we predicted.

In quantum mechanics (with $\hbar = 1$) the momentum $\mathbf{p}$ of a particle becomes the operator $-i\nabla$, and the energy becomes the operator $i\partial/\partial t$ [see Chapter IX, Section 4(d)]. Thus $p_\mu \rightarrow i\partial/\partial x^\mu$ and

$$p^\mu p_\mu = g^{\mu\nu}p_\mu p_\nu \rightarrow -g^{\mu\nu}\frac{\partial^2}{\partial x^\mu\,\partial x^\nu} = -\Box.$$

When the differential operators are supplied with wave (or field) functions to operate on, Eq. (32) then becomes

$$(\Box + m^2)\psi = 0.$$

This is why (76) is the quantum-mechanical equation of a particle and why λ is identified with the particle's mass.

The stress tensor is easily calculated:

$$T_\mu{}^\nu = \tfrac{1}{2}g^{\alpha\beta}(\delta^\nu_\alpha\psi_\beta + \psi_\alpha\,\delta^\nu_\beta)\psi_\mu - \delta^\nu_\mu\mathscr{L}$$

$$= g^{\nu\alpha}\psi_\alpha\psi_\mu - \tfrac{1}{2}\delta^\nu_\mu(g^{\alpha\beta}\psi_\alpha\psi_\beta - \lambda^2\psi^2).$$

Now we raise the index μ, obtaining

$$T^{\mu\nu} = \psi^\mu\psi^\nu - \tfrac{1}{2}g^{\mu\nu}(\psi^\beta\psi_\beta - \lambda^2\psi^2),$$

which is symmetric, as follows also from Problem 6.

The vector meson (uncharged). Let φ_σ be a vector and form the scalar Lagrangian density

$$\mathscr{L} = \frac{1}{2}\left(g^{\alpha\beta}\frac{\partial\varphi_\sigma}{\partial x^\alpha}\frac{\partial\varphi^\sigma}{\partial x^\beta} - \lambda^2\varphi_\sigma\varphi^\sigma\right). \tag{77}$$

Straightforward calculation yields the field equations

$$(\Box + \lambda^2)\varphi_\sigma = 0. \tag{78}$$

The similarity of (75) to (77) and of (76) to (78) is obvious. We leave it to the reader to find the stress tensor for this case. If we set $\lambda = 0$ we again obtain a simple wave equation. Observe the similarity of this zero-mass equation with Maxwell's equations in the Lorentz gauge; see Problem 12.

Remark. Note that so far all of our Lagrangian densities have been quadratic in the fields and their derivatives. In general this will be true if the field equations are to be linear in the fields.

The Hamiltonian formalism for this case yields the following results. The momentum conjugate to φ_σ is

$$\pi^\sigma = \frac{\partial \mathscr{L}}{\partial \varphi_{\sigma 0}} = \varphi_0^\sigma \equiv \frac{\partial \varphi^\sigma}{\partial x^0}.$$

The Hamiltonian density, written in terms of π^σ, $\varphi_{\sigma k}$ (and, equivalently, φ_k^σ) and φ^σ, is (note the sum on k)

$$\mathscr{H} = \tfrac{1}{2}\pi^\sigma\pi_\sigma + \tfrac{1}{2}\varphi_k^\sigma\varphi_{\sigma k} + \tfrac{1}{2}\lambda^2\varphi^\sigma\varphi_\sigma.$$

The canonical equations are

$$\varphi_{\sigma 0} = \pi_\sigma$$

(repeating the definitions of the π_σ), and

$$\begin{aligned}\pi_{\sigma 0} &= -\frac{\partial \mathscr{H}}{\partial \varphi^\sigma} + \frac{\partial}{\partial x^l}\frac{\partial \mathscr{H}}{\partial \varphi_l^\sigma} \\ &= -\lambda^2\varphi_\sigma + \frac{\partial}{\partial x^l}\varphi_{\sigma l}.\end{aligned}$$

These last are the field equations, for $\pi_{\sigma 0}$ is, according to the first set of canonical equations $\partial^2\varphi_\sigma/\partial t^2$. Thus we have

$$\frac{\partial^2 \varphi_\sigma}{\partial t^2} - \frac{\partial}{\partial x^l}\varphi_{\sigma l} = \Box\varphi_\sigma = -\lambda^2\varphi_\sigma,$$

in agreement with (78).

(c) Gauge Invariance, Charge, and Coupled Fields

Consider a complex-valued field, that is, one consisting of field functions ψ_A of the form

$$\psi_A = \varphi_A + i\chi_A,$$

where φ_A and χ_A are real. We could discuss this problem in terms of the φ_A and χ_A, and then the field would consist of $2N$ rather than N real functions.

When we construct ε-families of transformations, we can do so with these $2N$ real functions, but then it is clear that there will be $2N$ independent variations. This means that instead of the φ_A and χ_A we can vary the ψ_A and their complex conjugates ψ_A^* *independently* to arrive at the same result. The field equations, as shown in Problem 3, are then

$$\frac{\partial}{\partial x^\mu}\frac{\partial \mathscr{L}}{\partial \psi_{A\mu}} - \frac{\partial \mathscr{L}}{\partial \psi_A} = 0, \tag{79a}$$

and

$$\frac{\partial}{\partial x^\mu}\frac{\partial \mathscr{L}}{\partial \psi^*_{A_\mu}} - \frac{\partial \mathscr{L}}{\partial \psi^*_A} = 0. \tag{79b}$$

Let us assume then that we have a Lagrangian density $\mathscr{L}(\psi, \psi^*, \psi', \psi'^*, x)$ which depends, however, only in a special way on the field functions and their derivatives. Let us assume, namely, that $\mathscr{L}$ does not change under simultaneous replacements of the form

$$\psi_A(x, \varepsilon) = \psi_A(x)e^{i\varepsilon}, \qquad \psi^*_A(x, \varepsilon) = \psi^*_A(x)e^{-i\varepsilon},$$

which we shall call a *gauge transformation of the first kind.* An $\mathscr{L}$ which depends, for instance, only on the absolute values of the ψ_A and $\psi_{A\mu}$ is of this kind. The invariance of $\mathscr{L}$ under such transformations leads, as usual to a conserved current. Equations (14) reads

$$0 = i\frac{\partial}{\partial x^\alpha}\left[\frac{\partial \mathscr{L}}{\partial \psi_{A\alpha}}\psi_A - \frac{\partial \mathscr{L}}{\partial \psi^*_{A\alpha}}\psi^*_A\right],$$

so that the conserved current is

$$j^\alpha = i\left(\frac{\partial \mathscr{L}}{\partial \psi_{A\alpha}}\psi_A - \frac{\partial \mathscr{L}}{\partial \psi^*_{A\alpha}}\psi^*_A\right), \tag{80}$$

Consider, for instance, *the charged scalar meson*, namely, the first example of Section 4(b), but now with a complex field. We write[5]

$$\mathscr{L} = g^{\alpha\beta}\frac{\partial \psi}{\partial x^\alpha}\frac{\partial \psi^*}{\partial x^\beta} - \lambda^2\psi\psi^*. \tag{81}$$

The field equations, complex Klein-Gordon equations, are then

$$(\Box + \lambda^2)\psi = 0$$

and

$$(\Box + \lambda^2)\psi^* = 0.$$

These equations are complex conjugates of each other and they are obviously the same as Eq. (76) for the *uncharged* scalar meson. The (covariant)

[5] We might have defined $\mathscr{L}$ as half of what is given by (81), in analogy with Eq. (75), but that would give the field equations multiplied by $\frac{1}{2}$, which is equivalent, but perhaps not as pleasing esthetically.

conserved current is, according to Eq. (80),

$$j_\alpha = i\left(\psi \frac{\partial \psi^*}{\partial x^\alpha} - \frac{\partial \psi}{\partial x^\alpha}\psi^*\right). \tag{82}$$

Now this is the usual expression for the electric current four-vector of a particle satisfying the Klein-Gordon equation, or at least a multiple of it. This is why we call the meson "charged." In particular,

$$Q = e\int j_0 \, d^3x$$

is the total charge, where e is the elementary charge. Thus we may suspect that invariance of the Lagrangian density under gauge transformations of the first kind implies conservation of charge.

The way to check the suspected relation between gauge transformations and charge conservation is to try to couple the Klein-Gordon field to the Maxwell field so that several reasonable criteria are satisfied: *First*, the electromagnetic field should be described by Maxwell's equations with certain electric-current four-vector source functions for the right-hand side of (34) or (39). *Second*, the j^μ as defined by (82), or something reasonably similar, should be those source functions. *Third*, the Klein-Gordon equation should be modified in some reasonable way by the presence of the electromagnetic field. *Fourth*, the equations we obtain should be invariant under gauge transformations of the electromagnetic field.

The problem is to write an interaction Lagrangian density $\mathscr{L}_\mathrm{I}$ to be added to the Klein-Gordon density $\mathscr{L}_\mathrm{KG}$ of Eq. (81) and to the electromagnetic density $\mathscr{L}_\mathrm{em}$ of Eq. (40). But unfortunately there are no general rules for finding such interaction Lagrangians based on the kind of criteria we have stated, so we must proceed by educated guesswork. Recall that one can write down the Hamiltonian for an unquantized nonrelativistic particle in an electromagnetic field by replacing $\mathbf{p}$ by $\mathbf{p} - e\mathbf{A}$ and the Hamiltonian H of a free particle by $H + e\Phi$, as shown by Eq. (61) of Chapter VI. Relativistically this corresponds to replacing p^μ by $p^\mu - eA^\mu$. It is reasonable to suppose that this is how the Klein-Gordon equation should be modified by the presence of an electromagnetic field. That is, we can identify p_μ with $i\partial/\partial x^\mu$ [see the discussion following Eq. (76)] and make the replacement, so that wherever $\partial/\partial x^\mu$ appears in the free Klein-Gordon equation, $\partial/\partial x^\mu + ieA_\mu$ will appear in the presence of an electromagnetic field. What we shall do then instead of trying to find $\mathscr{L}_\mathrm{I}$ is to make this replacement in $\mathscr{L}_\mathrm{KG}$ and see what happens. We will find that this will satisfy three of our four criteria, and therefore we will amost have found an acceptable $\mathscr{L}_\mathrm{I}$. (Actually, we will have found $\mathscr{L}_\mathrm{I}$, but will also have to modify one of the criteria.)

We therefore write

$$\mathscr{L} = \mathscr{L}'_{\text{KG}} + \mathscr{L}_{\text{em}}, \tag{83}$$

where

$$\mathscr{L}'_{\text{KG}} = g^{\alpha\beta}\left(\frac{\partial\psi}{\partial x^\alpha} + ieA_\alpha\psi\right)\left(\frac{\partial\psi^*}{\partial x^\beta} - ieA_\beta\psi^*\right) - \lambda^2\psi\psi^* \tag{84}$$

and $\mathscr{L}_{\text{em}} = -\frac{1}{4}f_{\mu\nu}f^{\mu\nu}$. We have written $-i$ rather than $+i$ in the second factor of the first term of $\mathscr{L}'_{\text{KG}}$ in order to make $\mathscr{L}$ real. Let us now calculate the field equations obtained from this density. First we have the ψ-equation

$$\begin{aligned}\frac{\partial}{\partial x^\mu}\frac{\partial\mathscr{L}}{\partial\psi_\mu} - \frac{\partial\mathscr{L}}{\partial\psi} &= g^{\mu\beta}\frac{\partial}{\partial x^\mu}\left(\frac{\partial\psi^*}{\partial x^\beta} - ieA_\beta\psi^*\right)\\ &\quad - g^{\alpha\beta}ieA_\alpha\left(\frac{\partial\psi^*}{\partial x^\beta} - ieA_\beta\psi^*\right) + \lambda^2\psi^*\\ &= \left[g^{\alpha\beta}\left(\frac{\partial}{\partial x^\alpha} - ieA_\alpha\right)\left(\frac{\partial}{\partial x^\beta} - ieA_\beta\right) + \lambda^2\right]\psi^* = 0.\end{aligned} \tag{85}$$

The ψ^*-equation is simply the complex conjugate of this, namely

$$\left[g^{\alpha\beta}\left(\frac{\partial}{\partial x^\alpha} + ieA_\alpha\right)\left(\frac{\partial}{\partial x^\beta} + ieA_\beta\right) + \lambda^2\right]\psi = 0. \tag{86}$$

In view of what was written in the previous paragraph, Eqs. (85) and (86) satisfy the third criterion. Now we obtain the equations for the A_μ. We have

$$\frac{\partial}{\partial x^\nu}\frac{\partial\mathscr{L}}{\partial A_{\mu\nu}} - \frac{\partial\mathscr{L}}{\partial A_\mu} = \frac{\partial}{\partial x^\nu}\frac{\partial\mathscr{L}_{\text{em}}}{\partial A_{\mu\nu}} - \frac{\partial\mathscr{L}_{\text{em}}}{\partial A_\mu} + \frac{\partial}{\partial x^\nu}\frac{\mathscr{L}'_{\text{KG}}}{\partial A_{\mu\nu}} - \frac{\partial\mathscr{L}'_{\text{KG}}}{\partial A_\mu} = 0.$$

We have already calculated the first two terms on the right-hand side when we derived Eq. (43). The second two yield (let us tentatively call this j^μ)

$$j^\mu = -g^{\mu\beta}ie\left[\psi\left(\frac{\partial}{\partial x^\beta} - ieA_\beta\right)\psi^* - \psi^*\left(\frac{\partial}{\partial x^\beta} + ieA_\beta\right)\psi\right]. \tag{87}$$

Thus the A_μ-equations read

$$\frac{\partial f^{\mu\nu}}{\partial x^\nu} = j^\mu, \tag{88}$$

where j^μ is defined by Eq. (87). This satisfies the first criterion, and the second also if we can accept this new definition of j^μ. Now except for the factor $-e$ the new definition agrees with Eq. (82) in the limit $A_\mu \to 0$. The factor $-e$ is irrelevant for these considerations, for we never claimed that (82) defined more than a multiple of the current. As for the rest, we recall that $i(\partial/\partial x^\mu + ieA_\mu)$ corresponds to $p_\mu - eA_\mu$. Now Eq. (60) of Chapter VI shows that

$m\dot{\mathbf{x}} = \mathbf{p} - e\mathbf{A}$ (remember: $c = 1$), so that $\partial/\partial x^\mu + ieA_\mu$ rather than $\partial/\partial x^\mu$ is to be associated with m times the velocity of the particle. But the j^μ (or rather their space components) are in general proportional to the charge and to the velocity, and thus it is reasonable for the j^μ to be, in agreement with Eq. (87), linear in $\partial/\partial x^\mu + ieA_\mu$ rather than in $\partial/\partial x^\mu$.

But now two problems arise. The first is that Eqs. (85), (86), and (87) and in fact in a hidden way also (88), depend on the field functions A_μ themselves and are therefore not invariant under the electromagnetic gauge transformation $A_\mu \rightarrow \bar{A}_\mu = A_\mu + \partial\chi/\partial x^\mu$. Second, it is no longer obvious that invariance under gauge transformations of the first kind leads to conservation of the newly redefined current. It turns out that these problems are related. Let us see what happens, say, to $(\partial/\partial x^\alpha + ieA^\alpha)\psi$ under an electromagnetic gauge transformation $A_\mu \rightarrow \bar{A}_\mu$. We have

$$(\partial/\partial x_\alpha + ie\bar{A}_\alpha)\psi = (\partial/\partial x^\alpha + ieA_\alpha + ie\,\partial\chi/\partial x^\alpha)\psi = e^{-ie\chi}(\partial/\partial x^\alpha + ieA_\alpha)e^{ie\chi}\psi,$$

so that

$$(\partial/\partial x^\alpha + ie\bar{A}_\alpha)e^{-ie\chi}\psi = e^{-ie\chi}(\partial/\partial x^\alpha + ieA_\alpha)\psi.$$

This suggests that we extend the idea of a gauge transformation to the following:

$$\begin{aligned} A_\mu \rightarrow \bar{A}_\mu &= A_\mu + \partial\chi/\partial x^\mu, \\ \psi \rightarrow \bar{\psi} &= e^{-ie\chi}\psi, \\ \psi^* \rightarrow \bar{\psi}^* &= e^{ie\chi}\psi^*. \end{aligned} \tag{89}$$

Then Eqs. (85) and (86) remain valid for the barred field functions, the definition (87) of the current is the same in terms of the barred and unbarred functions, and therefore Eqs. (88) are gauge invariant in this extended sense. The Lagrangian density of Eq. (81) is also invariant under this *extended gauge transformation* or *gauge transformation of the second kind.* (Note, incidentally, that a gauge transformation of the first kind is the special case of this one with χ = const.) This then modifies our fourth criterion. In order, however, to accept this modification we must be convinced that the extended gauge transformation does not affect any physical results, as was previously true for an electromagnetic gauge transformation. This involves an interpretation of the ψ function, about which we say only this: all physical results depend only on combinations of the ψ function such as $\psi^*\psi$, in which any arbitrary *phase factor* like $e^{ie\chi}$ will drop out. Thus we can in fact accept the modification.

We leave to Problems 13 and 14 two remaining questions: *1.* What is the conserved quantity associated with the invariance of $\mathscr{L}$ under extended gauge transformations? Is the current defined in (87) conserved? *2.* What is $\mathscr{L}_\mathrm{I}$?

(d) Orthonormal Functions in Field Theory

The development of classical field theory given in Sections 1 and 3 started by considering the individual motions of a set of coupled particles and going to the limit as the number of particles becomes continuous. This leads us to interpret $\psi(x) = \psi(\mathbf{x}, t)$ as the coordinate of the "particle" normally located at $\mathbf{x}$. In Chapter IV, Section 5(b) it was shown that for some coupled systems it is often convenient to transform to a new set of coordinates, the *normal modes*. Since classical fields are similar to a set of coupled particles, the normal mode idea can be useful for them as well.

Let $\varphi_n(\mathbf{x})$, $n = \ldots, -1, 0, 1, 2, \ldots$ be an infinite complete set of orthonormal functions defined over three-space. By a complete set we mean that any function that goes to zero rapidly enough as $|\mathbf{x}| \to \infty$ can be expanded in a series of these functions. In particular, the field functions $\psi_A(x)$ and their time derivatives will have the expansions (remember $x^0 = t$)

$$\begin{aligned} \psi_A(x) &= \psi_A(\mathbf{x}, x^0) = \sum_n a_{An}(x^0)\varphi_n(\mathbf{x}), \\ \psi_{A0}(x) &= \sum_n \dot{a}_{An}(x^0)\varphi_n(\mathbf{x}). \end{aligned} \tag{90}$$

Functions that can be so expanded form an infinite-dimensional vector space in a rather obvious way. We want this vector space, moreover, to be unitary (see Chapter IV, Section 4 and some of the infinite-dimensional examples). For this purpose we define the inner product of any two functions $f(\mathbf{x})$ and $g(\mathbf{x})$ by

$$(f, g) = \int f^*(\mathbf{x})g(\mathbf{x})\, d^3x,$$

and we assume that for the functions we deal with all such integrals converge. Then since the φ_n are to be orthonormal we have

$$(\varphi_n, \varphi_m) = \int \varphi_n^*(\mathbf{x})\varphi_m(\mathbf{x})\, d^3x = \delta_{nm}.$$

The coefficients $a_{An}(x^0) = a_{An}(t)$ determine the field functions uniquely, and vice versa. They allow us to go from a field description involving continuous functions of $\mathbf{x}$ back to a particle-like description involving a discrete set of coordinates, although it is an infinite set. The equations of motion for the a_{An} are given, not surprisingly, by

$$\frac{d}{dt}\frac{\partial L}{\partial \dot{a}_{An}} - \frac{\partial L}{\partial a_{An}} = 0, \tag{91}$$

when $L = \int \mathscr{L}\, d^3x$ is written as a function of the a_{An}, $\dot{a}_{An}$ and t (the $\mathbf{x}$ dependence is integrated out).

This result can be proved by a straightforward calculation. We have

$$\frac{\partial L}{\partial a_{An}} = \frac{\partial}{\partial a_{An}} \int \mathscr{L}\, d^3x = \int \left[\frac{\partial \mathscr{L}}{\partial \psi_B}\frac{\partial \psi_B}{\partial a_{An}} + \frac{\partial \mathscr{L}}{\partial \psi_{Bk}}\frac{\partial \psi_{Bk}}{\partial a_{An}}\right] d^3x.$$

(We continue to use the summation convention for $A, B, \ldots$ and vector indices, but will use Σ's for sums on the orthonormal function indices $n, m, \ldots$.) Now

$$\frac{\partial \psi_B}{\partial a_{An}} = \delta_{AB}\varphi_n(\mathbf{x})$$

in accordance with (90). To calculate $\partial\psi_{Bk}/\partial a_{An}$ we must expand ψ_{Bk}:

$$\psi_{Bk} = \frac{\partial \psi_B}{\partial x^k} = \sum_m a_{Am}\frac{\partial \varphi_m}{\partial x^k}.$$

Then

$$\frac{\partial \psi_{Bk}}{\partial a_{An}} = \delta_{AB}\frac{\partial \varphi_n}{\partial x^k},$$

so that

$$\frac{\partial L}{\partial a_{An}} = \int \left[\frac{\partial \mathscr{L}}{\partial \psi_A}\varphi_n + \frac{\partial \mathscr{L}}{\partial \psi_{Ak}}\frac{\partial \varphi_n}{\partial x^k}\right] d^3x.$$

The second term in the integrand can be integrated by parts as in deriving Eq. (71), and we get

$$\frac{\partial L}{\partial a_{An}} = \int \left[\frac{\partial \mathscr{L}}{\partial \psi_A} - \frac{\partial}{\partial x^k}\frac{\partial \mathscr{L}}{\partial \psi_{Ak}}\right]\varphi_n\, d^3x.$$

Similarly we find

$$\frac{\partial L}{\partial \dot{a}_{An}} = \int \frac{\partial \mathscr{L}}{\partial \psi_{A0}}\varphi_n\, d^3x, \tag{92}$$

and therefore

$$\frac{d}{dt}\frac{\partial L}{\partial \dot{a}_{An}} - \frac{\partial L}{\partial a_{An}} = \int \left[\frac{\partial}{\partial x^0}\frac{\partial \mathscr{L}}{\partial \psi_{A0}} + \frac{\partial}{\partial x^k}\frac{\partial \mathscr{L}}{\partial \psi_{Ak}} - \frac{\partial \mathscr{L}}{\partial \psi_A}\right]\varphi_n\, d^3x$$

$$= \int \left[\frac{\partial}{\partial x^\mu}\frac{\partial \mathscr{L}}{\partial \psi_{A\mu}} - \frac{\partial \mathscr{L}}{\partial \psi_A}\right]\varphi_n\, d^3x.$$

This last equation shows that if the field equations are satisfied then so are Eqs. (91). Conversely, if (91) are satisfied, we have

$$\int \left[\frac{\partial}{\partial x^\mu}\frac{\partial \mathscr{L}}{\partial \psi_{A\mu}} - \frac{\partial \mathscr{L}}{\partial \psi_A}\right]\varphi_n\, d^3x = 0$$

for all n. But since the φ_n form a complete set, the equations

$$\int f(\mathbf{x})\varphi_n\, d^3x = (f^*, \varphi_n) = 0$$

(for all n) imply that $f^* = 0$. Thus we have shown that the complex conjugates of the field equations hold, and hence so do the field equations themselves.

As usual, we define the momentum b_n^A conjugate to the variable a_{An} by

$$b_n^A = \frac{\partial L}{\partial \dot{a}_{An}}. \tag{93}$$

It is then found that the momentum $\pi^A = \partial \mathscr{L}/\partial \psi_{A0}$ conjugate to the field, defined in Eq. (63), can be expanded in the form

$$\pi^A = \sum b_n^A \varphi_n^*, \tag{94}$$

using the complete orthonormal set φ_n^* rather than φ_n. This is easily verified by calculating (φ_m^*, π_A) and using Eq. (92).

The Hamiltonian is defined in the usual way:

$$H = \sum b_n^A \dot{a}_{An} - L. \tag{95}$$

It is shown in Problem 17 that then $H = \int \mathscr{H}\, d^3x$, as expected. The canonical equations are obtained also in the usual way, and we get

$$\dot{a}_{An} = \frac{\partial H}{\partial b_n^A} = [a_{An}, H],$$

$$\dot{b}_n^A = -\frac{\partial H}{\partial a_{An}} = [b_n^A, H],$$

where the Poisson brackets are now the ordinary ones that appear in particle theory (the a_{An} play the role of q's and the b_n^A the role of p's), except that the sums are infinite. As noted before, the introduction of the normal modes φ_n enables us to treat fields as a system consisting of an infinite number of particles. The fundamental Poisson brackets

$$\begin{aligned} [a_{An}, b_m^B] &= \delta_A^B\, \delta_{mn}, \\ [a_{An}, a_{Bm}] &= [b_n^A, b_m^B] = 0 \end{aligned} \tag{96}$$

are very important in the transition to quantum field theory. We leave it to the reader to show that if F is any function of the a_{An}, b_n^A, and t, then

$$\frac{dF}{dt} = \frac{\partial F}{\partial t} + [F, H],$$

as in the particle case.

As an example, we apply this normal-mode approach to the familiar elastic rod in one dimension. Since we know that sinusoidal (imaginary exponential) waves may be expected in the answer, we take the $\varphi_n(\mathbf{x})$ to be of the form $N \exp(ik_n x)$, where N is a real constant which we will adjust to normalize the φ_n. Since x is now a one-dimensional variable, orthonormality requires that

$$N^2 \int \exp(ik_n x) \exp(-ik_m x)\, dx = \delta_{nm}.$$

This condition can not be satisfied if the rod is of infinite length, for the integrals will not converge. We therefore consider a finite but very long rod of length V, and later we will say a few words about the limit as $V \to \infty$. Integrals over x then go from 0 to V, and in order for the functions to be orthogonal, we must restrict the values of k_n to

$$k_n = \frac{2\pi n}{V} = -k_{-n}, \tag{97}$$

where n is an integer. The normalized functions are then

$$\varphi_n = V^{-1/2} \exp(ik_n x).$$

In Section 1 the field function for the rod was called $q(x, t)$. Keeping this notation here, we write

$$\begin{aligned} q(x, t) &= V^{-1/2} \sum_{n=-\infty}^{\infty} a_n(t) \exp(ik_n x), \\ \frac{\partial q}{\partial t} &= V^{-1/2} \sum_{n=-\infty}^{n} \dot{a}_n(t) \exp(ik_n x), \end{aligned} \tag{98}$$

with the restriction that k_n satisfy (97). Clearly if q is to be real, we must have

$$a_{-n} = a_n^*. \tag{99}$$

To find L we first write $\mathscr{L}$ [the integrand of Eq. (5)] in terms of the φ_n:

$$\mathscr{L} = \frac{1}{2V} \sum_{n,m} (\mu \dot{a}_n \dot{a}_m - Y\{-k_n k_m\} a_n a_m) \exp\{i(k_n + k_m)x\}.$$

Then we integrate this over x and use (97) to obtain

$$\begin{aligned} L &= \tfrac{1}{2} \sum_{n.m} (\mu \dot{a}_n \dot{a}_m - Y\{-k_n k_m\} a_n a_m)\, \delta_{n,-m} \\ &= \frac{\mu}{2} \sum_n \dot{a}_n \dot{a}_{-n} - \frac{Y}{2} \sum_n k_n^2 a_n a_{-n}. \end{aligned} \tag{100}$$

By using (99) we can write this in the form

$$L = \frac{\mu}{2} \sum \dot{a}_n \dot{a}_n^* - \frac{Y}{2} \sum k_n^2 a_n a_n^*.$$

The Hamiltonian is easily obtained from L in the usual way. The conjugate momenta are

$$b_n = \frac{\partial L}{\partial \dot{a}_n} = \mu \dot{a}_{-n}$$

so that

$$H = \frac{1}{2\mu} \sum b_n b_{-n} + \frac{Y}{2} \sum k_n^2 a_n a_{-n}. \tag{101}$$

Again if we use (99), this can be written

$$H = \frac{1}{2\mu} \sum b_n b_n^* + \frac{Y}{2} \sum k_n^2 a_n a_n^*.$$

The field equations obtained from (100) are (for both positive and negative values of n)

$$\mu \ddot{a}_{-n} + Y k_n^2 a_{-n} = 0.$$

In spite of the fact that the Lagrangian is not the sum of simple harmonic oscillators (though it is very similar), we obtain simple-harmonic oscillator equations. The well-known solutions are

$$a_n = \alpha_n \exp(i\omega_n t) + \beta_n \exp(-i\omega_n t), \tag{102}$$

where

$$\omega_n = \omega_{-n} = \sqrt{Y k_n^2/\mu}. \tag{103}$$

In addition, in order to satisfy (99), we must have

$$\alpha_{-n} = \beta_n^*.$$

When these expressions for the a_n are inserted into (98), we arrive at the general solution for $q(x, t)$, namely

$$q(x, t) = V^{-1/2} \sum [\alpha_n \exp\{i(k_n x + \omega_n t)\} + \beta_n \exp\{i(k_n x - \omega_n t)\}]. \tag{104}$$

This solution represents a sum of waves propagating up and down the rod with velocity

$$v = \omega_n/|k_n| = \sqrt{Y/\mu}.$$

These are the normal modes of the system. The coefficients α_n and β_n are determined from the functions q and $\dot{q}$ at some initial time. Equation (104) is merely the Fourier expansion of the general solution of the wave equation.

Many Lagrangian densities are similar in form to this elastic-rod Lagrangian density. Because of this similarity, the simple harmonic oscillator plays an important role in field theory. To illustrate this, consider the slightly modified density

$$\mathscr{L} = \frac{\mu}{2}\left(\frac{\partial q}{\partial t}\right)^2 - \frac{Y}{2}\left(\frac{\partial q}{\partial x}\right)^2 - \frac{\lambda^2}{2}q^2.$$

The last term makes this look something like the Lagrangian density for the uncharged scalar meson of Section 4(b). In this case L becomes

$$L = \frac{\mu}{2}\sum \dot{a}_n\dot{a}_{-n} - \frac{Y}{2}\sum (k_n^2 + K^2)a_n a_{-n}, \tag{105}$$

where

$$K^2 = \lambda^2/Y.$$

We leave further discussion of this case to Problems 21–24.

The Lagrangian (100) and the Hamiltonian (101) are independent of V. The solutions (102) of the equations of motion are also independent of V, although the field functions (104), which are the solutions of the field equations, are not. This last fact makes it difficult to pass to the limit as $V \to \infty$ in a simple-minded way. It can be done, however, by going over from Fourier series to Fourier integrals, in which k takes on a continuum of values. This would take us far afield, and we will not describe it. See, however, Problem 20.

Problems

1. From the fact that $\partial T_\alpha{}^\mu/\partial x^\mu = 0$ and from the assumption that $T_\alpha{}^0$ is the energy-momentum density, prove that $T_\alpha{}^k$ is the energy-momentum flux in the k-direction. That is, show that if V is a volume in three-space, then

$$\oint_{\partial V} T_\alpha{}^k \, d\Sigma_k$$

is the flux of energy or momentum out of V, where $d\mathbf{\Sigma}$ is the outward surface element on ∂V. Equation (19b) may help.

2. (a) Show that if Φ is a scalar function, then $\partial\Phi/\partial x^\mu$ is a covariant vector. More generally, show that $\partial/\partial x^\mu$ is a covariant vector operator, increasing by one the covariant rank of any tensor it operates on. (b) Show that if f_μ and h_μ are vector functions, then $\Phi = f_\mu h^\mu$ is a scalar in the sense that $\Phi' = f'_\mu h'^\mu$ [see the Remark before Eq. (40)]. Show that $t_\mu^\nu = f_\mu h^\nu$ is a tensor of the kind indicated by its indices. (c) Show that if f_μ is an arbitrary vector and $\Phi = f_\mu h^\mu$ is a scalar, then h^μ is a vector.

3. Prove the following statement. If the field is complex (i.e., if the ψ_A are complex-valued functions) and the fields and their complex conjugates are varied independently as described at the beginning of Section 4(c), then the two sets of field equations (79) are obtained.

4. The general equations for the vibrations in a two-dimensional elastic medium are of the form

$$A\frac{\partial^2\psi_x}{\partial t^2} = 4BC\frac{\partial^2\psi_x}{\partial x^2} + B\frac{\partial^2\psi_x}{\partial y^2} + C\frac{\partial^2\psi_y}{\partial y\,\partial x},$$

$$A\frac{\partial^2\psi_y}{\partial t^2} = 4BC\frac{\partial^2\psi_y}{\partial y^2} + B\frac{\partial^2\psi_y}{\partial x^2} + C\frac{\partial^2\psi_x}{\partial y\,\partial x},$$

where A, B, and C are constants characteristic of the medium. Find a Lagrangian density that will yield these equations.

5. Show that if $R^{\mu\nu}$ is defined by (59) and (60), then $\bar{T}^{\mu\nu} = T^{\mu\nu} + R^{\mu\nu}$ is symmetric and leads to the same values of P^μ as does $T^{\mu\nu}$. Show further that $\bar{M}^{\mu\nu\lambda}$ defined in Eq. (61) leads to the same values of $m^{\mu\nu}$ as does $M^{\mu\nu\lambda}$.
6. Show that for a scalar relativistic field (i.e., if ψ_A is a scalar) $d_A^B = 0$ and $T^{\mu\nu} = \bar{T}^{\mu\nu}$.
7. Calculate the angular-momentum density of the electromagnetic field in terms of **E** and **B**.
8. Show that if $\mathscr{L}$ and $\mathscr{L}'$ differ by a *four-divergence*, that is, if there exists an $F^\mu(\psi, x)$ (independent of the derivatives of the ψ_A) such that

$$\mathscr{L} - \mathscr{L}' = (\partial/\partial x^\mu)F^\mu(\psi, x),$$

then $\mathscr{L}$ and $\mathscr{L}'$ will yield the same field equations. The converse is almost true: if $\mathscr{L}$ and $\mathscr{L}'$ yield the same field equations *term for term*, their difference is a four-divergence.

9. Calculate the (unsymmetrized) stress tensor for longitudinal waves in an elastic rod. Show explicitly that $\partial T_0{}^\alpha/\partial x^\alpha = 0$ as a result of the field equations. Calculate the energy density and energy flux in a traveling wave of arbitrary shape and show that they are related as you might have guessed. Calculate also the momentum density and flux. The *intensity* of a sinusoidal traveling wave is defined as the average energy crossing a point per unit time. Show that the intensity is proportional to the square of the amplitude.
10. The Lagrangian density for transverse waves on a stretched string is

$$\mathscr{L} = \frac{\mu}{2}\left[\left(\frac{\partial q_1}{\partial t}\right)^2 + \left(\frac{\partial q_2}{\partial t}\right)^2\right] - \frac{T}{2}\left[\left(\frac{\partial q_1}{\partial x}\right)^2 + \left(\frac{\partial q_2}{\partial x}\right)^2\right],$$

where q_1 and q_2 are the two components of transverse displacement, μ is the linear mass density, T is the (constant) tension, and x is position measured along the equilibrium string. Show that $\mathscr{L}$ is invariant [see the Remark preceding Eq. (40)] under rotations about the equilibrium string axis. Find the associated conserved current M^μ, $\mu = 0, 1$ and the conserved quantity $\int M^0\,dx$. Show that M^0 is the density of angular momentum about the string axis. What is M^1? Calculate the angular-momentum density and flux for sinusoidal traveling and standing waves as functions of the relative phase and amplitude of q_1 and q_2. Calculate also the momentum and energy density and flux and the intensity for sinusoidal traveling waves.

11. Obtain Eq. (74) by using (72) and (73) rather than (71). *Hint:* Show that

$$[F, H]^{\mathrm{f}} = \int dy\, dz \left[\int \mathscr{F}(x)\, \delta(y - x)\, dx, \int \mathscr{H}(x)\, \delta(z - x)\, dx\right]^{\mathrm{f}},$$

and proceed from the expressions for $\mathscr{F}$ and $\mathscr{H}$ and the rules for δ functions.

12. We have seen that if we set $\lambda = 0$ in the field equations (78) for the vector meson, we obtain Maxwell's equations in the Lorentz gauge. Prove that if you apply the Lorentz gauge condition to the $\mathscr{L}$ of Eq. (77) and set $\lambda = 0$, the $\mathscr{L}$ you obtain differs from Eq. (40) by a four-divergence.
13. Show that for the coupled Maxwell and Klein-Gordon fields $\mathscr{L}_{\mathrm{I}} = -j^\mu A_\mu$.
14. Show that the invariance of $\mathscr{L}$ under extended gauge transformation implies conservation of j^μ as defined in (87) for the coupled Klein-Gordon and Maxwell fields.
15. Calculate the stress tensor for the free charged scalar meson and for the charged scalar meson coupled to the electromagnetic field. Compare the energy densities for these two cases.
16. According to the definition of Eq. (65), $\mathscr{H}$ is a function of the φ_A, φ_{Ak}, π^A, and x^μ, while $\mathscr{L}$ is a function of the φ_A, $\varphi_{A\mu}$, and x^μ. Show that $\partial\mathscr{H}/\partial x^\mu = -\partial\mathscr{L}/\partial x^\mu$. Show also that $dH/dt = \int (\partial\mathscr{H}/\partial t)\, d^3x$. These are the field analogs of the particle equations $\partial H/\partial t = -\partial L/\partial t$ and $\partial H/\partial t = dH/dt$.
17. Let φ_n be a complete set of orthonormal functions, as defined in Section 4(d). Show that $\sum \varphi_n^*(\mathbf{x})\varphi_n(\mathbf{y}) = \delta(\mathbf{x} - \mathbf{y})$. [Find the coefficients of $\delta(\mathbf{x} - \mathbf{y})$ when expanded in a series of φ_n functions.] From this show that H as defined by (95) is $\int \mathscr{H}\, d^3x$. This would not be true if the φ_n rather than the φ_n^* appeared in (94).
18. (a) By summing the a_{An} and b_n^A of Eq. (96) with φ_n and φ_n^* functions we obtain a Poisson bracket for fields. Show that it agrees with the one defined in Eq. (72) (b) Conversely, let $F = \int \mathscr{F}\, d^3x$, where $\mathscr{F}$ is a function of the φ_A and their derivatives. Show that if F is written as a function of the a_{An} and b_n^A then it follows from (71) that $dF/dt = [F, H]$, where the Poisson bracket is the "particle" one of Eq. (96).
19. Equation (74) shows that a certain function F is a constant of the motion for the elastic rod. Obtain an expression for this F in terms of the a_n and b_n for the elastic rod, and show that F is a constant of the motion in accordance with the Hamiltonian of Eq. (101). What is the physical meaning of F?
20. An elastic rod is distorted, at time $t = 0$, into the shape given by

$$q(x, 0) = A \sin 2\pi x/l, \qquad -2l < x < 2l,$$
$$q(x, 0) = 0, \qquad |x| > 2l,$$

where the wavelength l is chosen to satisfy $l = V/(8N)$, N a positive integer. The rod is then released. (a) Find the resulting wave $q(x, t)$ by solving (6) and inserting the boundary conditions and initial conditions. Describe the motion in words. (b) Find the $a_n(t)$ and write the solution in the form of the Fourier series of Eq. (104). Can you sum the series and reproduce the solution of part (a)? (c) Discuss the limit $V \to \infty$ for the two forms of the solution. This illustrates the difficulty of going to this limit in the treatment of Section 4(d).

21. Obtain and solve the equations of motion from the Lagrangian of Eq. (105). Show that different frequencies propagate with different velocities, and that there is a minimum cut-off frequency ω_0 that will propagate at all. What is its velocity? Consider a wave packet of finite extent (e.g., something like the finite sine wave of Problem 20, or a Gaussian-shaped wave). Discuss its motion down the rod.
22. Derive the Lagrangian density $\mathscr{L}$ for transverse vibrations of a stretched string which, in addition to its elastic forces is subjected to a restoring force $\mathbf{F}$, on each length $\Delta\mathbf{x}$ of the string, which is (a) proportional to Δx in the limit as $\Delta x \to 0$, (b) perpendicular to the equilibrium string axis, and (c) proportional to the displacement $|q|$. In other words, $\mathbf{F} = -\lambda^2 \Delta x \mathbf{q}$, where λ is a constant. (It is as though each point of the string were attached by a tiny spring to its equilibrium position.) Derive L from $\mathscr{L}$ and the equations of motion from L. Compare the result with the uncharged vector meson of Section 4(b) and with Eq. (105). This derivation lends some meaning to the Lagrangian of (105): it is for an elastic rod with a similar restoring force.
23. The Lagrangians in Problems 21 and 22 yield minimum cutoff frequencies ω_0 for propagation. Yet the wave equations obtained from the $\mathscr{L}$'s have formal solutions for frequencies lower than ω_0. Find these solutions in the case of the rod and discuss their drawbacks from a physical point of view. That is, why do we reject these solutions? On the other hand, suppose that the point $x = 0$ of the rod is forced to vibrate at a frequency lower than ω_0. Describe the motion. (This corresponds to a bound state of the meson).
24. For a particle of mass m, energy E and momentum $\mathbf{p}$, Eq. (32) tells us that $E^2 - p^2 = m^2$. In quantum mechanics, an energy E is associated with a frequency given by $\hbar\omega$, and a momentum p is associated with a wavelength $2\pi/k$ given by $p = \hbar k$. Following Eq. (76) we identify λ with the meson mass. Show that this identification is consistent with the relativistic energy-momentum relation and the ω-k relation of Problems 21 and 22. Find the velocity of a particle in terms of E and p, and see how this is related to the wave velocity. (Consider this also in the limit as $\lambda^2 \to 0$.) Try to relate this to the last point of Problem 23.
25. Apply the approach of Section 4(d) to the uncharged scalar meson in a cubical box of volume V.
26. Extend the approach of Section 4(d) to the charged scalar meson in a cubical box of volume V.
27. Solve the canonical equations for the Hamiltonian of Eq. (101), thereby constructing $q(x, t)$ and its canonical conjugate $p(x, t)$. Of course, your

expression for $q(x, t)$ should agree with (104). Find the total energy in terms of the α_n and β_n. Find the α_n and β_n for a traveling sinusoidal wave.

28. Let (no sum on m)

$$Q_n = \frac{C_n}{\sqrt{2}}\left(\sqrt{Yk_n^2}\, a_n + \frac{i}{\sqrt{\mu}} b_n\right),$$

$$P_n = \frac{iC_n}{\sqrt{2}}\left(\sqrt{Yk_n^2}\, a_n - \frac{i}{\sqrt{\mu}} b_n\right).$$

Show that if $C_n = C_{-n} = (\omega_n)^{-1/2}$, this transformation is canonical, where ω_n is given by (103). Show that then the Hamiltonian of (101) can be written

$$H = -i \sum \omega_n Q_n P_{-n},$$

and that $P_{-n} = iQ_n^*$, so that

$$H = \sum \omega_n Q_n Q_n^*.$$

Solve for $Q_n(t)$ and $P_n(t)$. (Note that the first-order equations for Q_n and Q_{-n} are coupled for each n, and therefore that Q_n and Q_{-n} depend on a total of two constants of integrations. Similarly for P_n and P_{-n}.) Compare Chapter VI, Eq. (67).

29. Obtain the Hamiltonian for the Lagrangian of Eq. (105). Find a canonical transformation similar to the one of Problem 28 which puts the Hamiltonian in a simple form similar to the final form of that problem.

CHAPTER IX

The Group Theoretical Approach to Dynamics

In this chapter we give a brief introduction to the group theoretical or Lie algebraic view of dynamics.[1] Although it is not the last word in sophistication [see, for instance, Abraham 1967)], this is a relatively modern approach which has grown out of the successes of group theory in quantum mechanics. The achievements in classical mechanics are of current interest and provide some insight into the connection between classical and quantum mechanics.

We do not treat the subject matter of this chapter with all possible rigor; nor do we prove, or even justify sufficiently, all of our assertions. (Notable examples are that what we call throughout a Lie group is in reality a *local* Lie group and that we do not worry that some groups we consider are of infinite dimension.) However we hope that the material will give some feeling for what a Lie group is and how the concept can be applied in physics. In addition to canonical realizations, we discuss some representations, particularly of the rotation group, because of their importance in quantum mechanics.

IX-1 GROUPS AND ALGEBRAS

A *group* G is a set of elements $x, y, \ldots$ with a certain kind of operation defined on it. The operation, usually called *multiplication*, is binary (involves two elements at a time) and must have four properties, called the *group axioms*. If we denote the operation by a dot, so that x *multiplied by* y is written $x \cdot y$, these axioms are (compare Section 1 of Chapter IV)

G1. Closure. For every pair of elements $x, y \in G$ there exists in G a third element z such that[2]

$$z = x \cdot y.$$

[1] This chapter is based largely on work by Sudarshan (1963).

[2] Note the difference between this and G1 (see also G4) of Chapter IV, Section 1. If in general $x \cdot y = y \cdot x$, then G is said to be *Abelian*, and multiplication is sometimes called addition. The *identity* can then be called the *null* element and the *inverse*, the *negative*.

G2. Associativity. If $x, y, z \in G$, then

$$(x \cdot y) \cdot z = x \cdot (y \cdot z).$$

G3. Existence of identity element. There exists an element $e \in G$ such that for all $x \in G$

$$x \cdot e = e \cdot x = x.$$

Then e is called the group's *identity element.*

G4. Existence of the inverse. For each $x \in G$ there exists an *inverse* $y \in G$ with the property

$$x \cdot y = y \cdot x = e.$$

It can be shown (see Problem 1) that when x is given, its inverse is unique. We shall write x^{-1} for this element.

Remark. The dot to represent multiplication is usually omitted: $x \cdot y$ is written simply xy. We shall do so from now on.

Examples of groups are numerous. We have already seen that a vector space is a group *under* (meaning whose operation is) addition. Another simple example is the set of all positive numbers under ordinary multiplication. (However the set of all positive integers is not a group under multiplication since the inverse of an integer is not an integer.) Still another is the set of all nonsingular linear operators on a vector space under operator multiplication (successive application). We shall be interested, in fact, mostly in groups of this last kind, the so-called transformation groups.

A group may be *discrete* or *continuous*, *finite* or *infinite*. A finite group, necessarily discrete, has a finite number of elements. For instance the set $(1, -1)$ under ordinary multiplication is a (discrete) finite group. The set of positive and negative integers and zero under addition is an infinite discrete group, for it has an infinite number of elements and the elements are separate. The vectors of a vector space under addition, or all numbers of the form $e^{i\lambda}$ under multiplication (λ real) are continuous groups, for the elements form a continuum (the elements are infinitesimally close).[3] Continuous groups are necessarily infinite. For instance even though $e^{i(\lambda+2\pi)} = e^{i\lambda}$, so that λ need vary only over a finite range (e.g., from 0 to 2π), the number of elements in the group is infinite. A subset G' of a group G is a *subgroup* of G if G' itself is a group, that is, if the elements of G' satisfy properties G1–G4. A continuous group may have discrete, even finite sub-groups.

We will use as a running example the already mentioned *G*eneral *L*inear group in n dimensions over the *R*eal numbers $GL(n, R)$, consisting of all nonsingular linear operators on a real vector space of dimension n under ordinary operator multiplication. Although there are many real vector spaces of

[3] "Closeness" of elements is a topological concept. We do not want to go into it in this book with any rigor, for the intuitive idea will be enough.

dimension n, in a certain sense they are all the same, for they can all be represented by the space of n-dimensional column vectors. Similarly, the general linear groups of operators on these spaces are in a sense the same, for they can all be represented by the nonsingular n-dimensional square matrices. Thus $GL(n, R)$ stands for all these equivalent groups. We say these groups are *isomorphic*. More specifically, two groups G_1 and G_2 are isomorphic if there is a way to identify elements of G_1 with those of G_2 and vice versa so that group operations are preserved; that is, there exists a *mapping* φ from G_1 onto[4] G_2 with the properties:

1. If $x_1 \in G_1$ then $\varphi(x_1) \equiv x_2 \in G_2$.
2. $\varphi(x_1 y_1) = \varphi(x_1)\varphi(y_1) \equiv x_2 y_2$
3. φ^{-1} exists, mapping G_2 onto G_1 so that

$$\varphi^{-1}(\varphi(x_1)) = x_1 \qquad \text{for all } x_1 \in G_1.$$

(If no unique φ^{-1} exists, G_1 and G_2 are said to be *homomorphic*.) Then φ is called an *isomorphism* (or *homomorphism*). An isomorphism maps only one element of G_1 into each element of G_2; a homorphism maps more than one element of G_1 into a single element of G_2.

For example, consider the two groups $GL(n, R)$ and what we might call $\overleftarrow{GL}(n, R)$, which differs from $GL(n, R)$ in that we take multiplication in the opposite order. That is, if x and y are nonsingular operators in n dimensions, their product in $GL(n, R)$ is xy, while in $\overleftarrow{GL}(n, R)$ it is yx. Then $\varphi(x) = x^T$ is an isomorphism of $GL(n, R)$ and $\overleftarrow{GL}(n, R)$. We leave the proof to the reader. As an example of a homomorphism, consider $GO(3, R)$, the General Orthogonal group in three Real dimensions (with determinant ± 1, as we have seen in Chapter IV) and $SO(3, R)$ the Special (determinant 1) orthogonal group (what we have called the rotations). Let $\Delta(x)$ be the determinant of $x \in GO(3, R)$ and consider the mapping φ from $GO(3, R)$ to $SO(3, R)$ defined by

$$\varphi(x) = \Delta(x)x. \tag{1}$$

This is a homomorphism; the proof we leave again to the reader (Problem 5).

In application to mechanics we will be interested mostly in continuous groups, in particular those which are called *Lie groups.* Rather than define them with all available mathematical rigor and elegance, we give an intuitively simple definition. A Lie group is a continuous group in which one

[4] "Onto" means that every element of G_2 can be written in the form $\varphi(x_1)$ with $x_1 \in G_1$. These definitions can be extended to "into" mappings, when only some elements of G_2 can be written $\varphi(x_1)$, but these will not concern us.

can introduce a continuous coordinate system $\xi^1, \ldots, \xi^n$ of finite dimension n (called the dimension of the group) so that the identity element is at the origin [the coordinates of e are $(0, 0, \ldots, 0)$] and so that the multiplication law is given by analytic functions. That is, if the coordinates of x are ξ^α and the coordinates of y are η^α, then there exist n analytic functions $f^\alpha(\xi, \eta)$ of the $2n$ variables ξ^α, η^α such that the coordinates ζ^α of $z = xy$ are given by

$$\zeta^\alpha = f^\alpha(\xi, \eta). \tag{2}$$

All the continuous linear transformation groups, subgroups of $GL(n, R)$, are of this type, and it can be shown that essentially all Lie groups are isomorphic to subgroups of $GL(n, R)$. Thus we may think, when we want to, of Lie groups as such transformation groups. Consider for example $GL(n, R)$ itself, and let the coordinates of $x \in GL(n, R)$ be its matrix elements minus $\delta_{\alpha\beta}$.[5] That is $x \leftrightarrow \xi_{\alpha\beta} = x_{\alpha\beta} - \delta_{\alpha\beta}$ where $x_{\alpha\beta}$ are the matrix elements of x. Then if $xy = z$, the coordinates of z are

$$\begin{aligned}\zeta_{\alpha\beta} &= z_{\alpha\beta} - \delta_{\alpha\beta} = x_{\alpha\mu}y_{\mu\beta} - \delta_{\alpha\beta} = (\xi_{\alpha\mu} + \delta_{\alpha\mu})(\eta_{\mu\beta} + \delta_{\mu\beta}) - \delta_{\alpha\beta} \\ &= \xi_{\alpha\mu}\eta_{\mu\beta} + \xi_{\alpha\beta} + \eta_{\alpha\beta},\end{aligned}$$

which is certainly analytic in the coordinates of x and y. $GL(n, R)$ is of dimension n^2. Subgroups of it may be of lower dimension, and we would then choose other coordinates than the $\xi_{\alpha\beta}$. For instance, when dealing in Chapter V with the rotation group $SO(3, R)$, which is of dimension three, we used the Euler angles. Since every Lie group G has a coordinate system, curves may be defined in it. Specifically, a differentiable curve is a set of elements $x(\varepsilon) \in G$ whose coordinates $\xi^\alpha(\varepsilon)$ depend differentiably on a parameter ε; we will always choose the parameter so that $x(0) = e$. If $x(\varepsilon)$ is also a group, we shall write $g(\varepsilon)$ and call it a *one-parameter subgroup* of G. For example, rotations about the *1*-axis form a one-parameter subgroup of $GL(3, R)$. The parameter can in general be chosen in many ways. In the rotations about the *1*-axis, for instance, we may choose $\varepsilon = \theta$, the angle of rotation, or we may choose $\varepsilon' = \sin\theta$. The multiplication law is then differently expressed in these two parameterizations:

$$\begin{aligned}&\text{if} \quad \varepsilon = \theta, && g(\varepsilon_1)g(\varepsilon_2) = g(\varepsilon_1 + \varepsilon_2); \\ &\text{if} \quad \varepsilon' = \sin\theta, && g(\varepsilon_1')g(\varepsilon_2') = g(\varepsilon_1'[1 - \varepsilon_2'^2]^{1/2} + \varepsilon_2'[1 - \varepsilon_1'^2]^{1/2}).\end{aligned} \tag{3}$$

In general the multiplication law for a one-parameter subgroup (call the parameter ε') may be written

$$g(\varepsilon_1')g(\varepsilon_2') = g[\mu(\varepsilon_1', \varepsilon_2')], \tag{4}$$

where $\mu(\varepsilon_1', \varepsilon_2')$ is some function, assumed differentiable, such that

$$\mu(\varepsilon', 0) = \mu(0, \varepsilon') = \varepsilon',$$

[5] We subtract $\delta_{\alpha\beta}$ so that the coordinates of $e = \mathbb{1}$ become all zeros.

because $g(0) = e$. It is always possible, however, to switch to a new parameter ε in terms of which the multiplication law becomes

$$g(\varepsilon_1)g(\varepsilon_2) = g(\varepsilon_1 + \varepsilon_2). \tag{5}$$

This may be seen as follows. Let us write ε' as a function of ε: thus $\varepsilon' = \varepsilon'(\varepsilon)$. Then by combining (4) and (5) we have

$$\varepsilon'(\varepsilon_1 + \varepsilon_2) = \mu(\varepsilon'(\varepsilon_1), \varepsilon'(\varepsilon_2)).$$

Now take the derivative of this equation with respect to ε_2 at $\varepsilon_2 = 0$ and write $\varepsilon_1 \equiv \varepsilon$. Noting that $e = g(0)$ in both parameterization, so that $\varepsilon'(0) = 0$, we obtain

$$\frac{d\varepsilon'}{d\varepsilon} = \left[\frac{d\varepsilon'}{d\varepsilon}\right]_{\varepsilon=0}\left[\frac{\partial\mu}{\partial\varepsilon_2'}\right]_{\varepsilon_2'=0}. \tag{6}$$

But at $\varepsilon = 0$, $d\varepsilon'/d\varepsilon = k$ is some constant, and the second factor on the right is a known function of ε'. Thus this is a simple differential equation for $\varepsilon'(\varepsilon)$, and it will in general have solutions. These solutions can be inverted to give the new parameterization $\varepsilon(\varepsilon')$. See Problem 8. From now on we shall always take one-parameter subgroups in the *standard form* of Eq. (5).

Let $g(\varepsilon)$ be a one-parameter subgroup (in standard form) of some Lie group G. Remember that we can think of G as a transformation group, and hence of $g(\varepsilon)$ as a one-parameter group of matrices or operators. There is a very neat way of writing down this matrix function $g(\varepsilon)$, as we shall now show. From the fact that $gg^{-1} = e$, we have

$$\frac{dg}{d\varepsilon}g^{-1} + g\frac{dg^{-1}}{d\varepsilon} = 0,$$

or, multiplying on the right by g,

$$\frac{dg}{d\varepsilon} = -g\frac{dg^{-1}}{d\varepsilon}g. \tag{7}$$

Now Eq. (5) and $g(0) = e$ imply that $g^{-1}(\varepsilon) = g(-\varepsilon)$. Thus

$$\begin{aligned} g\frac{dg^{-1}}{d\varepsilon} &= g\lim_{\Delta\varepsilon\to 0}\left[\frac{g(-\varepsilon-\Delta\varepsilon) - g(-\varepsilon)}{\Delta\varepsilon}\right] \\ &= \lim_{\Delta\varepsilon\to 0}\left[\frac{g(\varepsilon)g(-\varepsilon-\Delta\varepsilon) - g(\varepsilon)g(-\varepsilon)}{\Delta\varepsilon}\right] \\ &= \lim_{\Delta\varepsilon\to 0}\left[\frac{g(-\Delta\varepsilon) - g(0)}{\Delta\varepsilon}\right] = -\dot{g}(0) \equiv -a, \end{aligned} \tag{8}$$

where we write $dg/d\varepsilon = \dot{g}$, and a is a matrix independent of ε. Let us insert this into (7) and include the "initial" condition. We then have

$$\frac{dg}{d\varepsilon} = ag, \qquad g(0) = e,$$

which is a problem in ordinary differential equations for the matrix function $g(\varepsilon)$. Its solution may be written

$$g(\varepsilon) = \exp(\varepsilon a), \tag{9}$$

where the exponential is defined by the power series. The operator a is called the *infinitesimal generator* of the one-parameter group $g(\varepsilon)$.

Remark. We are talking about transformation or operator groups. That is what allows us to write sums and thereby exponentials.

The converse of this result is also true in a sense: Let a be any linear operator on a real n-dimensional vector space. Then $g(\varepsilon) = \exp(\varepsilon a)$ is a one-parameter subgroup (in standard form) of $GL(n, R)$. Although in general a is not an element of $g(\varepsilon)$, in application to physics it will turn out that a and entities like it are the important ones. We have already seen something similar when, just before Eq. (12) of Chapter VII we remarked that dynamical variables are *infinitesimal generators* of translations in their canonical conjugates. Eventually we shall see the connection between these two kinds of infinitesimal generators.

As an example of an infinitesimal generator, consider the two-dimensional operator

$$a = \begin{vmatrix} 0 & 1 \\ -1 & 0 \end{vmatrix}.$$

The powers of a are

$$a^2 = -e,\ a^3 = -a,\ a^4 = e,\ a^5 = a, \ldots .$$

Then in complete analogy with the usual proof that $\exp(i\theta) = \cos\theta + i\sin\theta$, which depends only on the fact that $i^2 = -1$, we have

$$\begin{aligned} g(\varepsilon) = \exp(\varepsilon a) &= e\cos\varepsilon + a\sin\varepsilon \\ &= \begin{vmatrix} \cos\varepsilon & \sin\varepsilon \\ -\sin\varepsilon & \cos\varepsilon \end{vmatrix}. \end{aligned} \tag{10}$$

Thus the one-parameter subgroup of $GL(2, R)$ generated by a is the rotation group in two dimensions, $SO(2, R)$.

Associated with a Lie group G is the set of all of its infinitesimal generators, which are defined by the last equality in (8) with $g(\varepsilon)$ any one-parameter subgroup. Actually, it is easier to deal with a slightly broader definition of this set. Let $x(\varepsilon)$ be any differentiable curve in G [with $x(0) = e$] and let $\dot{x} = dx/d\varepsilon$. Then the *tangent vector* (actually an operator)

$$a = \dot{x}(0) \tag{11}$$

is the infinitesimal generator of some $g(\varepsilon)$ according to (9). [Actually this is not a definition, but a theorem stating that every tangent vector is an infinitesimal generator. But the proof that $\exp(\varepsilon a) \in G$ for every a defined by (11) is too complicated for us to go into, and we omit it.] The set of all such infinitesimal generators is called the *Lie algebra* $\mathfrak{G}$ of G [but see Remark 1 after Eq. (14)]. In an obvious way $\mathfrak{G}$ is a vector space, which means that it satisfies the nine properties listed at the beginning of Chapter IV. We shall prove only two of these.

1. Closure under addition. Let $x(\varepsilon)$ and $y(\varepsilon)$ be curves in G and let $\dot{x}(0) = a$, $\dot{y}(0) = b$. Form the curve $z(\varepsilon) = x(\varepsilon)y(\varepsilon)$, and call $\dot{z}(0) = c$. Then we have

$$c = \dot{z}(0) = \dot{x}(0)y(0) + x(0)\dot{y}(0) = ae + eb = a + b.$$

2. Closure under multiplication by a number. If $x(\varepsilon)$ is a curve in G, then clearly so is $y(\varepsilon) = x(\alpha\varepsilon)$ where α is a fixed number. Then if $\dot{y}(0) = b$ and $\dot{x}(0) = a$ we have

$$b = \dot{y}(0) = \alpha\dot{x}(0) = \alpha a.$$

The rest are left to the reader.

Since the elements of $\mathfrak{G}$ are operators, one might expect their products to be in $\mathfrak{G}$. In general, however, they are not. But if $a, b, \in \mathfrak{G}$, then their *commutator* or *Lie bracket*[6]

$$[a, b] = ab - ba \tag{12}$$

is in $\mathfrak{G}$. To prove this we again let $x(\varepsilon)$ and $y(\varepsilon)$ be curves in G with tangent vectors a and b, respectively. Let $\varepsilon' = \varepsilon^2$ and let $u(\varepsilon')$ be the curve

$$u(\varepsilon') = x(\varepsilon)y(\varepsilon)x^{-1}(\varepsilon)y^{-1}(\varepsilon) = [x, y]x^{-1}y^{-1} + e. \tag{13}$$

Then

$$\frac{du}{d\varepsilon'} = \frac{1}{2\varepsilon}([\dot{x}, y] + [x, \dot{y}])x^{-1}y^{-1} + \frac{1}{2\varepsilon}[x, y](\dot{x}^{-1}y^{-1} + x^{-1}\dot{y}^{-1}).$$

Now let $c = (du/d\varepsilon')_{\varepsilon'=0}$. To calculate c we cannot simply let ε go to zero in the expression for $du/d\varepsilon'$, for all three brackets that appear in that expression vanish in the limit, because $[z, e] = 0$ for all z, and $x(0) = y(0) = e$. Since

[6] We will use the same notation for commutator and Lie as for Poisson brackets as long as we do not have to distinguish them. Later we will put subscripts C, L, or P on them, as the case may be.

numerator and denominator both vanish, we may use L'Hôpital's rule; we then have

$$\lim_{\varepsilon\to 0} \frac{[\dot{x}, y]}{\varepsilon} = \lim_{\varepsilon\to 0} ([\ddot{x}, y] + [\dot{x}, \dot{y}]) = [a, b],$$

$$\lim_{\varepsilon\to 0} \frac{[x, \dot{y}]}{\varepsilon} = \lim_{\varepsilon\to 0} ([\dot{x}, \dot{y}] + [x, \ddot{y}]) = [a, b],$$

$$\lim_{\varepsilon\to 0} \frac{[x, y]}{\varepsilon} = \lim_{\varepsilon\to 0} ([\dot{x}, y] + [x, \dot{y}]) = 0.$$

Thus

$$c = [a, b], \tag{14}$$

which completes the proof.

Remarks. 1. The Lie algebra $\mathfrak{G}$ of G is actually the vector space of infinitesimal generators of G *with the algebraic bracket operation defined on it. 2.* The elements of $\mathfrak{G}$ are not in general those of G. Not only do the elements differ, but the operations differ radically. If $a,b \in \mathfrak{G}$ then ab is not in general in $\mathfrak{G}$. If $x, y \in G$, then $x + y$ and $[x, y]$ are not in general in G. Furthermore the Lie bracket is not even an associative operation! *3.* We see from the proof of closure under addition that multiplication in G is reflected by addition in $\mathfrak{G}$. Since $\mathfrak{G}$ has to do with the limit of G (or of curves in it) as one approaches the identity, this implies that in some sense the closer you get to e, the more nearly G is Abelian.

Since $\mathfrak{G}$ is a vector space, it has a basis $j_1, \ldots, j_n$. We state without proof that the dimension n of $\mathfrak{G}$ is the same as the dimension of G. Let the components of a in this basis be α^μ and those of b be β^μ. Then if the bracket relationships are known for the basis vectors (actually basis operators) they can be found for any two vectors in $\mathfrak{G}$ without using Eq. (12): Since $[j_\mu, j_\nu]$ is in $\mathfrak{G}$, it is a vector which can be expanded in the basis, its components depending on μ and ν. Thus there exist constants $c^\lambda_{\mu\nu}$ such that

$$[j_\mu, j_\nu] = c^\lambda_{\mu\nu} j_\lambda, \tag{15}$$

and then (12) can be written

$$\begin{aligned} c = \gamma^\lambda j_\lambda &= [\alpha^\mu j_\mu, \beta^\nu j_\nu] = \alpha^\mu \beta^\nu [j_\mu, j_\nu] \\ &= \alpha^\mu \beta^\nu c^\lambda_{\mu\nu} j_\lambda. \end{aligned}$$

In other words the components of c can be written down in terms of those of a and b:

$$\gamma^\lambda = \alpha^\mu \beta^\nu c^\lambda_{\mu\nu}. \tag{16}$$

The $c^\lambda_{\mu\nu}$ are called *structure constants* of the Lie algebra. Note that the structure constants depend on the basis chosen in $\mathfrak{G}$, as is clear from (15). The Lie algebra determines the $c^\lambda_{\mu\nu}$ only up to coordinate transformations.

Now, all that we have done in terms of operators can also be done for abstract groups, usually in terms of some coordinates on the group. Then $\mathfrak{G}$ becomes an abstract vector space in which the vectors are defined by their components obtained from the coordinates on G. For instance, if the coordinates on G of the curve $x(\varepsilon)$ are $\xi^\mu(\varepsilon)$, then we define its tangent vector a in the vector space $\mathfrak{G}$ by its components $\alpha^\mu = (d\xi^\mu/d\varepsilon)_{\varepsilon=0}$. The bracket is then defined as follows. The first equality of Eq. (13) is used to define the curve $u(\varepsilon')$, whose tangent vector is taken to be $[a, b]$, where b is the tangent vector to $y(\varepsilon)$. This will define the components γ^μ of $[a, b]$ in terms of the α^μ, β^μ and structure constants $c^\lambda_{\mu\nu}$ obtained from certain derivatives of the f^μ functions of Eq. (2). We state without proof that these structure constants will be just the ones obtained in the operator treatment, so that we are led in both ways to exactly the same Eq. (16). In other words, the multiplication law in G, as expressed by the f^μ, will lead to the same set (up to coordinate transformations) of structure constants $c^\lambda_{\mu\nu}$ as we have obtained in the operator-group treatment, and these define uniquely a certain Lie bracket and hence a Lie algebra. In general this Lie bracket has four important properties, easily proved in our operator-group procedure. (In fact we have already proved the first.) These are

A1. Closure. If $a, b \in \mathfrak{G}$ then

$$c = [a, b] \in \mathfrak{G}.$$

A2. Antisymmetry.[7] $[a, b] = -[b, a]$ for all $a, b \in \mathfrak{G}$.

A3. Linearity. $[\lambda a + \mu b, c] = \lambda[a, c] + \mu[b, c]$ for all $a, b, c \in \mathfrak{G}$ and all numbers λ, μ.

A4. Jacobi Identity.

$$[a, [b, c]] + [b, [c, a]] + [c, [a, b]] = 0 \tag{17}$$

for all $a, b, c \in \mathfrak{G}$.

It is now natural to ask whether the inverse construction is feasible. That is, given the Lie algebra, can one construct the group? The answer is a qualified yes. Let $\mathfrak{G}$ be an abstract Lie algebra, that is, a vector space with a Lie bracket or set of structure constants. Then there are only certain Lie groups with this Lie algebra, and close enough to the identity in these groups, or *locally*, these groups are all isomorphic. It is only when you start moving far away from the origin that they begin to differ [an example is given in Section 4(b)]. Furthermore, the local structure of these groups can be obtained by integrating certain differential equations involving the structure constants.

[7] Property A2 can be replaced by the equivalent property A2′: $[a, a] = 0$ for all $a \in \mathfrak{G}$. That A2′ is immediately implied by A2 is obvious. Linearity plus A2′ yield $0 = [a + b, a + b] = [a, b] + [b, a]$, which is A2.

Thus given a set of n^3 numbers $c^{\lambda}_{\mu\nu}$ satisfying properties A2 and A4 (see Problem 9), there exists at least one Lie group whose Lie algebra has the $c^{\lambda}_{\mu\nu}$ for structure constants. The mathematical importance of this result arises from the fact that it is much easier to study the Lie algebra than the group. The physical importance is that the group corresponds to symmetry transformations or invariances and the algebra to conserved quantities.

We give now an example of a Lie operator group and its Lie algebra which is not taken from any $GL(n, R)$. Consider differentiable functions of a single real variable q, and let G be defined as the group of all operators $x(k, \alpha, \beta)$ acting on the $f(q)$ according to

$$x(k, \alpha, \beta)f(q) = f(q + \beta) \exp (k + aq), \tag{18}$$

where k, α, β are real numbers. This is obviously a three-dimensional set of operators. That they form a group can be seen by calculating:

$$\begin{aligned} x(k_2, \alpha_2, \beta_2)[x(k_1, \alpha_1, \beta_1)f(q)] &= x(k_2, \alpha_2, \beta_2)[f(q + \beta_1) \exp (k_1 + \alpha_1 q)] \\ &= f(q + \beta_1 + \beta_2) \exp (k_1 + \alpha_1[q + \beta_2] + k_2 + \alpha_2 q) \\ &= f(q + \beta_1 + \beta_2) \exp (k_1 + k_2 + \alpha_1\beta_2 + [\alpha_1 + \alpha_2]q) \\ &= x(k_1 + k_2 + \alpha_1\beta_2, \alpha_1 + \alpha_2, \beta_1 + \beta_2)f(q). \end{aligned} \tag{19}$$

This proves closure. The rest is left to the reader: the x operators form a group. Now consider the curves

$$\begin{aligned} g_1(\varepsilon) &= x(\varepsilon, 0, 0), \\ g_2(\varepsilon) &= x(0, \varepsilon, 0), \\ g_3(\varepsilon) &= x(0, 0, \varepsilon). \end{aligned} \tag{20}$$

We have called them all $g_\alpha(\varepsilon)$ rather than $x_\alpha(\varepsilon)$ because they are all one-parameter subgroups in standard form, as is easily verified by using (19).

The tangent vectors to these curves are obtained by taking derivatives with respect to ε at $\varepsilon = 0$ (we leave off the subscript $\varepsilon = 0$):

$$\begin{aligned} a_1 f &= \frac{d}{d\varepsilon} g_1 f = \frac{d}{d\varepsilon} [f(q) \exp \varepsilon] = f, \\ a_2 f &= \frac{d}{d\varepsilon} g_2 f = \frac{d}{d\varepsilon} [f(q) \exp (\varepsilon q)] = qf, \\ a_3 f &= \frac{d}{d\varepsilon} g_3 f = \frac{d}{d\varepsilon} [f(q + \varepsilon)] = \frac{df}{d\varepsilon} = f'. \end{aligned} \tag{21}$$

It is seen that $\mathfrak{G}$ is an operator space that does not coincide with G, for neither a_2 or a_3 are elements of G. Now we can obtain the structure of the Lie algebra by taking commutator brackets. We have

$$[a_1, a_2]f = a_1a_2f - a_2a_1f = qf - qf = 0,$$

$$[a_2, a_3]f = a_2a_3f - a_3a_2f = qf' - \frac{d}{dq}(qf) = -f,$$

$$[a_3, a_1]f = a_3a_1f - a_1a_3f = f' - f' = 0.$$

Thus

$$\begin{aligned} [a_1, a_2] = [a_3, a_1] &= 0, \\ [a_2, a_3] &= -a_1. \end{aligned} \tag{22}$$

These commutation relations could also have been obtained by forming the curves $u(\varepsilon')$ as in Eq. (13). For instance, consider g_2 and g_3; then we have

$$\begin{aligned} u(\varepsilon')f &= g_2g_3g_2^{-1}g_3^{-1}f = g_2g_3[f(q-\varepsilon)\exp(-\varepsilon q)] \\ &= f(q)\exp(-\varepsilon^2) = f(q)\exp(-\varepsilon'). \end{aligned}$$

This yields

$$[a_2, a_3]f = \frac{d}{d\varepsilon'}uf = -f,$$

in agreement with the previous result. We can now write down the structure constants immediately:

$$c_{23}^1 = -c_{32}^1 = -1,$$

and all others are zero.

According to (18), g_3 is the group of translations in q, so that if this is to have physical application one would expect its infinitesimal generator a_3 to be related to the momentum. According to (21), a_2 is just q, and a_1 is simply 1. Thus the last of Eqs. (22) may be written

$$[q, a_3] = -1.$$

Except for sign this looks like the Poisson bracket of q with $a_3 = p$. In fact this relation is even deeper, for we see that a_3 is d/dq and in quantum mechanics p is simply a multiple of d/dq. We will return to this point in Section 4(d).

Finally, we have said that essentially all Lie groups are transformation groups. We shall show here only that a_1, a_2, a_3 can be thought of as operators on an infinite-dimensional vector space. This can be done through P_∞ (see Chapter IV). If $f(q) = \alpha_0 + \alpha_1 q + \alpha_2 q^2 + \cdots$, its components in P_∞ are

$\alpha_0, \alpha_1, \alpha_2, \ldots$. Then it is easily seen that

$$a_1 = \mathbb{1}, a_2 = \begin{vmatrix} 0 & 0 & 0 & \cdots \\ 1 & 0 & 0 & \cdots \\ 0 & 1 & 0 & \cdots \\ \cdot & \cdot & & \\ \cdot & & \cdot & \\ \cdot & & & \cdot \end{vmatrix}, \qquad a_3 = \begin{vmatrix} 0 & 1 & 0 & 0 & \cdots \\ 0 & 0 & 2 & 0 & \cdots \\ 0 & 0 & 0 & 3 & \cdots \\ \cdot & \cdot & \cdot & \cdot & \\ \cdot & \cdot & \cdot & & \cdot \\ \cdot & \cdot & \cdot & & \cdot \end{vmatrix}. \tag{23}$$

As with groups, we shall occasionally be interested in subsets of Lie algebras which are themselves Lie algebras; these are called *subalgebras.* For instance, in the example we just discussed, the subspace spanned by a_1 and a_2 is a subalgebra of $\mathfrak{G}$. But the subspace spanned by a_2 and a_3 is not (check properties A1–A4). We state without proof that in general if G_1 is a Lie subgroup of a Lie group G, then its Lie algebra $\mathfrak{G}_1$ is a subalgebra of the Lie algebra $\mathfrak{G}$ of G. Notice that the bracket in $\mathfrak{G}$ is the limit of elements in G of the form $xyx^{-1}y^{-1}$. If $\mathfrak{G}_1$ is an Abelian subgroup of G and $x, y \in \mathfrak{G}_1$, then $xyx^{-1}y^{-1} = e$, and in the algebra this will yield the null vector. Thus if $a, b \in \mathfrak{G}_1$, this will give $[a, b] = 0$. Such a Lie algebra, like the one spanned by a_1 and a_2 in the example, in which the bracket of any two elements in zero, is also called *Abelian.*

Also as with groups, we shall occasionally have to deal with *isomorphic* and *homomorphic* algebras. Let $\mathfrak{G}_1$ and $\mathfrak{G}_2$ be Lie algebras. They are called *homomorphic* if there exists a mapping φ of $\mathfrak{G}_1$ onto $\mathfrak{G}_2$ such that for all $a_1, b_1 \in \mathfrak{G}_1$ we have

$$\varphi(\alpha a_1 + \beta b_1) = \alpha\varphi(a_1) + \beta\varphi(b_1) \tag{24}$$

and

$$\varphi([a_1, b_1]) = [\varphi(a_1), \varphi(b_1)], \tag{25}$$

where the bracket on the left-hand side is in $\mathfrak{G}_1$ and that on the right-hand side is in $\mathfrak{G}_2$. If φ has an unique inverse mapping, then $\mathfrak{G}_1$ and $\mathfrak{G}_2$ are *isomorphic.* We state without proof that if G_1 and G_2 are isomorphic Lie groups, their Lie algebras $\mathfrak{G}_1$ and $\mathfrak{G}_2$ are isomorphic. (The situation with homomorphisms is more complicated.) The converse is true locally. In fact we have effectively stated it in the discussion following Eq. (17).

With this definition of isomorphism it is relatively simple to find all essentially different (i.e., up to isomorphisms) Lie algebras of low dimensions. For example, in one dimension there is only the Abelian one (implied by antisymmetry). In two dimensions, as we shall now show, there are two. First, the Abelian one. Second, the most general non-Abelian $\mathfrak{G}$ must be of the form $[a_1, a_2] = \alpha a_1 + \beta a_2$, where a_1 and a_2 form a basis, and α and β are not both zero (or it would be Abelian). Let $\mathfrak{G}_s$ be the *standard* non-Abelian one for which $[b_1, b_2] = b_2$ and b_1, b_2 form a basis. Every non-Abelian $\mathfrak{G}$ is isomorphic to $\mathfrak{G}_s$. For let $\varphi(b_1) = \lambda a_1 + \mu a_2$ and $\varphi(b_2) = \alpha a_1 + \beta a_2$, where λ and μ are chosen so that $\lambda\beta - \alpha\mu = 1$ (the reader should prove that this is always possible and that φ has a unique inverse). Then

$$[\varphi(b_1), \varphi(b_2)] = [\lambda a_1 + \mu a_2, \alpha a_1 + \beta a_2] = (\lambda\beta - \mu\alpha)(\alpha a_1 + \beta a_2)$$
$$= \alpha a_1 + \beta a_2 = \varphi(b_2) = \varphi([b_1, b_2]).$$

Thus every such $\mathfrak{G}$ is isomorphic to $\mathfrak{G}_s$. In three dimensions, already, there is an infinite number of nonisomorphic Lie algebras.

IX-2 REALIZATIONS AND REPRESENTATIONS

Almost always physicists are led to groups and algebras through sets of mathematical operations, sets that turn out to be homomorphic to certain groups and algebras. For instance, as is hinted in the example which led to Eq. (22), the fundamental Poisson-bracket relations for a system with one degree of freedom are isomorphic to a very simple abstract Lie algebra. By studying the abstract groups and algebras we come across in such ways and by understanding the homomorphisms between them and our sets of operations, physicists may learn about the underlying dynamical systems they represent.

A mathematical set of this type is called a *realization* of the group or algebra. More precisely, let G be a group and let G' be a set of operators. Then if G' is itself a group, and if there exists a homomorphism D from G onto G', then $G' = D(G)$ is called a realization of G. Similarly for algebras: we write $d(\mathfrak{G})$ instead of $D(\mathfrak{G})$, however. If D or d is an isomorphism, we say the realization is *faithful*. If G' consists of linear operators or matrices, the realization is called a *representation*.

In realizations of groups, multiplication is always represented by successive application, or operator multiplication. But for a Lie algebra $\mathfrak{G}$ we must also define the Lie bracket in $\mathfrak{G}' = d(\mathfrak{G})$. If d is a representation, the commutator bracket will do. We have seen, however, that the Poisson bracket has all of properties A2–A4 of the Lie bracket, and it follows that the Poisson bracket can be used in a realization of a Lie algebra (provided A1 is verified).

We have seen examples already. For instance Eq. (19) may be taken as the multiplication law defining a certain Lie group:

$$x(k_2, \alpha_2, \beta_2)x(k_1, \alpha_1, \beta_1) = x(k_1 + k_2 + \alpha_1\beta_2, \alpha_1 + \alpha_2, \beta_1 + \beta_2).$$

Then Eq. (18) is a realization of this group. Similarly, Eq. (22) may be taken as the definition of a certain abstract Lie algebra. Then (21) with the commutator bracket provides a realization; (23) with the commutator bracket provides an infinite dimensional representation; and

$$d(a_1) = 1, \qquad d(a_2) = q, \qquad d(a_3) = -p$$

with the Poisson bracket provides another realization.

In the group-theoretical approach to classical particle mechanics, representations, as opposed to realizations, play a minor role. On the other hand representations are important in classical field theory. For example, consider Eq. (48) of Chapter VIII. Recall that the D depend on the Lorentz transformation $a_\nu{}^\mu$; we will write $D_A^B(a)$. Now suppose $a_\nu{}^\mu$ is a transformation from the four-space coordinates x^μ to y^μ and $b_\nu{}^\mu$ is a transformation from y^μ to z^μ, and let $c_\nu{}^\mu = b_\lambda{}^\mu a_\nu{}^\lambda$ be the product transformation from x^μ to z^μ. Then if we go by successive steps, Eq. (48) of Chapter VIII reads

$$\begin{aligned}\psi_A''(z) &= D_A^B(b)\psi_B'(b^{-1}z) = D_A^B(b)D_B^C(a)\psi_C(a^{-1}b^{-1}z)\\ &= D_A^B(b)D_B^C(a)\psi_C(c^{-1}z),\end{aligned}$$

and if we go straight from x^μ to z^μ, it reads

$$\psi_A''(z) = D_A^C(c)\psi_C(c^{-1}z).$$

Since $c = ba$, we have

$$D_A^B(b)D_B^C(a) = D_A^C(ba). \tag{26}$$

This means that the D_A^B matrices form a representation of the Lorentz group. In particular, let $D(a)$ be the matrix of the $D_A^B(a)$. Then D is a mapping from the Lorentz group G onto some set G' of $N \times N$ matrices such that $D(b)D(a) = D(ba)$ for all $a, b \in G$. Thus D is a homomorphism if G' is a group, which we leave to the reader to verify.

In Chapter VIII we discussed the requirement that the Lagrangian density be Lorentz invariant. This requirement can be stated in terms of the D matrices, without specifying whether the ψ_A are vector or tensor components, or some combinations. The Lagrangian density $\mathscr{L}$ should be made up of combinations of the ψ_A such that the D matrices cancel out in the expression for the transformed $\mathscr{L}$. For instance the abstract vector space of ψ functions may have an inner product on it with respect to which the D matrices are unitary:

$$(\psi, \psi) = (D\psi, D\psi).$$

Then $\mathscr{L}$ could be made up of such inner products. This is a very general way of putting it, and does not require that ψ be a covariant quantity like a vector or tensor (we shall say that such covariant quantities transform by *covariant representations* of the Lorentz group). There exist, in fact, other representations of the Lorentz group, and invariant Lagrangians can be constructed out of ψ-functions transforming by such other representations. The Lagrangian of the Dirac equation for the spin-$\frac{1}{2}$ particles is of this type.

Now by studying the Lorentz group one can find all of its representations. Then with certain additional requirements one can write down all reasonable Lagrangian densities and obtain all reasonable wave equations. In addition, one would then know all possible forms for the $D(a)$ matrices. Their derivatives at $a = \mathbb{1}$, as established in Eq. (54) of Chapter VIII, are very important in calculating angular momentum densities and symmetric stress tensors (in quantum mechanics the connection is even more immediate). Thus group theoretical considerations yield a great deal of information about possible fields and about any given field. Furthermore, we see again that it is derivatives at the identity element, or actually elements of the Lie algebra, which carry this physical information.

Promising as these results seem, we will not go into them in this book [although see Sections 4(a) and 4(b), which discuss the rotation group]. They belong rather in a book on quantum field theory. What we will be more interested in is the group of canonical transformations and its relation to dynamical variables, and how, given a set of dynamical variables, one constructs the corresponding subgroup of the canonicals. To do this, we consider the more general problem of constructing a representation or realization of a Lie group from its Lie algebra.

Let K be an abstract Lie group and G a finite-dimensional Lie subgroup of K, and let $\mathfrak{K}$ and $\mathfrak{G}$, respectively, be their abstract Lie algebras. For $a \in \mathfrak{G}$ we define the linear operator $d(a)$ by its operation on any $k \in \mathfrak{K}$:

$$d(a)k = [a, k], \tag{27}$$

where the bracket is the *abstract* bracket given in $\mathfrak{K}$.

Then if $a, b \in \mathfrak{G}$ we may form $[a, b]$ in $\mathfrak{G}$ and we may ask for $d([a, b])$. From (27) we have, using the Jacobi identity,

$$\begin{aligned} d([a, b])k &= [[a, b], k] = [a, [b, k]] - [b, [a, k]] \\ &= d(a)[b, k] - d(b)[a, k] \\ &= \{d(a)\, d(b) - d(b)\, d(a)\}k = [d(a), d(b)]_{C}k, \end{aligned} \tag{28}$$

where the subscript C stands for "commutator." Thus Eq. (27) defining the $d(a)$ for $a \in \mathfrak{G}$, taken together with the commutator bracket, forms a representation $d(\mathfrak{G})$ of $\mathfrak{G}$ by linear operators on $\mathfrak{K}$.

From this we can construct a representation of G in the following way. We know [see Eq. (9)] that from an element of $d(\mathfrak{G})$, which is now an operator or matrix Lie algebra, we can construct a one-parameter subgroup by exponentiation: let $d(a) \in d(\mathfrak{G})$, and consider

$$\begin{aligned} X(\varepsilon, a) &= \exp \{\varepsilon\, d(a)\} \\ &= \mathbb{1} + \varepsilon\, d(a) + \tfrac{1}{2}\varepsilon^2\, d^2(a) + \cdots. \end{aligned} \tag{29}$$

Clearly

$$X(\varepsilon_1, a)X(\varepsilon_2, a) = X(\varepsilon_1 + \varepsilon_2, a),$$

so that $X(\varepsilon, a)$ is a one-parameter group of operators on $\mathfrak{K}$. Its action on an element $k \in \mathfrak{K}$ may be written down in the form

$$X(\varepsilon, a)k = k + \varepsilon[a, k] + \tfrac{1}{2}\varepsilon^2[a, [a, k]] + \cdots. \tag{30}$$

Now suppose $a, b, c, \ldots$ is a basis in $\mathfrak{G}$ and that we have formed the one-parameter groups $X(\varepsilon, a)$, $X(\varepsilon, b), \ldots$. We can further form products of the form $X(\varepsilon_1, a)X(\varepsilon_2, b)$, products of such products, etc., obtaining finally a group G' of operators on $\mathfrak{K}$.

The form of G' can be simplified a bit. If $a \in \mathfrak{G}$ then so is $\varepsilon a = a'$, and consequently every operator of the form $X(\varepsilon, a)$ can be written in the form $X(1, a')$. Thus we need concern ourselves only with products like $X(1, a)X(1, b)$, etc. Now there is a theorem, too difficult and lengthy to prove here, which shows that for each pair a, b close enough to the null vector in $\mathfrak{G}$ there exists a $c \in \mathfrak{G}$ such that

$$X(1, a)X(1, b) = X(1, c). \tag{31}$$

In fact a series expression can be written down for c:

$$c = a + b + \tfrac{1}{2}[a, b] + \tfrac{1}{12}([a, [a, b]] + [b, [b, a]]) + \cdots.$$

This is called the Baker-Hausdorff series (Magnus, 1954); its convergence determines how far from the null vector of $\mathfrak{G}$ we can go before (31) breaks down. Now for each $X(1, c) \in G'$ for which (31) is valid we can introduce coordinates in the following way: the coordinates of $X(1, c)$ in G' are the coordinates of c in $\mathfrak{G}$. The Baker-Hausdorff series gives the coordinates of $X(1, c)$ analytically[8] in terms of the coordinates of its factors $X(1, a)$ and $X(1, b)$, so that G' is a Lie group. Furthermore, its Lie algebra is just $d(\mathfrak{G})$, homomorphic to $\mathfrak{G}$. Thus G' is homomorphic to G and is therefore a representation $D(G)$ of G by operators on $\mathfrak{K}$.

Remarks. 1. This procedure will work even if $\mathfrak{G}$ is $\mathfrak{K}$ itself. In general, whether or not $\mathfrak{G}$ is a *proper subalgebra* (i.e., not $\mathfrak{K}$ itself), the representations $d(\mathfrak{G})$ and $D(G)$ so obtained may not be faithful. *2.* By this procedure we can construct a representation of G only for $a, b, \ldots$ close enough to the null vector in $\mathfrak{G}$, or only for $X(1, a)$, $X(1, b), \ldots$ close enough to the identity element in $D(G)$, where the Baker-Hausdorff series converges. But this is not surprising, for as was mentioned after Eq. (17) the Lie algebra determines the group in general only locally. Thus all that we can guarantee about $D(G)$ is that it represents G locally (although in any actual case, the representation may be complete).

[8] It is analytic because by means of the structure constants the coordinates of $[a, b]$ are expressed as quadratic functions of the coordinates of a and b.

Equation (30) shows that in forming the representation we use not the commutator bracket $[d(a), d(b)]_C$, but the abstract bracket $[a, b]$, or rather $[a, k]$. In fact, therefore, we never write down the linear operators $X(\varepsilon, a)$, and thus we end up not with a representation, but with a realization. To illustrate what this means, we give an example of constructing a realization (Problem 12 is an example of constructing a representation). We take $\mathfrak{K}$ to be the space of all reasonably well-behaved functions of q and p [that is, now $k = f(q, p)$], and the abstract bracket in $\mathfrak{K}$ to be the Poisson bracket. We shall construct a realization of G, the group of Eq. (18), whose Lie algebra $\mathfrak{G}$ is given by (22). [Note: the functions appearing around Eqs. (18) and (22) are functions of q alone. Now we are dealing with functions of q and p.] It follows from previous discussion that we may write $a_1 = 1$, $a_2 = q$, $a_3 = -p$, for (the subscript P stands for Poisson)

$$[1, q]_P = [1, -p]_P = 0, \qquad [q, -p]_P = -1.$$

Thus for every $f(q, p) \in \mathfrak{K}$ we define $d(\mathfrak{G})$ by

$$d(a_1)f = d(1)f = [1, f]_P,$$

$$d(a_2)f = d(q)f = [q, f]_P,$$

$$d(a_3)f = d(-p)f = [-p, f]_P.$$

We now form the realization $D(G)$, acting also on $\mathfrak{K}$. First we form realizations of the three one-parameter groups $g_1(\varepsilon)$, $g_2(\varepsilon)$, $g_3(\varepsilon)$ of Eq. (20) by exponentiating $d(a_1)$, $d(a_2)$, $d(a_3)$ according to Eq. (30). Let $f(q, p)$ be any function in $\mathfrak{K}$. Then for a_1 Eq. (30) reads

$$X(\varepsilon, a_1)f \equiv D(g_1)f = f + \varepsilon[1, f]_P + \tfrac{1}{2}\varepsilon^2[1, [1, f]_P]_P + \cdots = f. \quad (32)$$

The entire subgroup $g_1(\varepsilon)$ is realized by $\mathbb{1}$ in $D(G)$. For a_2 Eq. (30) reads

$$X(\varepsilon, a_2)f \equiv D(g_2)f = f + \varepsilon[q, f]_P + \tfrac{1}{2}\varepsilon^2[q, [q, f]_P]_P + \cdots$$
$$= f + \varepsilon\frac{\partial f}{\partial p} + \tfrac{1}{2}\varepsilon^2\frac{\partial^2 f}{\partial p^2} + \cdots = f(q, p + \varepsilon). \quad (33)$$

The subgroup $g_2(\varepsilon)$ is realized by translations in p. For a_3 Eq. (30) reads

$$X(\varepsilon, a_3)f \equiv D(a_3)f = f + \varepsilon[-p, f]_P + \tfrac{1}{2}\varepsilon^2[-p, [-p, f]_P]_P + \cdots$$
$$= f + \varepsilon\frac{\partial f}{\partial q} + \tfrac{1}{2}\varepsilon^2\frac{\partial^2 f}{\partial q^2} + \cdots = f(q + \varepsilon, p). \quad (34)$$

The subgroup $g_3(\varepsilon)$ is realized by translations in q. These are exactly the results we had in Chapter VII: p generates translations in q, and q generates translations in p. (Even the signs are the same, which we leave to the reader to verify.) We are beginning to make contact with physics. The realization of G which we have obtained is not faithful: recall that the general element of G depends on three parameters k, α, β. In $D(G)$ we have

$$D(x(k, \alpha, \beta))f(q, p) = f(q + \beta, p + \alpha); \tag{35}$$

k does not appear. Thus D is a homomorphism, not an isomorphism. This can be expected whenever $\mathfrak{G}$ contains a nonnull *center*, that is, any elements like a_1 that commute with all others.

Remark. We could, on the other hand, have started with the group of (35) and asked for the representation of its Lie algebra in terms of commutators of operators on functions of q and p. This would have led to $[d(q), d(p)] = 0$, which is clearly not the Lie algebra we started with. It turns out that what we are sometimes forced to deal with are not the simplest realizations of groups and algebras, but realizations *up to a factor*, in which we add to the Lie algebra of the realization a nonnull center, even though the original Lie algebra itself has none. We will avoid any detailed discussion of such subtleties, however.

This example points up strikingly the relation between Lie algebras and groups on the one hand and Poisson brackets and canonical transformations on the other. By exponentiating certain Poisson brackets we have obtained a realization of G by (restricted) canonical transformations. That the transformations are canonical in our example is clear, but it is important and interesting that they will always be canonical, as can be verified by checking that Poisson brackets are preserved. We state this result more generally (see Problem 14):

Let $D(G)$ be constructed as above from a realization $d(\mathfrak{G})$ on a Lie algebra $\mathfrak{K}$ through Eq. (30). Let $k_1, k_2 \in \mathfrak{K}$, so that $[k_1, k_2]$, $X(\varepsilon, a)k_1$ and $X(\varepsilon, a)k_2$ are in $\mathfrak{K}$ for any $a \in \mathfrak{G}$. Then

$$X(\varepsilon, a)[k_1, k_2] = [X(\varepsilon, a)k_1, X(\varepsilon, a)k_2]. \tag{36}$$

PROOF. This can be proved by expanding both sides of (36) according to (30) and seeing that the coefficients of each power of ε are equal. It is simpler, however, to show that both sides satisfy the same differential equation in ε with the same "initial" condition. The initial condition is trivial: at $\varepsilon = 0$ both sides are simply $[k_1, k_2]$. For the derivative, we have in general

$$\begin{aligned}
\frac{d}{d\varepsilon}\{X(\varepsilon, b)k\} &= \frac{d}{d\varepsilon}\{k + \varepsilon[b, k] + \tfrac{1}{2}\varepsilon^2[b, [b, k]] + \frac{1}{3!}\varepsilon^3[b, [b, [b, k]]] + \cdots\} \\
&= [b, k] + \varepsilon[b, [b, k]] + \tfrac{1}{2}\varepsilon^2[b, [b, [b, k]]] + \cdots \\
&= [b, (k + \varepsilon[b, k] + \tfrac{1}{2}\varepsilon^2[b, [b, k]] + \cdots)] \\
&= [b, X(\varepsilon, b)k].
\end{aligned} \tag{37}$$

Now let us write $X(\varepsilon, a) = X$ and apply (37) to both sides of (36). Let L be the left-hand side. Then

$$\frac{dL}{d\varepsilon} = \frac{d}{d\varepsilon}\{X[k_1, k_2]\} = [a, X[k_1, k_2]] = [a, L].$$

Now let R be the right-hand side of (36). Then[9] by using the Jacobi identity we have

$$\begin{aligned}\frac{dR}{d\varepsilon} = \frac{d}{d\varepsilon}[Xk_1, Xk_2] &= \left[\frac{d}{d\varepsilon}Xk_1, Xk_2\right] + \left[Xk_1, \frac{d}{d\varepsilon}Xk_2\right] \\ &= [[a, Xk_1], Xk_2] + [Xk_1, [a, Xk_2]] \\ &= [a, [Xk_1, Xk_2]] = [a, R].\end{aligned}$$

Thus both sides of (36) satisfy the same differential equation, as asserted.

Remarks. *1.* Note that the common differential equation satisfied by L and R is just (37). If, as in Eq. (27), we write $[b, k] = d(b)k$, then Eq. (37) may be treated as a differential equation for $X(\varepsilon, b)$:

$$\frac{d}{d\varepsilon}X(\varepsilon, b) = d(b)X(\varepsilon, b).$$

Its solution is well known: $X(\varepsilon, b) = \exp\{\varepsilon\, d(b)\}$. See Eq. (29). To the extent that there is any difference, it is only that now we are using a realization rather than a representation. *2.* Since for each pair ε, a the operator $X(\varepsilon, a)$ preserves brackets in $\mathfrak{K}$, this operator is an isomorphism (why not a homomorphism?) of $\mathfrak{K}$ onto itself, called an *automorphism*. Then $D(G)$ is a group of automorphisms of $\mathfrak{K}$.

IX-3 DYNAMICS AND CANONICAL TRANSFORMATIONS

The considerations we have been discussing can now be applied more generally to canonical transformations and functions of q, p in n degrees of freedom. Let $\mathfrak{K}$ now consist, as in the last example of Section 2, of all reasonably well behaved functions $f(q, p)$, with the Poisson bracket. Let $\mathfrak{G}$ be any subalgebra of $\mathfrak{K}$ (for example, the one consisting of $q, p, 1$; or the one consisting of l_1, l_2, l_3 as in Section 6(c) of Chapter VI). Then by proceeding just as we did in the previous example, we can construct a local realization of $D(G)$ by canonical transformations, what we shall call a *canonical realization*.

[9] The reader should satisfy himself that $d/d\varepsilon$ moves into the bracket according to the sort of product rule we are assuming is valid. Recall that this is an important point for d/dt and Poisson brackets (the Poisson Bracket Theorem of Chapter VI). It is valid here because of linearity and the way ε appears in Eq. (30).

Because it applies immediately to dynamics and to some other considerations of Chapter VII, we start with a particularly simple $d(\mathfrak{G})$. Consider a fixed function $h(q, p) \in \mathfrak{K}$. Together with its multiples αh it forms a canonical realization of the (Abelian) one-dimensional Lie algebra: $[\alpha h, h]_{\mathrm{P}} = 0$. This realization operates on any $f \in \mathfrak{K}$ according to

$$d(h)f = [h, f]_{\mathrm{P}}.$$

In particular it can operate on the functions q_α and p_α. Moreover, h generates a realization of a one-parameter group on $\mathfrak{K}$, converting every $f \in \mathfrak{K}$ into a family of new functions, one for each ε. We shall write (note the minus sign: see Problem 17)

$$f_\varepsilon(q, p) = X(-\varepsilon, h)f(q, p). \tag{38}$$

The realization of the Lie algebra is just the derivative of this at $\varepsilon = 0$. Indeed, from (37) we obtain

$$\begin{aligned}\frac{df_\varepsilon}{d\varepsilon} &= \frac{d}{d\varepsilon}\{X(-\varepsilon, h)f\} = -[h, X(-\varepsilon, h)f]_{\mathrm{P}} \\ &= [f_\varepsilon, h]_{\mathrm{P}} = -d(h)f_\varepsilon,\end{aligned} \tag{39}$$

which gives back the realization of the Lie algebra when we set $\varepsilon = 0$ [except for the choice of sign, which will become clear at Eq. (41)].

Equation (39) is a differential equation for f_ε as a function of ε. Its solution is (38). As mentioned in Remark 1 at the end of Section 2, the solution can always be written in the form

$$f_\varepsilon = \exp\{-\varepsilon\, d(h)\}f = f + \varepsilon[f, h]_{\mathrm{P}} + \tfrac{1}{2}\varepsilon^2[[f, h]_{\mathrm{P}}, h]_{\mathrm{P}} + \cdots \tag{40}$$

(inverting the order in the Poisson brackets takes care of the minus sign). This is a power-series solution of (39); all one need do is to calculate the brackets and sum the series.

As special cases of (40) we may take $f = q_\alpha$ and $f = p_\alpha$. Let us write $q_{\alpha,\varepsilon} = q_\alpha(\varepsilon)$ and $p_{\alpha,\varepsilon} = p_\alpha(\varepsilon)$ for these cases.[10] Given some other $f(q, p)$ consider the function $f^\varepsilon(q, p) = f(q(\varepsilon), p(\varepsilon))$. Then $f^\varepsilon = f_\varepsilon$, as we shall now show. First, $f^0 = f_0$ obviously, for both are just $f(q, p)$. Second,

$$\frac{df^\varepsilon}{d\varepsilon} = \frac{\partial f^\varepsilon}{\partial q_\alpha}\frac{dq_\alpha(\varepsilon)}{d\varepsilon} + \frac{df^\varepsilon}{\partial p_\alpha}\frac{dp_\alpha(\varepsilon)}{d\varepsilon}.$$

The derivatives of the $q_\alpha(\varepsilon)$ and $p_\alpha(\varepsilon)$ are obtained from (39), and then it follows immediately from the definition of the Poisson bracket that

$$\frac{df^\varepsilon}{d\varepsilon} = [f^\varepsilon, h]_{\mathrm{P}}.$$

[10] In functional form the $q_\alpha(\varepsilon)$ and $p_\alpha(\varepsilon)$ depend on the $q_\alpha(0)$ and $p_\alpha(0)$. See the example that follows Eq. (42).

As in the proof of (36), we see that f^{ε} and f_{ε} satisfy the same differential equation with the same initial condition: thus $f^{\varepsilon} = f_{\varepsilon}$. This means that the one-parameter group of transformations on $\mathfrak{R}$ generated by $h(q, p)$ transforms each phase point (q, p) in an ε-dependent continuous way to a set of new phase points $(q(\varepsilon), p(\varepsilon))$ sweeping out certain trajectories. At the same time every other function $f(q, p) \in \mathfrak{R}$ is "left alone." That is, in order to find its functional form $f_{\varepsilon}(q, p)$ you merely evaluate $f(q, p)$ at $(q(\varepsilon), p(\varepsilon))$.

This is all very similar to what we have done before, so the next step is obvious. We apply this to dynamics by replacing ε by the time t, and h by the Hamiltonian H. Then the old phase point (q, p) is what we have called elsewhere the initial point (q^0, p^0) and the new point $(q(\varepsilon), p(\varepsilon))$ becomes the moving point $(q(t), p(t))$. The function $f(q, p)$ becomes a dynamical variable, and Eq. (39) gives the familiar old result

$$\frac{df}{dt} = [f, H]_{\text{P}}. \tag{41}$$

Now, however, we have a solution to this equation in Eq. (40). In particular, for q_α and p_α we have

$$\begin{aligned} q_\alpha(t) &= q_\alpha + t[q_\alpha, H]_{\text{P}} + \tfrac{1}{2}t^2[[q_\alpha, H]_{\text{P}}, H]_{\text{P}} + \cdots, \\ p_\alpha(t) &= p_\alpha + t[p_\alpha, H]_{\text{P}} + \tfrac{1}{2}t^2[[p_\alpha, H]_{\text{P}}, H]_{\text{P}} + \cdots, \end{aligned} \tag{42}$$

Where $q_\alpha = q_\alpha(0)$ and $p_\alpha = p_\alpha(0)$ are the initial points.

We take the familiar dynamical example of the simple harmonic oscillator in one dimension: $H = \frac{1}{2}(q^2 + p^2)$. Then with $[q, H]_{\text{P}} = p$, the first of Eqs. (42) reads

$$q(t) = q + t \cdot p + \frac{t^2}{2!}[p, H]_{\text{P}} + \frac{t^3}{3!}[[p, H]_{\text{P}}, H]_{\text{P}} + \cdots.$$

Further, $[p, H]_{\text{P}} = -q$, so that

$$\begin{aligned} q(t) &= q + t \cdot p + \frac{t^2}{2!}(-q) + \frac{t^3}{3!}(-p) + \cdots \\ &= q\left(1 - \frac{t^2}{2!} + \cdots\right) + p\left(t - \frac{t^3}{3!} + \cdots\right) \\ &= q \cos t + p \sin t. \end{aligned}$$

Similarly,

$$\begin{aligned} p(t) &= p\left(1 - \frac{t^2}{2!} + \cdots\right) - q\left(t - \frac{t^3}{3!} + \cdots\right) \\ &= -q \sin t + p \cos t. \end{aligned}$$

This is the familiar solution, with q and p appearing as constants of integration.

We have arrived at a new expression of dynamics. Let $\xi(0) = (q, p)$ be the initial phase point. Then there is a one-parameter group of operators $U(t)$ in standard form [i.e., $U(t_1)U(t_2) = U(t_1 + t_2)$] such that at any later time, $\xi(t)$ is given by

$$\xi(t) = U(t)\xi(0).$$

The problem is to find $U(t)$, and the solution to this problem is Eq. (42): $U(t) = X(-t, H)$. This view is not very different from the Hamilton-Jacobi approach of Chapter VII. There we asked for the global generator, given the infinitesimal generator. The global generator $S(q, Q)$ transforms from the initial phase point (Q, P) to the moving point (q, p), although in any actual solution of the problem the Q and P were not the coordinates of the initial point, but some functions of them.

Suppose now that $h(q, p)$ and $f(q, p)$ form two separate one-dimensional Lie algebra realizations $d(\mathfrak{G}_h)$ and $d(\mathfrak{G}_f)$. Then we may ask how f varies along the ε-trajectories of $D(G_h)$ and how h varies along the similar δ-trajectories of $D(G_f)$. At any phase point (q, p), we may apply Eq. (39), dropping the subscript ε (we leave the justification to the reader), which gives

$$\frac{df}{d\varepsilon} = [f, h]_{\mathrm{P}} = -\frac{dh}{d\delta}. \tag{43}$$

In particular, f is constant along the trajectories generated by h if and only if h is constant along the trajectories generated by f. This is a familiar result as well [see the discussion following Eq. (11) of Chapter VII], and reestablishes the relation between symmetry and conservation laws.

The space $\mathfrak{R}$ of functions $f(q, p)$ is more than a Lie algebra; it has additional structure. Specifically two elements f and h in $\mathfrak{R}$ can be multiplied to give a third $g = fh$, and the bracket behaves as a derivative with respect to this multiplication: for any $f, g, h \in \mathfrak{R}$ we have

$$[fg, h]_{\mathrm{P}} = f[g, h]_{\mathrm{P}} + [f, h]_{\mathrm{P}}g. \tag{44}$$

Realizations and representations, incidentally, always have such additional structure. Physically this is a reflection of the fact that there is more to dynamics than can be learned by studying only the Lie algebraic structure, and for this reason expressions like (44) are very important in physical applications.

Certain groups are particularly important in physics. For instance, the rotation group, the Galilei group (see Chapter III, Section 4), and the Lorentz group. We have seen also that the realizations on $\mathfrak{R}$ of Lie algebras yield dynamical variables. We could therefore establish a program of finding canonical realizations of these important groups and their Lie algebras, and then of identifying dynamical systems in which the elements of the Lie algebra are actual dynamical variables. A similar program has been fruitful in relativistic quantum mechanics, as already mentioned. It has been less fruitful in classical mechanics, although it does provide some information. We shall not go into any such program in detail, but we give here a relatively simple example. Another is given in Section 4(c).

It is shown in Section 4(a) that the Lie algebra of the rotation group is

$$[a_1, a_2] = a_3, \quad [a_2, a_3] = a_1, \quad [a_3, a_1] = a_2. \tag{45}$$

Now we know that the components l_α of the angular momentum of a single particle have Poisson brackets that agree with Eq. (45).[11] Thus for a single-particle system we may take

$$d(a_\alpha) = l_\alpha = \epsilon_{\alpha\beta\gamma} x_\beta p_\gamma,$$

where x_β is the Cartesian coordinate of the particle. Because we can not write a single-particle Hamiltonian in terms of the l_α, this realization is not very useful. However for a system of particles the *total* angular momentum components l_α also satisfy the same Poisson-bracket relations as (45), and for some systems of particles the Hamiltonian can in fact be written down as a function of the l_α. For a free rigid body, for instance, the Hamiltonian in a body system is

$$H = \tfrac{1}{2} N_{\alpha\beta} l_\alpha l_\beta,$$

where $N_{\alpha\beta}$ is the inverse of the moment-of-inertia tensor: $N_{\alpha\beta} I_{\beta\mu} = \delta_{\alpha\mu}$. The l_α are not, incidentally, themselves part of a canonical system, for their mutual Poisson brackets are not zero (if they were, we would have $\dot{l}_\alpha = 0$). The equations of motion for the l_α are

$$\begin{aligned} \dot{l}_\alpha &= [l_\alpha, H]_{\mathrm{P}} = \tfrac{1}{2} N_{\mu\nu} [l_\alpha, l_\mu l_\nu]_{\mathrm{P}} \\ &= \varepsilon_{\alpha\mu\beta} N_{\mu\nu} l_\nu l_\beta, \end{aligned} \tag{46}$$

where we have used the symmetry of $N_{\mu\nu}$. These equations, as is shown in Problem 19, are Euler's equations for a rigid body with no applied torques.

A property of the rotation group is that in all of its realizations $l^2 = l_\alpha l_\alpha$ commutes with every element of $d(\mathfrak{G})$ and hence, by exponentiation, also with every element of $D(G)$. It follows that it commutes with every function of the l_α, including the Hamiltonian if H is a function only of the l_α. Therefore l^2 is a constant of the motion even though each l_α itself is not (in the

[11] For this and some other statements in the next few paragraphs the reader is referred back to Chapter VI, Section 6(c) and to Chapter V.

body system). This is true of Hamiltonians which depend in more complicated ways on the l_α, as well. See Problem 20.

IX-4 EXAMPLES, APPLICATIONS, AND EXTENSIONS

(a) The Three-Dimensional Rotation Group

We are already familiar with some aspects of the rotations in three dimensions from Chapter IV. Here we shall discuss some of the group properties, the Lie algebra, and some representations.

As we know, the rotations depend on three parameters, such as, for instance, the three Euler angles. Moreover, the Euler angles of the product of two rotations can be written down in terms of trigonometric, that is, analytic, functions of the Euler angles of the factors, so that $SO(3, R)$ is a Lie group. Each element of $SO(3, R)$, each rotation, can be expressed by a point in a three-dimensional *group space.* For instance if we specify each rotation not by its Euler angles, but by its axis $\mathbf{e}$ (a unit vector) and its angle θ (using the right-hand rule), the point corresponding to the rotation $(\mathbf{e}, \theta)$ may be chosen at a distance θ from the origin in the direction of $\mathbf{e}$ in some Cartesian *3*-space. Then all rotations lie within and on the surface of a sphere of radius π about the origin, which itself corresponds to the identity transformation. Since a rotation by π is the same as one by $-\pi$ (both physically and mathematically) opposite points on the surface of the sphere correspond to the same rotation. Thus we see that the group space of the rotation group is not simply a Euclidean space. If you go out far enough in one direction (keep $\mathbf{e}$ fixed and let θ grow from zero to 2π) you come back (to the origin) where you started. Thus it is not a trivial matter to parametrize $SO(3, R)$ continuously so that each rotation will be described uniquely.[12]

What are the one-parameter subgroups in $SO(3, R)$? To answer this question we first find the elements of its Lie algebra $so(3, R)$ and then exponentiate them. Let $R(\varepsilon)$ be a curve in $SO(3, R)$ with $R(0) = e = \mathbb{1}$, as usual. Then because $RR^T = \mathbb{1}$, we have

$$\frac{dR}{d\varepsilon} R^T + R \frac{dR^T}{d\varepsilon} = 0,$$

or, writing a for $dR/d\varepsilon$ at $\varepsilon = 0$,

$$a + a^T = 0.$$

[12] A similar problem arises in establishing a coordinate system on the surface of a sphere, say the Earth. For instance what is the longitude of the North Pole?

The elements of the Lie algebra are thus antisymmetric 3 × 3 matrices. The general antisymmetric matrix may be written in the form

$$\begin{Vmatrix} 0 & \alpha_3 & -\alpha_2 \\ -\alpha_3 & 0 & \alpha_1 \\ \alpha_2 & -\alpha_1 & 0 \end{Vmatrix} = \alpha_1 \begin{Vmatrix} 0 & 0 & 0 \\ 0 & 0 & 1 \\ 0 & -1 & 0 \end{Vmatrix} + \alpha_2 \begin{Vmatrix} 0 & 0 & -1 \\ 0 & 0 & 0 \\ 1 & 0 & 0 \end{Vmatrix}$$

$$+ \alpha_3 \begin{Vmatrix} 0 & 1 & 0 \\ -1 & 0 & 0 \\ 0 & 0 & 0 \end{Vmatrix} \tag{47}$$

$$= \alpha_1 a_1 + \alpha_2 a_2 + \alpha_3 a_3,$$

defining the a_α, which therefore form a basis in $so(3, R)$. The one-parameter subgroups they generate are easily found. For example, a_1 generates

$$g_1(\varepsilon) = \exp\{\varepsilon a_1\} = \mathbb{1} + \varepsilon a_1 - \tfrac{1}{2}\varepsilon^2 E_1 - \frac{1}{3!}\varepsilon^3 a_1 + \frac{1}{4!}\varepsilon^4 E_1 + \cdots,$$

where

$$E_1 = -(a_1)^2 = \begin{Vmatrix} 0 & 0 & 0 \\ 0 & 1 & 0 \\ 0 & 0 & 1 \end{Vmatrix}.$$

Therefore the one-parameter subgroup is of the form

$$g_1(\varepsilon) = \mathbb{1} - E_1 + E_1 \cos\varepsilon + a_1 \sin\varepsilon$$

$$= \begin{Vmatrix} 1 & 0 & 0 \\ 0 & \cos\varepsilon & \sin\varepsilon \\ 0 & -\sin\varepsilon & \cos\varepsilon \end{Vmatrix}, \tag{48}$$

which is a rotation through the angle ε about the *1*-axis. In general a_α generates such rotations about the α-axis [see the example at Eq. (10)]. There is nothing special about the way we chose the axes, so each rotation about any fixed axis lies on the one-parameter subgroup of all rotations about that axis, and there are no other such one-parameter subgroups.

The structure of the Lie algebra is found by calculating the commutator brackets of the a_α. It is a simple matter to show that

$$[a_\alpha, a_\beta] = \epsilon_{\alpha\beta\gamma} a_\gamma, \tag{49}$$

so that the structure constants are

$$c_{\alpha\beta}^{\gamma} = \epsilon_{\alpha\beta\gamma}.$$

We see that the a_α satisfy Lie bracket relations that are the same as the Poisson bracket relations satisfied by the components of angular momentum.

We want to proceed to a discussion of finite-dimensional representations of $SO(3, R)$. Recall that a representation is a realization in which the $d(a_\beta)$ are matrices and the bracket is the commutator. For simplicity we shall write $d(a_\beta) = d_\beta$. Then in every representation $d(\mathfrak{G})$ of $so(3, R)$ it will be found that[13]

$$d^2 \equiv d_\beta d_\beta \tag{50}$$

commutes with every element of $d(\mathfrak{G})$. Indeed, from (49) we have

$$[d_\alpha, d_\beta d_\beta] = d_\beta[d_\alpha, d_\beta] + [d_\alpha, d_\beta]d_\beta = \epsilon_{\alpha\beta\gamma}(d_\beta d_\gamma + d_\gamma d_\beta) = 0.$$

Since d^2 commutes with $d(\mathfrak{G})$, it commutes also with $D(G)$, where G is $SO(3, R)$, for all elements of $D(G)$ are obtained by exponentiating elements of $d(\mathfrak{G})$.

This property of every $d(\mathfrak{G})$ can be used in the following way. Any particular $d(\mathfrak{G})$ is a set of matrices acting on some finite-dimensional, column-vector space S. Let us assume that $d(\mathfrak{G})$ is *irreducible*, that is, that there is no subspace of S which is mapped into itself by every element of $d(\mathfrak{G})$. The rationale for this assumption is the following. If there were such a subspace, the matrices of $d(\mathfrak{G})$ operating on it alone would form a smaller representation of $\mathfrak{G} = so(3, R)$, and we could then study that one. Then any representation which is not irreducible (there is a theorem about this) can be built up out of such smaller irreducible ones, so we lose no generality by our assumption. Now it can be shown that any matrix which commutes with all the matrices of an irreducible representation must be a multiple of the unit matrix. This is a consequence of a theorem known as *Schur's Lemma* which we shall not prove (but see Problem 21). However we did prove essentially this result when we showed [Chapter III, Eq. (45); Chapter IV, Problem 32] that if K commutes with all of $SO(3, R)$, then $K = \lambda\mathbb{1}$. Thus for each $d(\mathfrak{G})$ in which we are interested there is a number λ such that

$$d^2 = \lambda\mathbb{1}. \tag{51}$$

We will see that λ characterizes the representation.

We define three new matrices in $d(\mathfrak{G})$:[14]

$$d^\uparrow = -d_2 + id_1, \qquad d^\downarrow = d_2 + id_1, \qquad d^z = id_3. \tag{52}$$

[13] We can form products of elements of $d(\mathfrak{G})$, as we have already mentioned [see Eq. (44)], only because it is a representation. These products are not themselves in $d(\mathfrak{G})$; thus d^2 is not in $d(\mathfrak{G})$.

[14] Although $\mathfrak{G}$ is a real vector space, it is convenient to allow complex coefficients in discussing $d(\mathfrak{G})$. This is because the matrices of $d(\mathfrak{G})$ themselves will be complex valued in general, although in the final analysis we will consider only linear combinations of them with real coefficients. We shall assume that the matrices of $D(G)$ are unitary, which it turns out involves no loss of generality for $SO(3, R)$. Then just as we showed that the a_α are antisymmetric, we can show that the d_α are anti-Hermitian: $d_\beta{}^\dagger = -d_\beta$. It follows that d^3 in Hermitian and that $(d^\uparrow)^\dagger = d^\downarrow$.

Then the Lie algebra becomes

$$[d^z, d^\uparrow] = d^\uparrow, \qquad [d^z, d^\downarrow] = -d^\downarrow, \qquad [d^\uparrow, d^\downarrow] = 2d^z, \tag{53}$$

and also

$$d^2 = -(d^z)^2 - \tfrac{1}{2}(d^\uparrow d^\downarrow + d^\downarrow d^\uparrow). \tag{54}$$

Let $x_\mu \in S$ be an eigenvector of d^z belonging to eigenvalue μ (necessarily real, for d^z is Hermitian). Then the first of Eqs. (53) yields

$$d^z(d^\uparrow x_\mu) = d^\uparrow x_\mu + d^\uparrow d^z x_\mu = (\mu + 1)\, d^\uparrow x_\mu.$$

Similarly,

$$d^z(d^\downarrow x_\mu) = -d^\downarrow x_\mu + d^\downarrow d^z x_\mu = (\mu - 1)\, d^\downarrow x_\mu.$$

This means that $d^\uparrow x_\mu$ is also an eigenvector of d^z but belonging to eigenvalue $\mu + 1$, and $d^\downarrow x_\mu$ is an eigenvector belonging to eigenvalue $\mu - 1$. In other words, there exist numbers $\alpha_{\mu+1}$ and β_μ such that (no sum!)

$$d^\uparrow x_\mu = \alpha_{\mu+1} x_{\mu+1}, \qquad d^\downarrow x_\mu = \beta_\mu x_{\mu-1}.$$

In fact since the eigenvectors of an operator can be multiplied by an arbitrary factor, we can choose all the eigenvectors x_ν of d^z so that $\alpha_\nu = \beta_\nu$. Then we have

$$d^\uparrow x_\mu = \alpha_{\mu+1} x_{\mu+1}, \qquad d^\downarrow x_\mu = \alpha_\mu x_{\mu-1}. \tag{55}$$

Because of these properties $d^\uparrow$ and $d^\downarrow$ are known as *raising* and *lowering* operators, respectively.

Since d^z is Hermitian, its eigenvectors form an orthogonal basis in S. We see that in this basis d^z is diagonal with eigenvalues (elements on the main diagonal) $l, l - 1, l - 2, \ldots, l - (n - 1) = k$, where l is the maximum eigenvalue, k the minimum, and n is the dimension of S. To find out more about d^z we must find l and n, and to find out about $d^\uparrow$ and $d^\downarrow$ we need only evaluate the α_μ. We now insert (55) into the third of Eqs. (53) and apply it to x_μ. This yields

$$\alpha_\mu^2 - \alpha_{\mu+1}^2 = 2\mu \tag{56}$$

for $\mu \neq l$ and $\mu \neq k$. Since there is no x_{l+1} and no x_{k-1}, it follows that $d^\uparrow x_l = 0$ and $d^\downarrow x_k = 0$. Inserting these into the third of (53) yields

$$\alpha_l^2 = 2l, \qquad \alpha_{k+1}^2 = -2k. \tag{57}$$

Recall that (sum on β)

$$d^2 = d_\beta d_\beta = -\{(d^z)^2 + \tfrac{1}{2}(d^\uparrow d^\downarrow + d^\downarrow d^\uparrow)\} = \lambda \mathbb{1}. \tag{58}$$

If we apply this to x_l and x_k we get

$$\alpha_l^2 = -2(\lambda + l^2), \qquad \alpha_{k+1}^2 = -2(\lambda + k^2).$$

Insert these into (57) and use $k = l - (n - 1)$. The results are

$$\lambda = -l(l + 1), \qquad \lambda = -k(k - 1) = -(l - n)(l - n + 1).$$

On eliminating λ, we arrive at

$$2l = n - 1.$$

Thus l is a half integer related to the dimension n of the representation (that is, of S), and $k = -l$. There is a representation of every dimension:

Dimension n	1	2	3	4	...
$l = \frac{1}{2}(n - 1)$	0	$\frac{1}{2}$	1	$\frac{3}{2}$	...
$-\lambda = l(l + 1)$	0	$\frac{3}{4}$	2	$\frac{15}{4}$	...

In each representation Eq. (57) gives α_l and Eq. (56) is a recursion relation for α_μ, $\mu < l$. Thus we now have all the matrix elements of $d^\uparrow$ and $d^\downarrow$ in the x_μ basis consisting of the eigenvectors $\{x_l, x_{l-1}, \ldots, x_{-l+1}, x_{-l}\}$. This completes the picture, for from the matrices d^z, $d^\uparrow$, $d^\downarrow$ we can construct the entire representation. The value of l is called the *spin* of the representation.

As examples we give the matrix elements of the first three representations.

Spin 0, $n = 1$

$$d^z = 0, \qquad d^\uparrow = 0, \qquad d^\downarrow = 0.$$

Every element of G is represented by $\mathbb{1}$.

Spin $\frac{1}{2}$, $n = 2$ (spinor representation)

$$d^z = \begin{vmatrix} \frac{1}{2} & 0 \\ 0 & -\frac{1}{2} \end{vmatrix}, \qquad d^\uparrow = \begin{vmatrix} 0 & 1 \\ 0 & 0 \end{vmatrix}, \qquad d^\downarrow = \begin{vmatrix} 0 & 0 \\ 1 & 0 \end{vmatrix}. \tag{59}$$

This representation is discussed more fully in Section 4(b).

Spin 1, $n = 3$ (vector representation)

$$d^z = \begin{vmatrix} 1 & 0 & 0 \\ 0 & 0 & 0 \\ 0 & 0 & -1 \end{vmatrix}, \qquad d^\uparrow = \begin{vmatrix} 0 & \sqrt{2} & 0 \\ 0 & 0 & \sqrt{2} \\ 0 & 0 & 0 \end{vmatrix}, \qquad d^\downarrow = \begin{vmatrix} 0 & 0 & 0 \\ \sqrt{2} & 0 & 0 \\ 0 & \sqrt{2} & 0 \end{vmatrix}.$$

To construct $D(G)$ we may first use (52) to find the d_α. These are

$$d_1 = \frac{1}{2i}\begin{vmatrix} 0 & \sqrt{2} & 0 \\ \sqrt{2} & 0 & \sqrt{2} \\ 0 & \sqrt{2} & 0 \end{vmatrix}, \qquad d_2 = \tfrac{1}{2}\begin{vmatrix} 0 & -\sqrt{2} & 0 \\ \sqrt{2} & 0 & -\sqrt{2} \\ 0 & \sqrt{2} & 0 \end{vmatrix},$$

$$d_3 = \begin{vmatrix} -i & 0 & 0 \\ 0 & 0 & 0 \\ 0 & 0 & i \end{vmatrix}. \tag{60}$$

The representation we obtain is $so(3, R)$ itself. Although these d_α look very different from the a_α of Eq. (47), they are not as removed as may seem. The one-parameter subgroup generated by d_3, for instance, is of the form

$$\begin{vmatrix} e^{i\theta} & 0 & 0 \\ 0 & 1 & 0 \\ 0 & 0 & e^{-i\theta} \end{vmatrix}.$$

Now we already saw in Chapter IV that a general rotation matrix cannot be diagonalized over the real numbers, and in agreement with that result we find that the rotations about the 3-axis, when diagonalized, involve complex numbers. There exists, in fact, a unitary coordinate transformation U on S [the three-dimensional space on which $D(G)$ acts] which brings each d_α to the form a_α. Thus we have obtained $SO(3, R)$ itself, but in a different and complex coordinate system. See Problem 22.

(b) The Spinor Group

The spin-$\frac{1}{2}$ representation of the rotation group is particularly interesting both mathematically and physically. It is a homomorphic rather than an isomorphic representation of the group [although, as is evident from Section 4(a), it represents the Lie algebra isomorphically]. It gives a way of finding the axis of the product of two rotations if the axes of its factors are known. It lies at the basis of the relativistic equation for the electron, the Dirac equation. For these reasons we discuss the spin-$\frac{1}{2}$ representation separately.

From Eq. (59) we can construct the d_α in the case of spin $\frac{1}{2}$. We have,

$$d_3 = \begin{vmatrix} -\frac{1}{2}i & 0 \\ 0 & \frac{1}{2}i \end{vmatrix}, \qquad d_2 = \begin{vmatrix} 0 & -\frac{1}{2} \\ \frac{1}{2} & 0 \end{vmatrix}, \qquad d_1 = \begin{vmatrix} 0 & -\frac{1}{2}i \\ -\frac{1}{2}i & 0 \end{vmatrix}. \tag{61}$$

It is usual to define the *Pauli spin matrices* σ_α by $\sigma_\alpha = 2id_\alpha$ so that

$$\sigma_1 = \begin{vmatrix} 0 & 1 \\ 1 & 0 \end{vmatrix}, \qquad \sigma_2 = \begin{vmatrix} 0 & -i \\ i & 0 \end{vmatrix}, \qquad \sigma_3 = \begin{vmatrix} 1 & 0 \\ 0 & -1 \end{vmatrix}. \tag{62}$$

Then the spin matrices satisfy the relations

$$\begin{aligned} \sigma_\alpha \sigma_\beta &= i\epsilon_{\alpha\beta\gamma}\sigma_\gamma + \delta_{\alpha\beta}\mathbb{1}, \\ [\sigma_\alpha, \sigma_\beta] &= 2i\epsilon_{\alpha\beta\gamma}\sigma_\gamma. \end{aligned} \tag{63}$$

Consider a rotation through an angle θ about the axis parallel to some arbitrary unit vector $\mathbf{e}$. As follows from Problem 23 (which, however, concerns the three-dimensional representation), in two dimensions this rotation is represented by

$$X(\mathbf{e}, \theta) = \exp\,(e_\alpha\, d_\alpha) = \exp\,(-\tfrac{1}{2}i\theta\mathbf{e} \cdot \boldsymbol{\sigma}), \tag{64}$$

where $\boldsymbol{\sigma}$ is the vector whose components are the σ_α matrices:

$$\boldsymbol{\sigma} = \mathbf{e}_\alpha \sigma_\alpha.$$

(Note: $\mathbf{e}_\alpha$ is the unit vector in the α-direction; e_α is the α-component of the arbitrary unit vector $\mathbf{e}$.) Since we will have to calculate powers of $\mathbf{e} \cdot \boldsymbol{\sigma}$, let us calculate $(\mathbf{v} \cdot \boldsymbol{\sigma})(\mathbf{u} \cdot \boldsymbol{\sigma})$ for general vectors $\mathbf{v}$ and $\mathbf{u}$.[15] Using Eq. (63), we obtain

$$\begin{aligned} (\mathbf{v} \cdot \boldsymbol{\sigma})(\mathbf{u} \cdot \boldsymbol{\sigma}) &= v_\alpha \sigma_\alpha u_\beta \sigma_\beta = v_\alpha u_\alpha \mathbb{1} + i v_\alpha u_\beta \epsilon_{\alpha\beta\gamma}\sigma_\gamma \\ &= \mathbf{v} \cdot \mathbf{u}\mathbb{1} + i(\mathbf{v} \times \mathbf{u}) \cdot \boldsymbol{\sigma}. \end{aligned} \tag{65}$$

Then Eq. (64) becomes

$$\begin{aligned} X(\mathbf{e}, \theta) &= \mathbb{1} - i\frac{\theta}{2}\mathbf{e} \cdot \boldsymbol{\sigma} - \frac{1}{2}\left(\frac{\theta}{2}\right)^2 \mathbb{1} + \frac{1}{3!}\left(\frac{\theta}{2}\right)^3 i\mathbf{e} \cdot \boldsymbol{\sigma} + \cdots \\ &= \mathbb{1} \cos\frac{\theta}{2} - i\mathbf{e} \cdot \boldsymbol{\sigma} \sin\frac{\theta}{2}. \end{aligned} \tag{66}$$

This is sometimes written in the form

$$X(\mathbf{e}, \theta) = \alpha_0 \mathbb{1} - i\boldsymbol{\alpha} \cdot \boldsymbol{\sigma}, \tag{67}$$

where $\alpha_0 = \cos\theta/2$ and $\boldsymbol{\alpha} = \mathbf{e}\sin\theta/2$. The four real numbers α_0, α_1, α_2, α_3 are called the *Cayley-Klein parameters* of the rotation group. Although there are four of them, only three are independent, for they satisfy the relation

$$\alpha_0^2 + \boldsymbol{\alpha} \cdot \boldsymbol{\alpha} = 1.$$

Equation (66) or (67) shows that if we know the two-dimensional representation of some rotation operator, we can find its axis immediately. Now

[15] If $\mathbf{v}$ and $\mathbf{u}$ are vector operators, as in Eq. (74), they must commute with each other and with $\boldsymbol{\sigma}$ in order for this calculation to apply.

consider a rotation whose Cayley-Klein parameters are α_0, $\boldsymbol{\alpha}$ followed by another whose Cayley-Klein parameters are β_0, $\boldsymbol{\beta}$. What are the Cayley-Klein parameters γ_0, $\boldsymbol{\gamma}$ of the resultant rotation? The two-dimensional representation of the resultant rotation is

$$\begin{aligned} X &= (\beta_0 \mathbb{1} - i\boldsymbol{\beta} \cdot \boldsymbol{\sigma})(\alpha_0 \mathbb{1} - i\boldsymbol{\alpha} \cdot \boldsymbol{\sigma}) \\ &= (\alpha_0\beta_0 - \boldsymbol{\alpha} \cdot \boldsymbol{\beta})\mathbb{1} - i(\beta_0\boldsymbol{\alpha} + \alpha_0\boldsymbol{\beta} + \boldsymbol{\beta} \times \boldsymbol{\alpha}) \cdot \boldsymbol{\sigma}, \end{aligned} \tag{68}$$

where we have used Eq. (65). Thus

$$\boldsymbol{\gamma} = \beta_0\boldsymbol{\alpha} + \alpha_0\boldsymbol{\beta} + \boldsymbol{\beta} \times \boldsymbol{\alpha},$$

which gives the axis of the new rotation (see Problem 24).

According to Eq. (62) the σ_α are all Hermitian. Therefore if $X = \alpha_0 \mathbb{1} - i\boldsymbol{\alpha} \cdot \boldsymbol{\sigma}$ then $X^\dagger = \alpha_0 \mathbb{1} + i\boldsymbol{\alpha} \cdot \boldsymbol{\sigma}$ and (68) shows that

$$XX^\dagger = (\alpha_0^2 + \boldsymbol{\alpha} \cdot \boldsymbol{\alpha})\mathbb{1} - i(\alpha_0\boldsymbol{\alpha} - \alpha_0\boldsymbol{\alpha} + \boldsymbol{\alpha} \times \boldsymbol{\alpha}) \cdot \boldsymbol{\sigma} = \mathbb{1}.$$

Thus every element of $D(G)$ is unitary. Moreover, its determinant is

$$\det \left(\left| \begin{matrix} \alpha_0 - i\alpha_3 & -i\alpha_1 + \alpha_2 \\ -i\alpha_1 - \alpha_2 & \alpha_0 + i\alpha_3 \end{matrix} \right| \right) = \alpha_0^2 + \alpha_3^2 + \alpha_2^2 + \alpha_1^2 = 1,$$

so that $D(G)$ is contained in the *S*pecial (determinant $= 1$) *U*nitary group in two dimensions over the *C*omplex numbers $SU(2, C)$. To show that $D(G)$ actually is $SU(2, C)$ is left as a problem for the reader.

Notice that when θ goes from 0 to 2π, $X(\mathbf{e}, \theta)$ goes from $\mathbb{1}$ to $-\mathbb{1}$. Thus the identity element of $SO(3, R)$ is represented by two matrices in $SU(2, C)$ and so, consequently is every other element of $SO(3, R)$. The representation is hence not faithful, or is only homomorphic. The Lie algebras of $SO(3, R)$ and $SU(2, C)$ are, however, isomorphic.

The spin-$\frac{1}{2}$ representation of the rotation group finds little application in classical mechanics. In relativistic quantum mechanics, however, it is intimately involved in the Dirac equation, the field equation for the electron and certain other particles, among them the mass-zero neutrino. Here we present a similar equation in nonrelativistic classical field theory.

We have mentioned that there exist other than covariant representations of the Lorentz group. The situation is similar for the rotation group. For example, under a rotation R a scalar transforms under the covariant spin-zero representation according to

$$\psi'(x, t) = \psi(R^{-1}x, t),$$

in analogy with Eq. (48) of Chapter VIII, where $x = (x_1, x_2, x_3)$. The components of a vector $\psi = (\psi_1, \psi_2, \psi_3)$ transform under the covariant spin-one representation [see Eq. (60) and Problem 22] according to

$$\psi'(x, t) = R\psi(R^{-1}x, t).$$

Corresponding to these are the scalar and vector analogs of Eq. (6) of Chapter VIII, namely

$$\mu \frac{\partial^2 \psi}{\partial t^2} - Y\nabla^2 \psi = 0, \tag{69}$$

where ψ may be either the scalar or any component of the vector.

We want now to deal with *spinor* fields, two-component fields that transform under rotation according to the noncovariant two-dimensional spin-$\frac{1}{2}$ representation. Thus we will write

$$\psi = \begin{pmatrix} \psi_1 \\ \psi_2 \end{pmatrix},$$

and require that under a rotation R the components of the field spinor ψ transform according to

$$\psi'(x, t) = X(R)\psi(R^{-1}x, t),$$

or as we shall write, simply

$$\psi' = X(R)\psi, \tag{70}$$

where $X(R)$ is the representative of R in $SU(2, C)$. In addition to ψ we shall define its *Hermitian adjoint* $\psi^\dagger$ as the *row* spinor

$$\psi^\dagger = (\psi^\dagger_1, \psi^\dagger_2),$$

where $\psi^\dagger_A = (\psi_A)^*$, $A = 1, 2$. Then a spinor and its adjoint can be combined by matrix multiplication to yield a number:

$$\psi^\dagger \psi = \sum_A \psi^\dagger_A \psi_A = \sum_A |\psi_A|^2.$$

It is left to the reader to show that under a rotation, $\psi^\dagger$ changes to $\psi^{\dagger\prime}$ not in accordance with (70), but given by

$$\psi^{\dagger\prime} = \psi^\dagger X^\dagger(R). \tag{71}$$

To obtain a spinor wave equation we form the Lagrangian density

$$\mathscr{L} = \frac{i}{2}\left(\sqrt{\mu}\,\psi^\dagger \frac{\partial \psi}{\partial t} + \sqrt{Y}\,\psi^\dagger \boldsymbol{\sigma} \cdot \nabla \psi\right) - \frac{i}{2}\left(\sqrt{\mu}\frac{\partial \psi^\dagger}{\partial t}\psi + \sqrt{Y}\,\nabla \psi^\dagger \cdot \boldsymbol{\sigma}\psi\right). \tag{72}$$

The second term is the Hermitian adjoint of the first, for $\boldsymbol{\sigma}^\dagger = \boldsymbol{\sigma}$ according to Eq. (62). The field equations obtained from this Lagrangian are

$$i\left(\sqrt{\mu}\frac{\partial \psi^\dagger}{\partial t} + \sqrt{Y}\,\nabla \psi^\dagger \cdot \boldsymbol{\sigma}\right) = 0 \tag{73a}$$

and

$$-i\left(\sqrt{\mu}\frac{\partial \psi}{\partial t} + \sqrt{Y}\,\boldsymbol{\sigma} \cdot \nabla \psi\right) = 0. \tag{73b}$$

We need deal only with the second of these, for the two are merely Hermitian adjoints.

To solve Eq. (73*b*) we first show that every solution is also a solution of the ordinary wave equation. Indeed, dividing by $-i$ and applying the operator $\sqrt{\mu}\,\partial/\partial t$, we have

$$\mu \frac{\partial^2 \psi}{\partial t^2} + \sqrt{Y}\boldsymbol{\sigma} \cdot \nabla \left(\sqrt{\mu}\frac{\partial \psi}{\partial t}\right) = 0.$$

If ψ is a solution of (73*b*), this becomes

$$\mu \frac{\partial^2 \psi}{\partial t^2} = Y(\boldsymbol{\sigma} \cdot \nabla)^2 \psi = Y\nabla^2 \psi, \tag{74}$$

where we have used Eq. (65). Thus each component of ψ satisfies Eq. (69). Given any spinor solution of (74), we can insert it into (73*b*) to obtain a relation between its two components.

For instance, consider a plane-wave solution propagating in the positive 3-direction. Then, as is well known, the solution of (74) is of the form $\psi = \psi(x_3 - vt)$, where $v = \sqrt{Y/\mu}$. If we write $x_3 - vt = y$, Eq. (73*b*) becomes

$$i\left(v\sqrt{\mu}\frac{d\psi}{dy} - \sqrt{Y}\,\sigma_3 \frac{d\psi}{dy}\right) = 0,$$

or

$$\frac{d\psi}{dy} = \sigma_3 \frac{d\psi}{dy}.$$

In other words, $d\psi/dy$ is an *eigenspinor* of σ_3 belonging to eigenvalue 1, and then a glance at Eq. (62) shows that $d\psi_2/dy = 0$. Integrating and setting the constant of integration equal to zero (why?), we obtain

$$\psi_1 = \psi_1(x_3 - vt), \qquad \psi_2 = 0.$$

which means that ψ itself is an eigenspinor of σ_3:

$$\sigma_3 \psi(x_3 - vt) = \psi(x_3 - vt). \tag{75}$$

Remark. If in the Lagrangian density and the field equations we had taken the negative root of either μ or Y, that is if $\sqrt{Y}$ had been replaced everywhere by $-\sqrt{Y}$, instead of (75) we would have had

$$\sigma_3 \psi(x_3 - vt) = -\psi(x_3 - vt).$$

Such a field is said to have *negative chirality*, and one satisfying (75) is said to have *positive chirality*.

There is nothing special about our choice of 3-direction, so that (75) implies that every plane wave traveling in the direction of $\mathbf{e}$ is an eigenspinor of $\mathbf{e} \cdot \boldsymbol{\sigma}$. In classical field theory this is a result difficult to interpret. In quantum mechanics it means that the 3-component of the spin angular momentum of such a plane wave is always positive. This is because in quantum mechanics it is not realizations but representations of Lie algebras that carry physical significance. Then it turns out that $i\, d_\alpha = \frac{1}{2}\sigma_\alpha$ is an operator which gives the angular momentum, and if $i\, d_3\psi = \frac{1}{2}\psi$, as is implied by Eq. (75), the system is in a state which has 3-component of angular momentum equal to $\frac{1}{2}$. This is a well-known result for mass-zero spin-$\frac{1}{2}$ particles (neutrinos).

(c) The Galilei Group

In this section we consider the group of Galilei transformations (Chapter III, Section 4) and two canonical realizations (see Section 3) of its Lie algebra. The general Galilei transformation $\Gamma(R, v, \xi, \tau)$ (in Chapter III we wrote D instead of R) takes a point x_1, x_2, x_3, t in space-time to another point x_1', x_2', x_3', t' given by

$$x' = Rx + vt + \xi,$$
$$t' = t + \tau,$$

where $R \in SO(3, R)$, ξ and v are fixed real three-vectors, and τ is a real constant. Then the group multiplication law is

$$\Gamma(R_2, v_2, \xi_2, \tau_2)\Gamma(R_1, v_1, \xi_1, \tau_1) = \Gamma(R_2R_1, R_2v_1 + v_2, R_2\xi_1 + \xi_2 + v_2\tau_1, \tau_2 + \tau_1). \quad (76)$$

The one-parameter subgroups are obvious: rotations about a fixed axis, transformations to frames moving in a fixed direction (*pure Galilei* transformation), displacements of the origin in a fixed direction, and time displacements. These correspond to fixed axes of rotation for R (three subgroups), fixed directions of v (three subgroups), fixed directions of ξ (three subgroups) and time displacements (one subgroup). Thus the Galilei group is ten-dimensional. The commutation relations for the Lie algebra are

$$\begin{aligned}
[M_\alpha, M_\beta] &= \epsilon_{\alpha\beta\gamma}M_\gamma, \\
[M_\alpha, P_\beta] &= \epsilon_{\alpha\beta\gamma}P_\gamma, \\
[M_\alpha, G_\beta] &= \epsilon_{\alpha\beta\gamma}G_\gamma, \\
[H, G_\alpha] &= -P_\alpha, \\
[M_\alpha, H] &= [P_\alpha, G_\beta] = [G_\alpha, G_\beta] = [P_\alpha, P_\beta] = [P_\alpha, H] = 0,
\end{aligned} \quad (77)$$

where the M_α is the infinitesimal generator of rotations about the α axis, G_α of pure Galilei transformations in the α direction, P_α of displacements of the origin in the α direction, and H of time displacements. The significance of the notation H and P_α should be obvious.

Let us now try to find a canonical realization of this Lie algebra on the space $\mathfrak{R}$ of functions $f(q, p)$, assuming three degrees of freedom, that is, $q = (q_1, q_2, q_3)$ and $p = (p_1, p_2, p_3)$. We start with the observation that the six-dimensional subspace spanned by the M_α and P_β forms a subalgebra, as is immediately evident from (77). This subalgebra clearly generates the group of translations and rotations, called the *Euclidean group in three dimensions*. Not surprisingly, since they are the infinitesimal generators of rotations, the M_α satisfy the commutation relations of the components of angular momentum. Let us therefore realize them by these components:

$$M_\alpha \rightarrow \epsilon_{\alpha\beta\gamma} q_\beta p_\gamma = l_\alpha. \tag{78}$$

According to Section 6(c) of Chapter VI, then, the P_α may be the components of any vector in $\mathfrak{R}$. Because they are the infinitesimal generators of translations, we try the realization

$$P_\alpha \rightarrow p_\alpha. \tag{79}$$

Then the first two of Eqs. (77) are satisfied, and the realization of $[P_\alpha, P_\beta]$ is zero, as it should be.

Now let us move on to the entire Galilei Lie algebra. Since H commutes with all the M_α, we take its realization to be a scalar function $h(q, p)$. But then the realization of $[P_\alpha, H]$ is

$$[p_\alpha, h]_{\text{P}} = -\partial h/\partial q_\alpha.$$

If this is to vanish, h must be a scalar function of p alone. The simplest such function is

$$H \rightarrow h = \kappa p_\alpha p_\alpha, \tag{80}$$

where κ is a constant. For the moment we put off the question of more complicated $h(p)$, and go on to the G_α. First, we take the g_α, realizations of the G_α, to form a vector in accordance with the third of Eqs. (77). Then the fourth tells us that

$$-p_\beta = \kappa[p_\alpha p_\alpha, g_\beta]_{\text{P}} = -2\kappa p_\alpha\, \partial g_\beta/\partial q_\alpha,$$

so that $\partial g_\beta/\partial q_\alpha = \frac{1}{2}\delta_{\alpha\beta}/\kappa$, or

$$g_\beta = \frac{q_\beta}{2\kappa} + f_\beta(p).$$

From $[g_\alpha, g_\beta]_{\text{P}} = 0$ we learn that $\partial f_\beta/\partial p_\alpha = \partial f_\alpha/\partial p_\beta$, so that there exists a function $F(p)$ such that $f_\alpha = \partial F/\partial p_\alpha$, thus

$$G_\beta \rightarrow g_\beta = \frac{q_\beta}{2\kappa} + \frac{\partial F}{\partial p_\beta}. \tag{81}$$

Only one of Eqs. (77) remains, namely $[P_\alpha, G_\beta] = 0$. We have, using (79) and (81),

$$0 = [p_\alpha, g_\beta]_\mathrm{P} = -\frac{1}{2\kappa}\,\delta_{\alpha\beta},$$

which is impossible with any finite κ. Yet we must accept the second equality, and we are thus forced to add to our realization of the Galilei group a nonnull center consisting of multiples of 1, as mentioned in the Remark following Eq. (35). In fact we are forced to do this precisely at the same point as in that Remark, in the bracket of q and p. Notice that F does not appear in this Poisson bracket. Again, we make the simplest choice, namely $F = 0$. Then (81) becomes

$$G_\beta \to g_\beta = \frac{q_\beta}{2\kappa}, \tag{82}$$

and

$$[p_\alpha, g_\beta]_\mathrm{P} = -\frac{1}{2\kappa}\,\delta_{\alpha\beta}, \tag{83}$$

This completes the realization. What is the dynamical system we have obtained? Obviously it is the free particle of mass $m = (2\kappa)^{-1}$. Notice that the realization is characterized by the value of m, and that $m \neq 0$. The infinitesimal generator of time-translations is the Hamiltonian, as it should be, and the infinitesimal generators of pure Galilei transformations, or "velocity translations," are the $g_\alpha = mq_\alpha$. The reader may check that this means that q_α is the infinitesimal generator of translations in p_α, as it should be. Not surprisingly, the Galilei group has led us to the free particle, as in Chapter III.

Remarks 1. When we chose the simplest h we lost really no generality. Any other function of $p_\alpha p_\alpha$ would still lead to the free particle with $\dot{p}_\alpha = 0$ and $\dot{q}_\alpha =$ const. But see Remark 4 below. *2.* Similarly, even if we had chosen $F \neq 0$ we would have ended up with the free particle. *3.* More general free-particle canonical realizations exist, corresponding to particles with finite spin, or intrinsic angular momentum. *4.* Our procedure has been somewhat arbitrary, and many choices were made for convenience and simplicity. However the result can be obtained in a more rigorous way; the interested reader is referred to the literature (Pauri, 1968).

The representation we have obtained is up to a factor, and in fact it is known that in general this cannot be avoided for canonical realizations of the Galilei group in three degrees of freedom. A nonfactor realization in six degrees of freedom is the following.[16] Let q and p carry two indices $q_{i\alpha}$ and $p_{i\alpha}$, where $i = 1, 2$, $\alpha = 1, 2, 3$ (you may think of two particles in three

[16] This is taken from G. Sudarshan (1963).

dimensions). Then, using the summation convention on both indices, write

$$M_\alpha \to \epsilon_{\alpha\beta\gamma} q_{i\beta} p_{i\gamma},$$
$$P_\alpha \to p_{1\alpha},$$
$$G_\alpha \to q_{2\alpha},$$
$$H \to p_{1\alpha} p_{2\alpha}.$$

To verify that this is a realization is left to the reader. It can be interpreted as a dynamical system, but rather artificially. Again, the interested reader is referred to the literature.

(d) Some Remarks about the Transition to Quantum Mechanics

Quantum mechanical equations for particles, such as those we came across in Chapter VIII, are equations for a field function $\psi(\mathbf{x}, t)$ from which one derives physical information about the system. Without going into detail, we mention several properties of these functions.

The ψ functions lie in an infinite-dimensional unitary space (a *Hilbert space*) and are all of unit norm. Thus as time progresses and ψ changes, its norm remains fixed, and therefore we may write (suppressing the $\mathbf{x}$ dependence)

$$\psi(t) = U(t)\psi_0, \tag{84}$$

where $\psi_0 = \psi(0)$ and $U(t)$ is a unitary time-translation operator on the Hilbert space. To say that U is a time-translation operator is to say that it will give the time development of the system starting at any time, and therefore that

$$U(t_2 + t_1) = U(t_2)U(t_1).$$

This and the fact that U is a unitary implies for simple enough systems[17] that there exists a Hermitian operator H on the Hilbert space such that (we take Planck's constant $\hbar = 1$)

$$U = \exp(-itH). \tag{85}$$

It then follows that $\psi(t)$ satisfies the differential-operator equation

$$\frac{\partial \psi}{\partial t} = -iH\psi. \tag{86}$$

Physical information is obtained by applying time-independent operators to $\psi(t)$. Let F_0 be such an operator, and let us write

$$\varphi(t) = F_0\psi(t), \tag{87}$$

[17] The system must be such that its classical Hamiltonian is time independent.

where $\varphi(t)$ is also in the Hilbert space. The physical information provided by Eq. (87) can be obtained also by using the time-independent functions φ_0 and ψ_0 and shifting the time dependence to the operator F. From Eq. (84) we obtain

$$\varphi_0 = \exp(itH)F_0 \exp(-itH)\psi_0 = F(t)\psi_0, \tag{88}$$

defining $F(t)$. It is a simple matter to calculate dF/dt:

$$\dot{F} = -i(FH - HF) = -i[F, H], \tag{89}$$

where the bracket is the commutator.

Now $\dot{F}(t)$ itself is an operator on our Hilbert space, and just as (87) or (88) provides information about whatever physical property F represents, so $\dot{F}(t)\psi_0$ provides information about the time derivative of that physical property. We may then ask for the time-independent form of $\dot{F}(t)$, which, according to (88) and (89) is

$$\begin{aligned} \dot{F}_0 &= \exp(-itH)\dot{F}\exp(itH) \\ &= -i(F_0H - HF_0) = -i[F_0, H], \end{aligned} \tag{90}$$

where we have used the fact that H commutes with $\exp(itH)$, so that $H_0 = H(t) = H$. Note that $\dot{F}_0$ is not itself the time derivative of F_0, for F_0 is time independent. Nevertheless $\dot{F}_0\psi(t)$ provides the same information as does $\dot{F}(t)\psi_0$, namely, information about the time derivative of whatever property $F(t)$ (or F_0) represents. In this sense $\dot{F}_0$ gives the time rate of change of F_0.

Let us now assume that among the operators on our Hilbert space are the *position operator* $\mathbf{x}$ (corresponding, as usual, to the q's) and the *momentum operator* $\mathbf{p}$, which by their action on $\psi(t)$ provide information about the position and momentum of the particle. Then according to Eq. (90), the operator which corresponds to the time rate of change of the α-component of position is

$$\dot{x}_\alpha = -i[x_\alpha, H], \tag{91a}$$

and similarly for momentum

$$\dot{p}_\alpha = -i[p_\alpha, H]. \tag{91b}$$

Equations (91) look very much like Hamilton's canonical equations if we interpret H as a *Hamiltonian operator* and $-i$ times the commutator as the Poisson bracket. In fact this is essentially what one does in making the transition from classical to quantum mechanics: the q_α and p_α are thought of as operators on Hilbert space, other dynamical variables $f(q, p)$ as functions of these operators, and Poisson brackets are replaced by $-i$ times the commutator bracket.

Now we see that quantum mechanics may be thought of as a representation, by linear operators on Hilbert space, of the infinite dimensional Lie algebra $\mathfrak{K}$ (consisting of functions of p and q with the Poisson bracket) which we discussed in Sections 2 and 3. In particular, the operators representing position and momentum have the commutation relations

$$[x_\alpha, p_\beta] = i\,\delta_{\alpha\beta}, \tag{92}$$

and the commutators of all other functions of these operators can be deduced from these relations. Let us assume that the operator x_α changes the wave function $\psi(\mathbf{x}, t)$ to $x_\alpha\psi(\mathbf{x}, t)$. Then the discussion above Eq. (23) shows that we may take p_α to be a multiple of $\partial/\partial x_\alpha$, and in order to obtain the correct factor of i in (92), we must choose

$$p_\alpha = -i\,\partial/\partial x_\alpha. \tag{93}$$

Finally, according to our identification of operators with functions in $\mathfrak{K}$, the Hamiltonian operator of (91), and hence also of (86), is obtained from the Hamiltonian function by interpreting the p_α everywhere in accordance with (93). Then Eq. (86) becomes a differential equation, the Schrödinger equation, for the wave function $\psi(\mathbf{x}, t)$. Its solutions together with the operators of the representation of $\mathfrak{K}$ provide all the physical information.

This simple view of the relation between classical and quantum mechanics may leave one with the impression that they are equivalent. Actually, of course, they are not. The difference, from our point of view, lies mainly in the multiplicative structure of $\mathfrak{K}$, which is mentioned at Eq. (44). In quantum mechanics the product of two dynamical variables depends on the order in which they are multiplied, and the difference between the two products is just the commutator bracket. Thus the quantum commutator bracket is involved in both the Lie algebraic and the multiplicative structures, whereas the classical Poisson bracket is involved in only the first. Thus, in spite of their Lie algebraic similarities, quantum and classical mechanics are far from equivalent. If this were not the case, in fact, there would be no need for quantum mechanics. Nevertheless they are also deeply related, and an understanding of this relationship will aid in an understanding of quantum mechanics, the natural next step in the study of dynamical systems.

Problems

1. Prove the uniqueness of the identity element and of the inverse from properties G1–G4.
2. Check whether the following are or are not subgroups of $GL(3, R)$. (a) The set of all rotations. (b) The three transformations that take every vector $\mathbf{x}$ into $\mathbf{e}_i \times \mathbf{x}$, where the $\mathbf{e}_i$ are three fixed orthonormal basis

vectors. (c) (Let A be

$$A = \begin{vmatrix} 1 & 0 & 0 \\ 0 & a & b \\ 0 & -b & a \end{vmatrix}, \qquad a^2 + b^2 = 1$$

for fixed a, b.) The set of all A^n, $n = 0, \pm 1, \pm 2, \ldots$. (If this is a subgroup, is it finite, infinite, discrete, continuous? For what values of a, b do the A^n, $n = 0, 1, 2, \ldots$ form a subgroup?) (d) The set of six matrices

$$e = \begin{vmatrix} 1 & 0 & 0 \\ 0 & 1 & 0 \\ 0 & 0 & 1 \end{vmatrix}, \quad a = \begin{vmatrix} 1 & 0 & 0 \\ 0 & -1 & 0 \\ 0 & 0 & 1 \end{vmatrix}, \quad b = \tfrac{1}{2}\begin{vmatrix} -1 & -\sqrt{3} & 0 \\ \sqrt{3} & -1 & 0 \\ 0 & 0 & 2 \end{vmatrix},$$

$$c = \tfrac{1}{2}\begin{vmatrix} -1 & \sqrt{3} & 0 \\ -\sqrt{3} & -1 & 0 \\ 0 & 0 & 2 \end{vmatrix}, \quad d = \tfrac{1}{2}\begin{vmatrix} -1 & \sqrt{3} & 0 \\ \sqrt{3} & 1 & 0 \\ 0 & 0 & 2 \end{vmatrix},$$

$$f = \tfrac{1}{2}\begin{vmatrix} -1 & -\sqrt{3} & 0 \\ -\sqrt{3} & 1 & 0 \\ 0 & 0 & 2 \end{vmatrix}.$$

(e) All operators in $GL(3, R)$ with positive determinant. (f) All operators in $GL(3, R)$ with negative determinant.

3. Show that the groups $\{1, -1, i, -i$ under ordinary multiplication$\}$ and

$$\left\{\begin{vmatrix} 1 & 0 \\ 0 & 1 \end{vmatrix}, \begin{vmatrix} -1 & 0 \\ 0 & -1 \end{vmatrix}, \begin{vmatrix} 0 & -1 \\ 1 & 0 \end{vmatrix}, \begin{vmatrix} 0 & 1 \\ -1 & 0 \end{vmatrix} \text{ under matrix multiplication}\right\}$$

are isomorphic. Find two isomorphisms between them.

4. Let $GL^+(n, R)$ be the subgroup of all elements of $GL(n, R)$ with positive determinant [see Problem 2(e)]. Let $\Delta(x)$ be the determinant of $x \in GL(n, R)$ and form the operator $x' = [\Delta(x)]^{-1/n}x$ for $x \in GL^+(n, R)$, taking the positive root. Find $\Delta(x')$. The set of all such x' is called the *S*pecial *L*inear group $SL(n, R)$. Show that the mapping $\varphi(x) = x'$ from $GL^+(n, R)$ onto $SL(n, R)$ is a homomorphism. How many elements of $GL^+(n, R)$ are mapped by φ into a given element of $SL(n, R)$?

5. Prove that the mapping of Eq. (1), namely $\varphi(x) = \Delta(x)x$ is from $GO(3, R)$ to $SO(3, R)$ and is a homomorphism. How many elements of $GO(3, R)$ are mapped by φ into a given element of $SO(3, R)$?

6. Let φ be a homomorphism from G_1 to G_2. Show that $\varphi(e_1) = e_2$, where e_i is the identity element of G_i. If $\varphi(x_1) = x_2$ show that $\varphi(x_1^{-1}) = x_2^{-1}$. Show that all $a_1 \in G_1$ such that $\varphi(a_1) = e_2$ form a subgroup $K_1 \subset G_1$. Prove that $a_1 \in K_1$ implies $x_1 a_1 x_1^{-1} \in K_1$ for all $x_1 \in G_1$ (a subgroup with this property is called an *invariant* or *normal* subgroup).

7. Find the dimensions of $SO(2, R)$, $SL(2, R)$, and $SU(3, C)$ (*S*pecial [determinant $= 1$], *U*nitary, *C*omplex numbers).

8. Suppose you were given the second of Eqs. (3) as a definition of a one-parameter subgroup. Use (6) to find the reparametrization that brings it

into the standard form of the first of Eqs. (3). You will not quite get $\varepsilon' = \sin \varepsilon$. Discuss the ambiguity and the meaning of $[d\varepsilon'/d\varepsilon]_0 = k$ in (6). What if $k = 0$?

9. Prove that $[a, b]$, as defined in Eq. (12) satisfies the algebraic properties A2–A4 at Eq. (17). State A2 and A4 in terms of structure constants.
10. Since every matrix in $GL(2, R)$ has four elements, this is a four-dimensional group. Consider the four curves

$$x_1(\varepsilon) = \begin{vmatrix} 1+\varepsilon & 0 \\ 0 & 1 \end{vmatrix}, \qquad x_2(\varepsilon) = \begin{vmatrix} 1 & 0 \\ 0 & 1+\varepsilon \end{vmatrix}$$

$$x_3(\varepsilon) = \begin{vmatrix} 1 & \varepsilon \\ 0 & 1 \end{vmatrix}, \qquad x_4(\varepsilon) = \begin{vmatrix} 1 & 0 \\ \varepsilon & 1 \end{vmatrix}.$$

Which of these are one-parameter subgroups in standard form? Find all four tangent vectors a_α and show that $\mathfrak{G}$ contains all 2×2 matrices. (Show that the a_α you found form a basis in $\mathfrak{G}$.) Find the structure constants in the a_α basis. Transform to the basis (is it a basis?)

$$j_0 = \begin{vmatrix} 1 & 0 \\ 0 & 1 \end{vmatrix}, \qquad j_1 = \begin{vmatrix} 1 & 0 \\ 0 & -1 \end{vmatrix}, \qquad j_2 = \begin{vmatrix} 0 & 1 \\ 1 & 0 \end{vmatrix}, \qquad j_3 = \begin{vmatrix} 0 & 1 \\ -1 & 0 \end{vmatrix}.$$

Find the new structure constants. Calculate the subgroups $\exp(\varepsilon a_\alpha)$ and $\exp(\varepsilon j_\alpha)$.

11. Let the Lie algebra of Problem 10 be called $gl(2, R)$. (a) Find two non-isomorphic two-dimensional subalgebras of $gl(2, R)$. (b) Let $\mathfrak{G}$ be the three-dimensional Lie algebra defined by $[a, b] = c$, $[b, c] = a$, $[c, a] = -b$. Find a homomorphism from $gl(2, R)$ to $\mathfrak{G}$.
12. Use the procedure of Eqs. (27) to (29) to construct a representation of G from the $\mathfrak{G}$ [not $gl(2, R)$] of Problem 11. Let $\mathfrak{G}$ coincide with $\mathfrak{K}$ in this case. [You should obtain the Lorentz group in one time and two space dimensions: b generates the rotations in space, and a and c generate "pure" Lorentz translations in mutually perpendicular directions.]
13. According to Eqs. (21), (19), and (10),

$$\exp\left(\varepsilon \frac{d}{dq}\right) \cdot f(q) = f(q + \varepsilon).$$

Check this and show that it is just Taylor's formula. Show that (42) is just Taylor's formula.

14. Specialize Eq. (36) to $\mathfrak{K}$ being the space of all functions $f(q, p)$ with the Poisson bracket. Show that (36) then states simply that $X(\varepsilon, a)$ is a canonical transformation. Expand both sides of (36) in powers of ε and show explicitly that the coefficients of these powers agree up to terms in ε^2.
15. Solve the problem of a freely falling body using the technique of Eq. (42).
16. Solve for the motion in Problem 2 of Chapter VI using the technique of Eq. (42). (Sum the series obtained.)

17. Suppose $d(\mathfrak{G})$ with a certain bracket $[\ \ ]_R$ is a realization of $\mathfrak{G}$. Show that $d^-(\mathfrak{G})$ is also a realization, where $d^-(\mathfrak{G})$ differs from $d(\mathfrak{G})$ only in that $[\ \ ]_R$ is replaced by $[\ \ ]_{-R} = -[\ \ ]_R$. This is to make sure that the minus sign in (38) and (39) is legitimate.
18. Consider the plane pendulum, whose Hamiltonian is

$$H = \frac{p^2}{2ml^2} - mgl\cos\theta,$$

and let it start *from rest* at some value of $\theta_0 = \theta(0) < \pi$. Use Eq. (42) to develop a power series in t for $\theta(t)$, and carry it out to terms of order t^6. Check that in the limit $\sin\theta_0 \to \theta_0$, $\cos\theta_0 \to 1$ this goes to $\theta = \theta_0 \cos[(g/l)^{1/2}t]$.
19. Show that $H = \frac{1}{2}N_{\alpha\beta}l_\alpha l_\beta$ is the Hamiltonian for a torque-free rigid body in the body system. Write out Eq. (46) in vector notation and show that it agrees with Eq. (23) of Chapter V with $n = 0$.
20. Consider a nonconducting charged rigid sphere in a uniform magnetic field $\mathbf{B}$. Show that in a space system its Hamiltonian is

$$H = \tfrac{1}{2}Nl^2 + \mu B_\alpha l_\alpha,$$

where μ depends on the charge distribution. Solve for the $l_\alpha(t)$, and show that the angular momentum has constant magnitude and precesses about $\mathbf{B}$ with constant angular velocity $\omega = \mu B$.
21. Let $D(G)$ be an irreducible representation of G, and let C be a matrix which commutes with all elements of $D(G)$. Let x be any eigenvector of C belonging to eigenvalue c: $Cx = cx$. Show that the subspace consisting of all eigenvectors of C belonging to eigenvalue c is the entire space, and thus that $C = c\mathbb{1}$. This is part of Schurr's Lemma.
22. Find a unitary 3×3 matrix U such that $Ud_\alpha U^\dagger = a_\alpha$, where the d_α are given by Eq. (60) and the a_α by (47). Exponentiate the d_α and show that $U\{\exp(\varepsilon d_\alpha)\}U^\dagger$ is in the usual form.
23. Let $\mathbf{e}$ be any unit vector in 3-space. Show that, in the three-dimensional representation, $D(\mathbf{e}, \theta) = \exp(\theta e_\alpha a_\alpha)$ is a rotation about the axis along $\mathbf{e}$ through the angle θ, where the a_α are given by Eq. (47). [*Hint:* Apply $e_\alpha a_\alpha$, rather than $D(\mathbf{e}, \theta)$, to a vector parallel to $\mathbf{e}$ and to another perpendicular to $\mathbf{e}$.]
24. Show that

$$\frac{i}{2}\operatorname{tr}(\sigma_\alpha D) = e_\alpha \sin\frac{\theta}{2},$$

where $D = D(\mathbf{e}, \theta)$ is the two-dimensional representation of a rotation about the axis along $\mathbf{e}$ through the angle θ, thus showing how to calculate the axis and angle of a rotation when given its two-dimensional representation.
25. With every real three-vector $\mathbf{v}$ we may associate the trace-zero Hermitian 2×2 matrix

$$V = \begin{vmatrix} v_3 & v_1 - iv_2 \\ v_1 + iv_2 & -v_3 \end{vmatrix} = \mathbf{v} \cdot \boldsymbol{\sigma}.$$

Prove that this association is one-to-one. Let $U \in SU(2, C)$ and show that $V' = UVU^\dagger$ is Hermitian and has trace zero, and is associated with a vector $\mathbf{v}'$ obtained from $\mathbf{v}$ by rotation. [*Hint*: How do you find the magnitude of $\mathbf{v}$ when you know V?] Let us write $\mathbf{v}' = \mathbf{R}\mathbf{v}$ and $U = U(\mathbf{R})$, where $\mathbf{R}$ is the rotation operator. Show that $U(\mathbf{R}_1)U(\mathbf{R}_2) = U(\mathbf{R}_1\mathbf{R}_2)$, so that $SU(2, C)$ forms a two-dimensional representation of $SO(3, R)$. Complete the argument to prove that the spin-representation we used in Section 4(b) is in fact all of $SU(2, C)$. Check this result explicitly by using $X(\mathbf{e}, \theta)$ as given by Eq. (66) to calculate $V' = XVX^\dagger$ and seeing that the vector $\mathbf{v}$ associated with V is properly transformed. [*Hint:* Calculate $dV'/d\theta$ at $\theta = 0$.]

26. If ψ transforms according to (70) under rotations and $\mathbf{v}$ is a vector, show that $\psi^\dagger \boldsymbol{\sigma} \cdot \mathbf{v} \psi$ is a scalar, that is, that in any two coordinate systems obtained from each other by rotation

$$\psi^{\dagger\prime} \sigma_\alpha v'_\alpha \psi' = \psi^\dagger \sigma_\alpha v_\alpha \psi$$

where v_α and v'_α are the components of $\mathbf{v}$ in the two systems. Show therefore that the Lagrangian density of Eq. (72) is invariant under rotations. [*Hint:* Some of the calculations of Problem 25 may help.]

27. Show explicitly that those solutions of Eq. (73*b*) which represent plane waves traveling in the 2-direction satisfy $\sigma_2\psi = \psi$.
28. Derive Eqs. (76) and (77).
29. Derive the quantum-mechanical expression for the angular momentum of a particle, and show that its components satisfy the correct commutation relations. Show explicitly that l_3 has the proper commutation relations with the azimuth angle φ.
30. Write down the Schrödinger equation for a free particle of mass m in one dimension. Solve it by assuming a solution of the form $\psi(x, t) = X(x)T(t)$. Show that ψ is then an eigenfunction of H and p (this is interpreted to mean that the particle has well-defined energy and momentum equal to the respective eigenvalues). Show that $\dot{x} = p/m$ and that $\dot{p} = 0$.
31. (a) By studying its action on an arbitrary wave function $\psi(\mathbf{x}, t)$, verify that ip_α as given by (93) generates translation in x_α. (b) By studying its action on a solution of Problem 30, verify that a multiple (what multiple?) of x_α generates translations in p_α.
32. Consider Eqs. (67) of Chapter VI, interpreted quantum mechanically. (a) Show that if $[q, p] = i$, then $[Q, P] = i$. Write down $H = \frac{1}{2}(q^2 + p^2)$ in terms of Q and P and use the commutation relations to put Q and P in the same order (i.e., Q to the left of P or vice versa) in each term. (b) Show that if ψ is an eigenfunction of H belonging to eigenvalue E, then $Q\psi$ is an eigenfunction belonging to eigenvalue $E - 1$, and $P\psi$ is an eigenfunction belonging to eigenvalue $E + 1$. (c) Assuming that every eigenvalue E of H (that is, the energy) is positive, find the lowest possible value of E. (d) Solve the Schrödinger equation for $\psi(Q, t)$ in the representation $P = -i\partial/\partial Q$, separating variables as in Problem 30. Verify part (b) explicitly.

APPENDIX A

Vectors in 3-Space

Although we expect the reader to be familiar with the properties of vectors in 3-space, we have summarized some of the important ideas about them.

Fields. A field will be, for our purposes, a point function in 3-space. A field may be a *scalar* or a *vector* or even something else. Being a function, a field is a map from its *domain*, the points of 3-space, to its *range*. The range of a scalar field ϕ is the set of numbers (real or complex). The range of a vector field $\mathbf{A}$ is the set of vectors in 3-space.

Let a point in space have coordinates $x = (x_1, x_2, x_3)$. As usual, we write $\mathbf{x} = x_1\mathbf{e}_1 + x_2\mathbf{e}_2 + x_3\mathbf{e}_3$, where the *basis vector* $\mathbf{e}_\alpha$ is the unit vector in the αth coordinate direction, $\alpha = 1, 2, 3$. With the summation convention (see Chapter II) this may be written

$$\mathbf{x} = x_\alpha \mathbf{e}_\alpha.$$

A scalar field may be written in several equivalent ways:

$$\phi = \phi(x_1, x_2, x_3) = \phi(\mathbf{x}) = \phi(x).$$

In the last form, we use x to stand for the three coordinates x_1, x_2, x_3. Similarly, a vector field may be written

$$A_\alpha = A_\alpha(x_1, x_2, x_3) = A_\alpha(\mathbf{x}) = A_\alpha(x),$$

where A_α is the αth component of $\mathbf{A}$, so that

$$\mathbf{A} = A_\alpha \mathbf{e}_\alpha.$$

The three component equations defining the field can then be combined, for instance, in the form $\mathbf{A} = \mathbf{A}(x_1, x_2, x_3)$. Often, in fact most of the time, we shall call vector and scalar fields merely vectors and scalars.

Products of Vectors. We define $\mathbf{A} \cdot \mathbf{B}$ the *dot product*, *scalar product*, or *inner product* of every pair of vectors $\mathbf{A}$ and $\mathbf{B}$ as follows. For the coordinate vectors

$$\mathbf{e}_\alpha \cdot \mathbf{e}_\beta = \delta_{\alpha\beta}, \tag{A1}$$

where $\delta_{\alpha\beta}$, called the Kronecker delta symbol, is simply shorthand for "zero if $\alpha \neq \beta$, and 1 if $\alpha = \beta$." For any other pair of vectors, the dot product is given by linearity:

$$\mathbf{A} \cdot (\beta \mathbf{B} + \gamma \mathbf{C}) = \beta \mathbf{A} \cdot \mathbf{B} + \gamma \mathbf{A} \cdot \mathbf{C}, \tag{A2}$$

where β and γ are numbers.

It follows that (remember we are using the summation convention)

$$\mathbf{A} \cdot \mathbf{B} = A_\alpha B_\alpha = AB \cos \theta, \tag{A3}$$

where $A = (A_1^2 + A_2^2 + A_3^2)^{1/2}$ is the *magnitude* or *norm* of **A**, similarly for **B**, and θ is the angle between **A** and **B**. *Note:* The dot product of two vector fields is a scalar field.

We define **A** × **B**, the *cross product* of every pair of vectors, as follows. For the coordinate vectors

$$\mathbf{e}_\alpha \times \mathbf{e}_\beta = \epsilon_{\alpha\beta\gamma} \mathbf{e}_\gamma, \tag{A4}$$

where $\epsilon_{\alpha\beta\gamma}$ is simply shorthand for "zero if α, β, γ are not all different (i.e., if there is any repetition), 1 if α, β, γ is an *even* or *cyclic permutation* of 1, 2, 3 (i.e., if it is 1, 2, 3 or 2, 3, 1 or 3, 1, 2 in that order), and -1 if α, β, γ is an *odd permutation* of 1, 2, 3 (i.e., what is left)." For any other pair of vectors, the cross product is given by

$$\mathbf{A} \times (\beta \mathbf{B} + \gamma \mathbf{C}) = \beta \mathbf{A} \times \mathbf{B} + \gamma \mathbf{A} \times \mathbf{C}, \tag{A5}$$

where β and γ are numbers. It follows that

$$(\mathbf{A} \times \mathbf{B})_\alpha = -(\mathbf{B} \times \mathbf{A})_\alpha = \epsilon_{\alpha\beta\gamma} A_\beta B_\gamma \tag{A6}$$

or

$$\mathbf{A} \times \mathbf{B} = -\mathbf{B} \times \mathbf{A} = \mathbf{e}_\alpha \epsilon_{\alpha\beta\gamma} A_\beta B_\gamma,$$

and

$$|\mathbf{A} \times \mathbf{B}| = AB \sin \theta, \tag{A7}$$

where $|\mathbf{A} \times \mathbf{B}|$ is the magnitude of **A** × **B**. *Note:* the cross product of two vector field is a vector field (almost!—see Problem 3, Chapter V). The reader should prove that

$$\epsilon_{\alpha\beta\gamma} \epsilon_{\alpha\mu\nu} = \delta_{\beta\mu}\, \delta_{\gamma\nu} - \delta_{\gamma\mu}\, \delta_{\beta\nu}. \tag{A8}$$

This is useful in proving relations involving two or more cross products. By using it the reader can prove the second of the following two identities:

$$\mathbf{C} \cdot (\mathbf{A} \times \mathbf{B}) = \mathbf{A} \cdot (\mathbf{B} \times \mathbf{C}) = \mathbf{B} \cdot (\mathbf{C} \times \mathbf{A}); \tag{A9}$$

$$\mathbf{A} \times (\mathbf{B} \times \mathbf{C}) = (\mathbf{C} \times \mathbf{B}) \times \mathbf{A} = (\mathbf{A} \cdot \mathbf{C})\mathbf{B} - (\mathbf{A} \cdot \mathbf{B})\mathbf{C}. \tag{A10}$$

DIFFERENTIAL CALCULUS

There should be no problem about the (partial) derivatives of a scalar field. We may define the derivative of a vector with respect to a parameter t

(which may be one of the coordinates) either by

$$\frac{\partial \mathbf{A}}{\partial t} = \lim_{\Delta t \to 0} \frac{1}{\Delta t} [\mathbf{A}(t + \Delta t) - \mathbf{A}(t)]$$

or by (Cartesian coordinates only)

$$\left(\frac{\partial \mathbf{A}}{\partial t}\right)_\alpha = \frac{\partial A_\alpha}{\partial t}.$$

(The reader should prove the equivalence of these two definitions.)

We further define the vector operator ∇ (called *del* or *nabla*) in Cartesian coordinates by

$$\nabla = \mathbf{e}_\alpha \frac{\partial}{\partial x_\alpha}. \tag{A11}$$

By operating with ∇ on fields, we can form new vector and scalar fields. For instance if ϕ is a scalar, $\nabla\phi = \mathbf{e}_\alpha\, \partial\phi/\partial x_\alpha$ is a vector, called the *gradient* of ϕ. Similarly, if $\mathbf{A}$ is a vector, $\nabla \cdot \mathbf{A} = \partial A_\alpha/\partial x_\alpha$ (summation convention) is a scalar (called the *divergence* of $\mathbf{A}$), and $\nabla \times \mathbf{A} = \mathbf{e}_\alpha \epsilon_{\alpha\beta\gamma}\, \partial A_\gamma/\partial x_\beta$ is a vector (almost!—see Problem 3, Chapter V) called the *curl* or *rotation* of $\mathbf{A}$.

Since ∇ is a sum of partial derivatives, it satisfies most of the rules of differentiation. For instance, in an obvious way $\nabla(\phi\psi) = \phi\nabla\psi + \psi\nabla\phi$, where ϕ and ψ are scalars. Perhaps less obviously $\nabla \cdot (\mathbf{A}\phi) = \mathbf{A} \cdot \nabla\phi + \phi\nabla \cdot \mathbf{A}$, and $\nabla \times (\mathbf{A}\phi) = -\mathbf{A} \times \nabla\phi + \phi\nabla \times \mathbf{A}$.

The reader should prove the following identities:

$$\nabla(\mathbf{A} \cdot \mathbf{B}) = (\mathbf{A} \cdot \nabla)\mathbf{B} + (\mathbf{B} \cdot \nabla)\mathbf{A} + \mathbf{A} \times (\nabla \times \mathbf{B}) + \mathbf{B} \times (\nabla \times \mathbf{A}), \tag{A12}$$

where $\mathbf{A} \cdot \nabla = A_\alpha\, \partial/\partial x_\alpha$,

$$\nabla \cdot (\mathbf{A} \times \mathbf{B}) = \mathbf{B} \cdot (\nabla \times \mathbf{A}) - \mathbf{A} \cdot (\nabla \times \mathbf{B}), \tag{A13}$$

$$\nabla \times (\mathbf{A} \times \mathbf{B}) = \mathbf{A}(\nabla \cdot \mathbf{B}) - \mathbf{B}(\nabla \cdot \mathbf{A}) + \mathbf{A} \times (\nabla \times \mathbf{B}) + \mathbf{B} \times (\nabla \times \mathbf{A}), \tag{A14}$$

$$\nabla \times \nabla\phi = 0, \qquad \nabla \cdot (\nabla \times \mathbf{A}) = 0, \tag{A15}$$

$$\nabla \times (\nabla \times \mathbf{A}) = \nabla(\nabla \cdot \mathbf{A}) + (\nabla \cdot \nabla)\mathbf{A}. \tag{A16}$$

Remark. This last formula does not extend simply to curvilinear coordinates.

We write ∇^2 for

$$\nabla^2 \equiv \nabla \cdot \nabla = \frac{\partial^2}{\partial x_\alpha\, \partial x_\alpha} = \frac{\partial^2}{\partial x_1^2} + \frac{\partial^2}{\partial x_2^2} + \frac{\partial^2}{\partial x_3^2}, \tag{A17}$$

which is a scalar operator. For instance, the expression $\nabla^2\phi = \nabla \cdot \nabla\phi$ is a scalar.

INTEGRAL CALCULUS

There should be no problem about the (triple) integral of a scalar over a bounded volume V. Let the smooth closed surface bounding the volume be called ∂V. Then it can be shown (*Gauss's divergence theorem*) that

$$\int_V \nabla \cdot \mathbf{A}\, d^3x = \oint_{\partial V} \mathbf{A} \cdot dS. \qquad \text{(A18)}$$

Here d^3x is the element of volume (say $dx\, dy\, dz$), and $\int$ represents a triple integral over V. The *surface integral* on the right-hand side is defined as follows. Approximate the surface ∂V by a finite number n of small planes, each of area ΔS_i, $i = 1, \ldots, n$, each plane tangent to ∂V at some point $\mathbf{x}_i$. Define the area vector $\Delta \mathbf{S}_i$ at $\mathbf{x}_i$ as having magnitude ΔS_i and pointing in the outward normal direction (out of V and perpendicular to ∂V). Let $\mathbf{A}_i = \mathbf{A}(\mathbf{x}_i)$ and form the sum

$$\sum_{i=1}^{n} \mathbf{A}_i \cdot \Delta \mathbf{S}_i.$$

Then the integral is defined as

$$\int_{\partial V} \mathbf{A} \cdot d\mathbf{S} = \lim_{\substack{\Delta S i \to 0 \\ n \to \infty}} \sum \mathbf{A}_i \cdot \Delta \mathbf{S}_i,$$

provided this limit exists and is independent of the choice of $\mathbf{x}_i$. *Note:* The divergence theorem is sometimes taken as the definition of the divergence.

The divergence theorem involves only scalars formed out of vectors, such as $\nabla \cdot \mathbf{A}$. This is partly a reflection of the fact that integrals of vectors are somewhat more complicated. We will seldom have to deal with them, but when we do, we will stick to Cartesian coordinates, in which all is relatively simple: $(\int \mathbf{A}\, d^3x)_\alpha \equiv \int A_\alpha\, d^3x$.

Another scalar integral theorem involving vectors is Stokes's theorem. Let Σ be a smooth surface bounded by a smooth closed curve $\partial\Sigma$. Then Stokes's theorem is

$$\int_\Sigma (\nabla \times \mathbf{A}) \cdot d\mathbf{S} = \oint_{\partial\Sigma} \mathbf{A} \cdot d\mathbf{l}. \qquad \text{(A19)}$$

The integral on the left-hand side is defined by a limiting process similar to the one we used for Gauss's theorem; the only difference is that Σ is not a closed surface. The *line integral* on the right-hand side is defined as follows. Approximate the curve $\partial\Sigma$ by a finite number n of directed line segments $\Delta \mathbf{l}_i$, $i = 1, \ldots, n$, each one tangent to $\partial\Sigma$ at some point $\mathbf{x}_i$. The direction of the $\Delta \mathbf{l}_i$ is chosen according to the direction of $d\mathbf{S}$. When the fingers of the right hand are laid along $\Delta \mathbf{l}_i$, the thumb points in the direction of $d\mathbf{S}$ (like the relation between a magnetic field **B** and the current **i** producing it). Let $\mathbf{A}_i = \mathbf{A}(\mathbf{x}_i)$ and form the sum

$$\sum_{i=1}^{n} \mathbf{A}_i \cdot \Delta \mathbf{l}_i.$$

Then the integral is defined as

$$\oint_{\partial \Sigma} \mathbf{A} \cdot d\mathbf{l} = \lim_{\substack{\Delta l_i \to 0 \\ n \to \infty}} \sum \mathbf{A}_i \cdot \Delta \mathbf{l}_i,$$

provided this limit exists and is independent of the choice of $\mathbf{x}_i$. *Note:* Stokes's theorem is sometimes taken as the definition of the curl.

These theorems tell us that we can know something, albeit not much, about a field inside a volume (or on a surface) without looking inside. We need only look on the boundary surface of the volume (or boundary curve of the surface). A similar theorem in one dimension states that

$$\int_a^b \frac{df}{dx}\, dx = f(b) - f(a).$$

In fact, Gauss's and Stokes's theorems are, like this transparent formula, essentially statements that integration and differentiation are inverse operations. For instance, in the divergence theorem, the derivative in the integral reduces the integral from a triple (volume) to a double (surface) integral.

An example. One of Maxwell's equations states that $\nabla \cdot \mathbf{E} = 4\pi\rho$, where **E** is the electric field and ρ is the charge density. The total charge in a volume V is, of course $Q = \int_V \rho\, d^3x$. We then have

$$Q = \frac{1}{4\pi} \int_V \nabla \cdot \mathbf{E}\, d^3x = \frac{1}{4\pi} \oint_{\partial V} \mathbf{E} \cdot d\mathbf{S},$$

which is known as Gauss's law.

APPENDIX B

Some Integrals Arising in the Hamilton-Jacobi Treatment of the Central Force Problem

Let us write down the central force Hamiltonian in spherical polar coordinates. Then θ is the colatitude, ϕ the azimuth angle, and r the radius. This notation differs slightly from the two-dimensional form of the problem, in which we have written θ for what is here the azimuth angle. The Hamiltonian is

$$H = \frac{1}{2m}\left[p_r^2 + \frac{1}{r^2}\,p_\theta^2 + \frac{1}{r^2\sin^2\theta}\,p_\phi^2\right] + V(r),$$

and Hamilton's principal function can be found by separation. It is

$$S = -Q_1 t + Q_3\phi + \int\sqrt{Q_2 - Q_3^2/\sin^2\theta}\;d\theta + \int\sqrt{2mQ_1 - 2mV(r) - Q_2/r^2}\;dr.$$

From this expression for S one can obtain integral expressions for the J_α and in this appendix we discuss a method for performing those integrals when $V = -k/r$ is the gravitational potential.

With this potential the expressions for J_θ and J_r are

$$\begin{aligned} J_\theta &= \oint\sqrt{Q_2 - Q_3^2/\sin^2\theta}\;d\theta, \\ J_r &= \oint\sqrt{2mQ_1 + 2mk/r - Q_2/r^2}\;dr. \end{aligned} \tag{B1}$$

To calculate J_θ we note first that the limits $\theta_{\min}$ and $\theta_{\max}$ are given by $\sin^2\theta = Q_3^2/Q_2$. Now θ itself always lies between 0 and π, so that $\sin\theta > 0$ and hence $\sin\theta_{\min} = \sin\theta_{\max} = +\sqrt{Q_3^2/Q_2}$. The integral goes from $\theta_{\min} < \frac{\pi}{2}$ to $\frac{\pi}{2}$ to $\theta_{\max} > \frac{\pi}{2}$ and back again to $\theta_{\min}$. The sine goes from $\sqrt{Q_3^2/Q_2}$ to

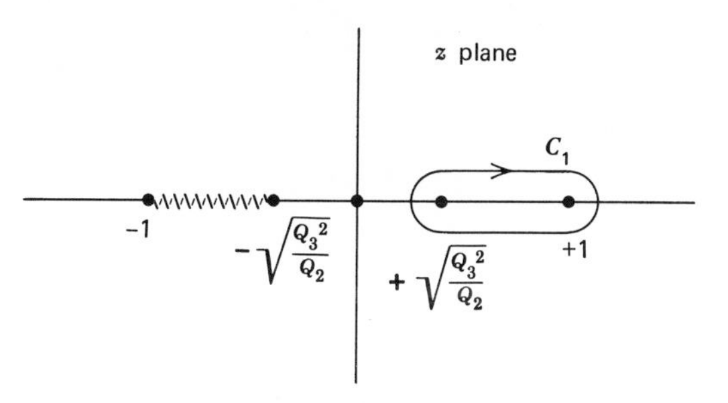

Figure 39. Analytic properties of the integrand in Eq. (B2).

1 to $\sqrt{Q_3^2/Q_2}$. While we integrate from $\theta_{\min}$ to $\theta_{\max}$, the integrand is positive, and on the way back it is negative.

Let us write $\sin\theta = z$, so that the integral becomes

$$J_\theta = \oint \sqrt{\frac{Q_2 z^2 - Q_3^2}{1 - z^2}}\,\frac{dz}{z}. \tag{B2}$$

We perform the integration in the complex z plane. The integrand of (B2) has a pole at $z = 0$ and branch points at $z = \pm 1$ and $z = \pm\sqrt{Q_3^2/Q_2}$. (Note that $Q_3^2 < Q_2$ in order that $\sin\theta < 1$.) These poles, branch points, and conveniently chosen branch cuts are illustrated in Fig. 39. The integral for J_θ is twice (see the discussion of limits) the contour integral around C_1. (There is no contribution to the integral from the singularity at $z = 1$ because the residue there is zero.) Now C_1 can be distorted, as shown in Fig. 40 into C_1', C_0 and C_∞, where C_∞ is a contour of infinite radius:

$$\int_{C_1} = \int_{C_1'} + \int_{C_0} + \int_{C_\infty}.$$

But it is seen from Eq. (B2) that under the interchange $z \to -z$ the integrand remains fixed. Under the same interchange, the contour C_1 changes to $-C_1'$. Thus

$$\int_{C_1'} = -\int_{C_1},$$

and we have

$$J_\theta = 2\int_{C_1} = \int_{C_0} + \int_{C_\infty}.$$

The remaining integrals are easy to calculate:

$$\int_{C_0} \sqrt{\frac{Q_2 z^2 - Q_3^2}{1 - z^2}}\,\frac{dz}{z} = 2\pi i\sqrt{-Q_3^2} = -2\pi Q_3,$$

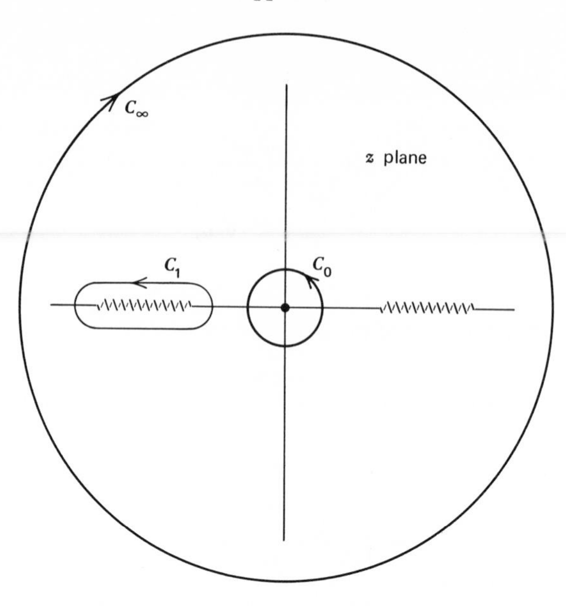

Figure 40. Deformed contour of Fig. 39.

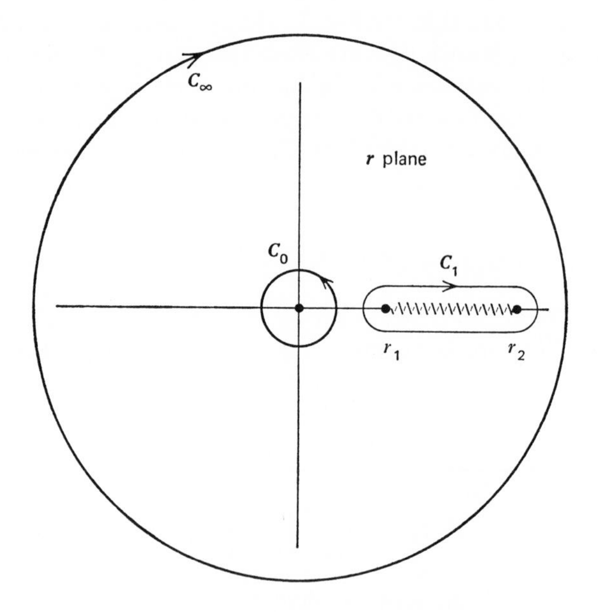

Figure 41. Deformed contour for the integral of Eq. (B4).

for $\sqrt{-Q_3^2}$ is the residue at zero. To calculate the integral at infinity, we write $z = \frac{1}{u}$ and integrate about $u = 0$. Then (note: the sense is reversed)

$$\int_{C_\infty} = \int_{C_0} \sqrt{\frac{Q_2 - Q_3^2 u^2}{u^2 - 1}} \left(-\frac{du}{u}\right) = -2\pi i \sqrt{-Q_2}.$$

Thus we arrive at

$$J_\theta = 2\pi(\sqrt{Q_2} - Q_3). \tag{B3}$$

Now let us calculate J_r. We have

$$J_r = \oint \sqrt{2mQ_1 + 2mk/r - Q_2/r^2}\, dr. \tag{B4}$$

We move to the complex r plane and note that integral is between the roots r_1 and r_2 of the radical in the integrand. (The roots are positive if $Q_1 < 0$ for they are the radii of the turning points.) These are branch points of the integrand, and again the integral along C_1 can be changed to integrals along C_0 and C_∞ of Fig. 41. The residue at the origin is $\sqrt{-Q_2}$. The integral at infinity is calculated by writing $u = 1/r$ and integrating around $u = 0$. The integrand becomes

$$\frac{1}{u^2}\sqrt{2mQ_1 + 2mku - Q_2u^2} = \frac{1}{u^2}\left[\sqrt{2mQ_1} + \frac{mk}{\sqrt{2mQ_1}}\, u + \cdots\right],$$

so that the residue is $mk/\sqrt{2mQ_1}$. Combining these with appropriate signs, we arrive at

$$J_r = 2\pi\left(-\sqrt{Q_2} + \frac{mk}{\sqrt{-2mQ_1}}\right). \tag{B5}$$

(As you may verify, the fact that r_1 and r_2 are real implies that J_r is positive.)

Equation (B4) for J_r is very similar to Eq. (70) of Chapter VII. Our result (B5) can be used to calculate that integral.

Bibliography

Abraham, Ralph, *Foundations of Classical Mechanics*, Benjamin, Inc., New York, 1967.

Bergman, Peter G., *Introduction to the Theory of Relativity*, Prentice-Hall, New York, 1948.

Brillouin, L., *Wave Propagation in Periodic Structures*, Dover Publications, Inc., New York, 1953.

Carathéodory, C., *Calculus of Variations and Partial Differential Equations of the First Order*. Part 1: *Partial Differential Equations of the First Order*, Holden-Day, Inc., New York, 1965.

Coffey, Timothy P., Jr., "Invariants to All Orders in Classical Perturbation Theory," *J. Math. Phys.*, **10,** 426 (1969).

Corben, H. C., and Stehle, Philip, *Classical Mechanics*, John Wiley and Sons, Inc., New York, 1960.

Courant, R., *Differential and Integral Calculus*, Vols. I and II, Interscience Publishers, Inc., 1950.

Currie, D., and Saletan, E., "Canonical Transformations and Quadratic Hamiltonians," *Nuovo Cimento*, (1971).

Dirac, P. A. M., *Principles of Quantum Mechanics*, 4th Ed., Oxford University Press, Oxford, 1958.

Eisenbud, L., "On the Classical Laws of Motion," *Am. J. Phys.*, **26,** 144 (1958).

Feynman, Richard P., Leighton, Robert B., and Sands, Matthew, *Lectures on Physics*, Addison-Wesley Publishing Co., Inc., Reading, Mass., 1963.

Fock, V., *The Theory of Space, Time, and Gravitation*, Macmillan, New York, 1964.

Goldstein, Herbert, *Classical Mechanics*, Addison-Wesley Publishing Co., Inc., Reading, Mass., 1950.

Gossick, Benjamin R., *Hamilton's Principle and Physical Systems*, Academic Press, New York, 1967.

Halmos, Paul R., *Finite Dimensional Vector Spaces*, 2nd Ed., Van Nostrand, New York, 1958.

Hauser, Walter, *Introduction to the Principles of Mechanics*, Addison-Wesley Publishing Co., Inc., Reading, Mass., 1965.

Hill, E. L., "Hamilton's Principle and the Conservation Theorems of Mathematical Physics," *Rev. Modern Phys.*, **23,** 253 (1951).

Kilmister, C. W., and Pirani, F. A. E., "Ignorable Coordinates and Steady Motion in Classical Mechanics," *Proc. Camb. Phil. Soc.*, **61,** 211 (1965).

Kittel, Charles, *Mechanics*, McGraw-Hill Book Company, Inc., New York, 1965.

Lanczos, Cornelius, *The Variational Principles of Mechanics*, 3rd Ed., University of Toronto Press, Toronto, 1966.

Landau, L., and Lifshitz, E., *The Classical Theory of Fields*, Addison-Wesley Publishing Co., Inc., Reading, Mass., 1951.

———, *Mechanics*, Addison-Wesley Publishing Co., Inc., Reading, Mass., 1960.

Mach, E., *Science of Mechanics*, Open Court, New York, 1942.

Mackey, George W., *Mathematical Foundations of Quantum Mechanics*, W. A. Benjamin, Inc., New York, 1963.

Magnus, W., "On the Exponential Solution of Differential Equations for a Linear Operator," Comm. Pure and Appl. Math., **7,** 649 (1954).

Marion, Jerry B., *Classical Dynamics of Particles and Systems*, Academic Press New York, 1965.

Møller, C., "On the Definition of the Centre of Gravity of an Arbitrary Closed System in the Theory of Relativity," Comm. Dublin Inst. for Adv. Studies, Ser. A, No. 5 (1949).

Mukunda, N., "Realizations of Lie Algebras in Classical Mechanics," *J. Math. Phys.*, **8,** 1069 (1967).

Noether, E., "Invariante Variationsprobleme," Nachr. Ges. Wiss., Göttingen, 235 (1918).

Northrop, Theodore G., *The Adiabatic Motion of Charged Particles*, Interscience Publishers, New York, 1963.

Palmieri, C., and Vitale, B., "On the Inversion of Noether's Theorem in the Lagrangian Formalism," *Nuovo Cimento*, **66A,** 299 (1970).

Pars, L. A., *A Treatise on Analytical Dynamics*, John Wiley and Sons, Inc., New York, 1965.

Pauri, M., and Prosperi, G. M., "Canonical Realizations of Lie Symmetry Groups," *J. Math. Phys.*, **7,** 366 (1966).

———, "Canonical Realizations of the Rotation Group," *J. Math. Phys.*, **8,** 2256 (1967).

———, "Canonical Realizations of the Galilei Group," *J. Math. Phys.*, **9,** 1146 (1968).

Pryce, M. H. L., "The Mass-Centre in the Restricted Theory of Relativity and its Connection with the Quantum Theory of Elementary Particles," Proc. Roy. Soc. (London) **195A,** 62 (1948).

Rund, Hanno, *The Hamilton-Jacobi Theory in the Calculus of Variations*, Van Nostrand, New York, 1966.

Saletan, E. J., and Cromer, A. H., "A Variational Principle for Nonholonomic Constraints," *Am. J. Phys.*, **38,** 892 (1970).

Shockley, William, and Gong, Walter A., *Mechanics*, Merrill, Columbus, 1966.

Slater, John C., and Frank, Nathaniel H., *Mechanics*, McGraw-Hill Book Co., Inc., New York, 1947.

Sommerfeld, A., *Mechanics; Lectures on Theoretical Physics*, Vol. 1, Academic Press, New York, 1964.

Sudarshan, E., *Principles of Classical Dynamics*, NYO 10250, University of Rochester, Department of Physics and Astronomy, Rochester, 1963.

Symon, Keith R., *Mechanics*, Addison-Wesley Publishing Co., Inc., Reading, Mass., 1953.

Taylor, Edwin F., and Wheeler, John Archibald, *Spacetime Physics*, W. H. Freeman and Co., San Francisco, 1966.

Ter Haar, D., *Elements of Hamiltonian Mechanics*, North-Holland Publishing Co., Amsterdam, 1964.

von Neumann, John, *Mathematical Foundations of Quantum Mechanics*, Princeton University Press, Princeton, N.J., 1955.

Whiteman, K. J., and MacNamara, B., "Invariants of Nearly Periodic Systems. II," *J. Math. Phys.*, **9**, 1385 (1968).

Whittaker, E. T., *A Treatise on the Analytical Dynamics of Particles and Rigid Bodies*, Dover Publishing Co., New York, 1944.

Yourgrau, W., and Mandlestam, S., *Variational Principles in Dynamics and Quantum Theory*, Pitman Publishing Co., New York, 1960.

Index

n denotes a footnote, p denotes a problem.